# The Maple Handbook

Maple V Release 3

Darren Redfern

# The Maple Handbook

Maple V Release 3

With 16 illustrations

Springer-Verlag
New York Berlin Heidelberg London
Paris Tokyo HongKong Barcelona Budapest

Darren Redfern
Practical Approach
P.O. Box 1007
Stratford, ON N5A 6W4
Canada

Library of Congress Cataloging-in-Publication Data
Redfern, Darren.
    The Maple handbook / Darren Redfern. - - 2nd ed.
        p.    cm.
    Includes bibliographical references and index.
    ISBN 0-387-94331-5 (New York : acid-free). - - ISBN 3-540-94331-5
    (Berlin : acid-free)
    1. Maple (Computer file)    2. Mathematics - - Data processing.
I. Title.
QA76.95.R43 1994
$510'.285'53$ - - dc20                      93-25796

Maple is a registered trademark of Waterloo Maple Software.

Printed on acid-free paper.

© 1993, 1994 Springer-Verlag New York, Inc.
All rights reserved. This work may not be translated or copied in whole or in part without the written permission of the publisher (Springer-Verlag New York Inc., 175 Fifth Avenue, New York, NY 10010, USA), except for brief excerpts in connection with reviews or scholarly analysis. Use in connection with any form of information storage and retrieval, electronic adaptation, computer software, or by similar or dissimilar methodology now known or hereafter developed is forbidden.
The use of descriptive names, trade names, trademarks, etc., in this publication, even if the former are not especially identified, is not to be taken as a sign that such names, as understood by the Trade Marks and Merchandise Marks Act, may accordingly be used freely by anyone.

Production managed by Karen Phillips; manufacturing supervised by Vincent Scelta.
Photocomposed copy prepared from the author's LaTeX file.
Printed and bound by R.R. Donnelley and Sons, Crawfordsville, IN.
Printed in the United States of America.

9 8 7 6 5 4 3 2 1

ISBN 0-387-94331-5 Springer-Verlag New York Berlin Heidelberg
ISBN 3-540-94331-5 Springer-Verlag Berlin Heidelberg New York

# Contents

Introduction .................................................... 1

Getting Started with Maple ........................... 9

Calculus ........................................................ 35

Linear Algebra ............................................. 69

Solving Equations ...................................... 111

Polynomials and Common Transformations ...... 145

Geometry ................................................... 189

Combinatorics and Graph Theory ................ 231

Number Theory .......................................... 265

Statistics .................................................... 299

Standard Functions and Constants ............. 331

Expression Manipulation ........................... 363

Plotting ...................................................... 383

Programming and System Commands ........... 415

Miscellaneous ............................................ 461

Index .......................................................... 515

# Introduction

## How to Use This Handbook

*The Maple Handbook* is a complete *reference* tool for the Maple language, and is written for all Maple users, regardless of their discipline or field(s) of interest. All the built-in mathematical, graphic, and system-based commands available in Maple V Release 3 are detailed herein.

Please note that *The Maple Handbook* does not teach about the mathematics behind Maple commands. If you do not know the meaning of such concepts as *definite integral*, *identity matrix*, or *prime integer*, do not expect to learn them here. As well, while the introductory sections to each chapter taken together do provide a basic overview of the capabilities of Maple, it is highly recommended that you also read a more thorough tutorial such as *Introduction to Maple* by André Heck or *First Leaves: A Tutorial Introduction to Maple V*.

## Overall Organization

One of the main premises of *The Maple Handbook* is that most Maple users approach the system to solve a particular problem (or set of problems) in a specific subject area. Therefore, all commands are organized in logical subsets that reflect these different categories (e.g., calculus, algebra, data manipulation, etc.) and the commands within a subset are explained in a similar language, creating a tool that allows you quick and confident access to the information necessary to complete the problem you have brought to the system.

This design goes hand in hand with the fact that there are many Maple commands that behave differently when faced with the varying data types and constructs inherent in different subject areas. When this is the case, the commands have separate entries in each relevant subset. A good example of this is the op command (which deals with the internal structure of data objects). The results of the op command are sufficiently different when dealing with expressions, matrices, plots, and procedures to warrant an entry in each corresponding section, as well as one in the section on manipulating data structures. The overall philosophy is to make you travel as little as possible to access information necessary to your work.

In addition, because there is much information about Maple that is very difficult to express on a purely command-by-command basis, each subject is prefaced with a short introductory section, written to represent an active Maple session where input, output, text, and graphics are combined. There is also an introductory session, titled *Getting Started with Maple*, which is intended to get those readers not already familiar with Maple off to a quick start.

## Cross Referencing

One of the most important goals of *The Maple Handbook* is to provide pointers to appropriate information so that you are able to solve your problems quickly and efficiently. *The Maple Handbook* is rife with valuable references, presented in such a way to be accessible yet not clutter the information to which they are attached.

There are three types of references present within *The Maple Handbook*. Firstly, each command listing has a *See also* section that points you to commands within that section, or in other sections, that contain related information. Secondly, in case you know the name of a command but are unsure to which subject area it belongs, there is a complete alphabetical index of all Maple commands at the end of the handbook. Finally, most command listings contain essential page references to supplementary information or illuminating examples contained in the official Maple manuals (*First Leaves: A Tutorial Introduction to Maple V, Maple V Language Reference Manual, Maple V Library Reference Manual*).

## Individual Command Entries

While the information contained in the entries for each individual command is unique, the format in which the information is presented is identical across all entries. The following fictitious example illustrates the various elements that are used throughout this book.

### acommand(expr, var)

Performs a command on expr with respect to var.

**Output:** If expr contains only numeric values, an expression sequence of integers is returned. Otherwise, a list of symbolic values is returned.

**Argument options:** (expr, var=a..b) to limit the computation to var ranging from a to b. a must be greater than b. ♣ ([$expr_1$, ..., $expr_n$], [$var_1$, ..., $var_n$]) to perform the calculation on $expr_1$ through $expr_n$ with respect to $var_1$ through $var_n$, respectively. An expression sequence of n lists is returned.

**Additional information:** If var is previously assigned to anything other than its own name, an error message is returned.

**See also:** bcommand, cpackage[dcommand], *anothercommand*

**FL = 12−14** *LA = 43* **LI = 386**

The *command call*, acommand(expr, var), gives the command name as well as its most common type of parameter sequence. The command name can be of several different forms but always represents how the command can be typed into Maple to be immediately executed:

acommand—a command that is in the standard library

apackage[acommand]—a command that is in a specialty package. To access these commands they must either be entered in their long form (as given) or they must follow an appropriate with command. Unfortunately, in Maple V Release 3, there are still a few packages where the long form (i.e., apackage[acommand]) does not provide access to the command. In these cases, the acommand call must be preceded by with(apackage) or with(apackage, acommand). The individual command entries always list the long form of the command name. If that doesn't appear to work, consult the command entry for the entire package that precedes the individual entries.

acommand(aparameter)—a command that has a unique function when combined with a particular type of parameter (including special optional arguments).

'acommand/bcommand'—used for specialized second-level commands that combine the functionality of acommand and bcommand. Backquotes ensure that special character / is not translated as a division sign. While there are many, many such second-level commands in Maple, very few of them should be used *directly* by you and are therefore not detailed in this book.

The command call's parameter sequence (as well as the parameter sequences found in other elements of a command entry) normally represents placeholders for the actual input you would use when calling the command. For example, a placeholder of int could be replaced with the integer 2, 754, or −11. When a parameter sequence contains an element that appears in italics, for example *numeric*, it means that the word *numeric* is to be entered as is in the command, not replaced with some other value. Such elements most frequently occur with predefined options and input values.

Whenever possible, the expected data type of a parameter is specified with one of the following abbreviations.

| | |
|---|---|
| A | an array data type |
| boolean | *true* or *false* |
| complex | a complex value or expression |
| eqn | an equation |
| expr | an expression |
| exprseq | an expression sequence |
| float, # | a floating-point value |
| filename | a file name |
| fnc | a function |
| fnc(var) | a function in variable var |
| ineq | an inequation |
| int | an integer |
| list, [] | a list data type |
| list[$type_1$] | a list with elements of type $type_1$ |
| n, m, i, j, posint | a positive integer |
| name | a name to be assigned a value |
| num, a, b | a numeric value |
| M | a two-dimensional array, matrix |
| option | one of a set of predefined options |
| poly | a polynomial expression |
| proc | a procedure |
| pt | a point (e.g., [a,b]) |
| rat | a rational expression or value |
| ratpoly | a rational polynomial |
| RootOf | a root notation |
| s, series | a series expression |
| set, {} | a set data type |
| set{$type_1$} | a set with elements of type $type_1$ |
| subexpr | a subexpression |
| T | a table data structure |
| V | a one-dimensional array, vector |
| var | an unassigned variable |

# Introduction 5

The above abbreviations deal with parameter types that are encountered across all disciplines covered by *The Maple Handbook*. There are also dozens of other data types that are specific to individual areas; these are detailed in the introductions to each chapter. [Note: when a sequence of parameters is represented with ..., for example, $expr_1$, ..., $expr_n$, do not confuse this with Maple's ellipsis operator ..—One is a short-hand representation, while the other is a language structure.]

Following the command call is a short description of how the command works on the given parameter sequence. This is meant to give you enough information to use the command in most instances. If more information is needed, the *Argument options* and *Additional information* sections should be read.

The *Output* listing gives some idea of what type of output (i.e., what data types) to expect from the most common calling structure. This is extremely helpful when you either want to dissect the answer for further use or correctly include the command within another command or a procedure.

The *Argument options* listing provides valid variations to the parameter sequence and brief explanations of their functioning. If any alternate parameter sequence is of paramount importance, there is an individual command listing to discuss it. ♣ characters appear in this section to separate multiple parameter sequences.

The *Additional information* section lists just that—additional information about the command. This could include, among other things, special pointers to other command entries, brief descriptions of algorithms used to compute the command, or warnings about dangerous combinations of parameters. ♣ characters appear in this section to distinguish separate items.

The *See also* section gives pointers to other command listings within that section (in normal typeface) or in other sections (in *italic* typeface) which contain related information or work in conjunction with the initial command. When searching for a command in another section, it is best to consult the index at the back of this book for an exact page location.

The Manual Cross References at the end of each command listing provide page numbers from the Maple manuals where more information related to the command is available. **FL** = *First Leaves*, *LA* = *Language Reference Manual*, **LI** = *Library Reference Manual*.

## Maple's On-line Help System

All versions of Maple come with an exhaustive on-line help facility containing hundreds of pages of detailed descriptions. These help pages have been updated during the many years of Maple development and contain much information that is valuable and much that is esoteric. On many platforms, there are on-line help topic browsers that allow you to navigate easily between various help pages.

*The Maple Handbook* adds to this facility. By trimming down the information presented to you, it allows you to save many hours of needless reading. The on-line help pages for Maple were written by many different researchers/developers and lack a common voice and a completely unifying style. By putting all the information in one voice, with consistent handling of similar situations throughout, *The Maple Handbook* furthers comprehension.

Please keep in mind that on-line help is always available to you when you are running Maple. *The Maple Handbook* is meant to offer you fast and efficient access to information about Maple— the on-line help pages can be used to broaden that knowledge at your leisure. To view the on-line help for any Maple command, simply enter the command name prepended with a question mark character. For example, the command ?factor displays the on-line help page for the factor command.

## Where to Go for More Information

The range of Maple books, courseware, and third-party applications is constantly growing. Apart from the standard Maple manuals (*First Leaves*, etc.) there are books on using Maple in subject areas from linear algebra to calculus to differential equations; and there are several more volumes currently being written by authors in the academic and commercial fields. The next few years will see an explosion in the number and variety of Maple materials.

In addition, there have been many scholarly papers, reports, and theses written around the Maple system. For more information on these or other Maple materials, contact the vendor who sold you Maple. They have access to much information that will be of interest to you.

In the meantime, for learning the intricacies of Maple there is no teacher like practical use. Take your copy of *The Maple Handbook* and go at it!

# Acknowledgements

I would like to thank all the people at Waterloo Maple Software and Springer-Verlag (New York) for their consistent support. Thanks go out to those friends who helped directly with the creation of this book, including Tom Casselman, Rüdiger Gebauer, Ken Dreyhaupt, Karen Kosztolnyik, Ron Neumann, David Clark, Tim Tyhurst, Greg Fee, Stefan Vorkoetter, Bruce Char, Jérôme Lang, David Doherty, Jacques Carette, and Kate Atherley.

As well, I would like to thank Ric Asselstine for the uplifting and unfailing encouragement he has given me from the first day I started delving into the mysteries of Maple.

A special thanks goes to my wife, Tamara Harbar. Without her love and inspiration, none of my dreams would be possible.

# Getting Started With Maple

## What is Maple?

In simplest terms, Maple is a computer environment for doing mathematics.

Symbolical, numerical, and graphical computations can all be done with Maple. While simple problems can be solved with Maple, its real power shows when given calculations that are extremely cumbersome or tediously repetitive to do by hand. Maple is a procedural language, combining an efficient programming language with a bevy of predefined mathematical commands.

The breadth of Maple's functionality is wide—topics covered range from calculus and linear algebra to differential equations, geometry, and logic. As well, Maple's coverage of these topics is deep—each subject area has a wealth of commands covering aspects both fundamental and far-reaching.

## Symbolics

The heart of Maple is its symbolical routines, and they afford you the greatest freedom. By allowing variables to remain unknown (i.e., without numerical values) and in exact form (e.g., $1/3$ as opposed to $0.333...$) throughout consecutive steps of a calculation, Maple provides *exact* answers with more accuracy than numerical approximation methods. If a floating-point result is needed, it can be calculated at the end of the computation, eliminating roundoff errors normally carried through numerical computations. Some examples of symbolic calculations follow.

```
> solve({a*x+b*y=1, 2*c*x+d*y=3}, {x,y});
```

$$\left\{ x = \frac{d-3b}{da-2bc}, y = \frac{-2c+3a}{da-2bc} \right\}$$

```
> int(exp(-x^2)*ln(x), x=0..infinity);
```

$$-\frac{1}{4}\sqrt{\pi}\,\gamma - \frac{1}{2}\sqrt{\pi}\,ln(2)$$

## Numerics

Besides solving problems that are inherently numeric, numerical routines provide alternative methods when a symbolical method either does not exist or is too slow. Symbolic constants and rational numbers can be evaluated numerically. Maple's *infinite precision* allows numerical calculations to be done to an accuracy of any number of digits. The following are numerical calculations.

```
> 5.0^(1/3);
```

$$1.709975947$$

```
> sum((-1)^i*1.0/i!, i=1..20);
```

$$-.6321205588$$

```
> evalf(Pi, 25);
```

$$3.141592653589793238462643$$

## Graphics

Functions in both one and two unknowns, as well as parametric equations, can be represented graphically. There are over twenty types of specialty plots as well as many available options for customizing the way plots are displayed. The following is an example of a three-dimensional plot.

```
> plot3d((x^2-y^2)/(x^2+y^2), x=-2..2, y=-2..2);
```

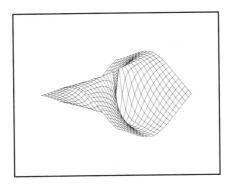

## Maple's Internal Structure

Internally, Maple consists of three component parts: the *kernel*, the *library*, and the *interface*.

The kernel is the "mathematical engine" behind Maple's calculations. This compact, highly optimized set of routines is written and compiled in the C programming language, and performs the majority of the basic computations done by the system.

Most of Maple's built-in commands reside in the Maple library and are written in its own programming language. Code written in Maple is not compiled, but *interpreted* as it is read or entered, allowing you to create Maple commands interactively within a session. By programming your own procedures and adding them to the standard library, you can increase the functionality of the entire system.

The interface is Maple's eyes to the world and defines, to a large extent, how you interact with commands and procedures. Depending on the quality of your terminal, and the version of Maple you are running, the interface may fluctuate between a TTY terminal version to a sophisticated interface supporting Maple documents combining input, output, text, and graphics. While interfaces to Maple on various platforms are becoming more and more similar, *The Maple Handbook* does not dwell further on them, but concentrates on aspects of the Maple *language*, which does not change from platform to platform or from interface to interface.

## Starting Maple

On systems which support command line interfaces (e.g., UNIX, DOS), Maple is typically started by entering the command maple at the prompt. Systems with more advanced graphical user interfaces (e.g., Macintosh, NeXT, Windows, Amiga) have a Maple program icon that activates the application.

Once Maple is started, you are presented with a Maple input prompt, that typically looks like:

>

## Basic Maple Syntax

Input to Maple is typed in at the input prompt. A *complete* input statement can consist of Maple commands, expressions, procedures, etc., and must be properly terminated. There are two ways to terminate a Maple statement. A semicolon (;) is the most commonly used Maple terminator. The resulting output is displayed directly below the input. A colon (:) terminator tells Maple to suppress the output and is particularly useful when the output is an inconsequential intermediate result or is expected to be too large to view conveniently.

When you have typed all the input that you wish to enter on any individual line, the *Enter* key sends it to Maple's mathematical engine to be interpreted. If your input contains a *complete* Maple statement (or completes a previously incomplete one) then that complete statement is executed and the result displayed in an output region directly below. If your input consists of an *incomplete* Maple statement, then the system simply holds the incomplete statement in a buffer and gives you another input prompt, patiently waiting for you to complete that statement.

Some common problems can arise when beginners (or even experts) enter input lines:

As detailed above, if the input is an incomplete statement, Maple does not perform any calculations, but merely provides you with another input prompt directly below the line just entered. This behavior happens naturally when an input is so long that it stretches over several lines, but most often it is the case that a semicolon or colon was omitted. Beginning users often mistake lack of activity for inability to perform the calculation. Just immediately add a terminator on the new line and hit *Enter*.

```
> int(x, x)
> ;
```

$$\frac{1}{2} x^2$$

A common reaction to forgetting a terminator is to retype the entire command on the second line. This only leads to a syntax error, as Maple concatenates the two input lines together and tries to execute the resulting expression.

```
> int(x, x)
> int(x, x);
syntax error:
int(x, x);
```

When there is a syntax error in the input, Maple prints a syntax error message. A caret (^) marks where Maple's interpreter first ran into trouble—it is then up to you to decide how to correct the error.

If your input is syntactically correct but contains another type of error, Maple issues an error message. Be warned, though, that Maple's error messages can be rather obtuse and fail to indicate exactly what you did wrong or how you should go about fixing it.

```
> diff(x^2);
Error, wrong number (or type) of parameters in function
diff
> assign([1,2,3,4]);
Error, (in assign) invalid arguments
```

Another basic consideration is the addition of blank spaces in Maple input. For the most part, blank spaces can be added at will and are automatically removed by Maple's parser if redundant; but there are a few rules and exceptions that are detailed in the appropriate places throughout this handbook.

# Basic Maple Objects

Numbers, constants, strings, and names are the simplest objects in Maple. The following sections give you a basic understanding of how to recognize, create, and use these objects.

## Integers and Rationals

There are different ways of specifying explicit values in Maple. Because Maple works in the symbolic realm, numbers need not always be given in decimal representation (though you certainly may do so if you wish). Integers are the most simple exact numbers to specify, while rational numbers use the division operator (/) to separate numerator and denominator.

> 31;

$$31$$

> 3/7;

$$\frac{3}{7}$$

> -39/13;

$$-3$$

As you can see, rationals are automatically put into their lowest terms.

## Floating-point Numbers

Decimal representations of exact values are expressed as you would expect, and appear as the results of many of Maple's numerical procedures. As well, numerical values can be represented in base 10 or scientific notation.

> 2.3;

$$2.3$$

> -123.45678;

$$-123.45678$$

> .143 * 10^(-44);

$$.1430000000\, 10^{-44}$$

```
> Float(3141, -3);
```

$$3.141$$

```
> -.12345678e3;
```

$$-123.45678$$

Note: blank spaces can be added before or after numbers; do *not* place them in the middle of a number.

## Mathematical Constants

While integers, rationals, and floating-point numbers can be thought of as constants of a sort, Maple also supports many other common mathematical constants. The list includes:

| | |
|---|---|
| Pi | $\pi$, 3.1415926535... |
| exp(1) | $e$, natural log base |
| I | $\sqrt{-1}$ |
| infinity | $\infty$ |
| -infinity | $-\infty$ |
| gamma | Euler's constant |
| Catalan | Catalan's constant |
| true, false | boolean values |

Maple is case sensitive, so be sure to use proper capitalization when stating these constants. The names Pi and pi are not equivalent!

## Global Variables

Another type of predefined values *can* be changed by the user. Global variables are set to initial values when Maple is started and they can be updated by either you or the system during operation. Some of the more common global variables include:

| | |
|---|---|
| Digits | number of digits carried |
| Order | truncation order for series |
| constants | the currently defined constants |
| libname | location of the standard library |
| printlevel | how much debugging information is displayed |
| lasterror | last ERROR message encountered |
| status | status of system variables |

# Mixing and Matching Different Number Types

As was discussed before, being able to leave values in their exact representation (e.g., Pi not 3.14159...) is part of the beauty of symbolic algebra. Normally, Maple allows you to retain values in their exact form throughout many calculations. One situation when exact values do get converted to approximations is when you mix and match types in an expression. The following examples should illustrate this idea. (For more information on Maple expressions see the next section.)

> 1/3 + 2;

$$\frac{7}{3}$$

> 1/3 + 2.0;

$$2.333333333$$

Placing even one floating-point value in a large exact expression causes complete conversion to floats.

> 1/2 + 2/3 + 3/5 + 5/7 + 7/11 + 1.2;

$$4.317316017$$

# Strings

A *string* in Maple consists of a number of characters of any sort surrounded by backquote characters ('). Following are some examples of Maple strings.

> 'This is a Maple string';

*This is a Maple string*

> '123abc';

$123abc$

```
> `invert.src`;
```

$$invert.src$$

As you can see, special characters (+, ., /, etc.) and blank spaces may be included anywhere in a string, so long as the backquote characters are present. If the backquote characters are *not* present, then these special characters are treated as normal Maple operators.

```
> 3+abc+4;
```

$$7 + abc$$

```
> directory/filename;
```

$$\frac{directory}{filename}$$

```
> invert.src;
```

$$invertsrc$$

A *name* in Maple is a special type of string, which typically contains a letter (a-z, A-Z) followed by zero or more letters, digits (0-9), and underscores (_). Names are case sensitive. The name My-Name is distinct from myname. One difference between names and strings is that names do not need to be enclosed with backquotes—unless the name contains special characters that you do not wish to be evaluated. The following are some examples of valid Maple names,

```
> MyVariable;
```

$$MyVariable$$

```
> hello;
```

$$hello$$

```
> `greatest common divisor`;
```

$$greatest\ common\ divisor$$

while these are examples of invalid Maple names.

```
> /thequotient;
syntax error:
/thequotient;
^

> ...etc;
syntax error:
...etc;
  ^

> `no backquotes `round me!;
syntax error:
`no backquotes `round me!;
                      ^
```

## Concatenation Operator

A handy tool for constructing Maple strings and names is the concatenation operator. Use the period character (.) for concatenation, but keep clear the difference between a period as a decimal point and a period as a concatenation operator. The rule to remember is that for the period to be processed as a concatenation operator, it must have a *name* as its left element.

```
> vname.2;
```

$$vname2$$

```
> Fred.` is a friend of mine`;
```

$$Fred\ is\ a\ friend\ of\ mine$$

```
> 3.amigos;
syntax error:
3.amigos;
  ^
```

# Maple Expressions

*Expressions* are extremely important structures in Maple. All Maple objects are, at one level or another, made up entirely of expressions. At its most basic level, an expression consists of a single value, unknown, or string. Conversely, Maple expressions can consist of thousands upon thousands of values, unknowns, and strings strung together with various arithmetic operators.

Maple's arithmetic operators include:

| | |
|---|---|
| + | addition |
| - | subtraction |
| * | multiplication |
| / | division |
| ^ | exponentiation |
| ! | factorial |
| abs() | absolute value |
| iqou() | integer quotient |
| irem() | integer remainder |

When necessary, use blank spaces between terms to keep your expressions readable. Following are some examples of simple Maple expressions.

> a+b+c;

$$a + b + c$$

> 3*x^3 - 4*x^2 + x - 7;

$$3x^3 - 4x^2 + x - 7$$

> x^2/25 + y^2/36;

$$\frac{1}{25}x^2 + \frac{1}{36}y^2$$

Maple echoes these expressions in a "pretty" form, the quality of which depends upon the capabilities of your monitor.

## Order of Operations

In expressions, the precedence of operators follows the standards found in most other areas of computation. If there are any ambiguities, use parentheses, (), to specify the order of operations.

> 2+3*4-5;

$$9$$

> (2+3)*4-5;

$$15$$

```
> (2+3)*(4-5);
```

$$-5$$

It is a good idea to use parentheses whenever there is any chance of ambiguity. If a set of parentheses is redundant, Maple's parser eliminates it during computation.

## Expression Sequences

Another often used data structure in Maple is the *expression sequence*. An expression sequence is one or more Maple expressions separated by commas. As you will see throughout this handbook, most Maple commands require an expression sequence as input, and many return a result that includes an expression sequence.

The simplest way to create an expression sequence is simply to enter it as such.

```
> 1, 2, 3, 4, 5;
```

$$1, 2, 3, 4, 5$$

```
> a+b, b+c, c+d, e+f, f+g;
```

$$a+b, b+c, c+d, e+f, f+g$$

Alternatively, there are two automatic ways to generate an *implicit* expression sequence. First, the $ operator can be used alone to create sequences containing multiples of one element or, in conjunction with the ellipsis operator .., to create well-ordered sequences.

```
> a$6;
```

$$a, a, a, a, a, a$$

```
> $1..6;
```

$$1, 2, 3, 4, 5, 6$$

```
> i^2 $ i=1..6;
```

$$1, 4, 9, 16, 25, 36$$

## Getting Started 21

As well, there is a Maple command, seq, that allows even more control in the creation of expression sequences.

> seq(i!/i^2, i=1..7);

$$1, \frac{1}{2}, \frac{2}{3}, \frac{3}{2}, \frac{24}{5}, 20, \frac{720}{7}$$

> seq(k*(k+1)/(k+2), k=-1..4);

$$0, 0, \frac{2}{3}, \frac{3}{2}, \frac{12}{5}, \frac{10}{3}$$

Note: the seq command operates very fast, and can be used in many situations to increase the speed of your Maple calculations.

## Sets and Lists

Now that you have learned how the create expressions and expression sequences, it is time to put those skills to use creating the next level of Maple objects: sets and lists. These two data types lend organization to Maple objects, though the exact manner in which they do so varies.

### Sets

A *set* is a *non-ordered* collection of expressions. Any valid Maple expression can be contained in a set. Sets are often used as input to Maple commands and are frequently contained in Maple output. A set is written as an expression sequence surrounded by braces, {}. One important thing to remember about sets is that repetitive elements are automatically removed. The first of the following three examples demonstrates this "non-repetitive" rule.

> {1, 1, 2, 3, 2};

$$\{1, 2, 3\}$$

> {a*x, blue, -234.456, 'Maple Tutorial'};

$$\{Maple\ Tutorial, blue, -234.456, a\,x\}$$

> {red, white, blue};

$$\{white, blue, red\}$$

As you can see from the last example, the order in which you list the elements of a set is not necessarily how Maple sees them internally. The ordering in which the elements are displayed depends on in what order the elements are stored in Maple's internal memory.

There are four basic operators that work on sets: the union operator combines the elements of two sets into one (eliminating any repetitive elements, of course), the intersect operator creates a set that contains any elements common to the two initial sets, the minus operator removes from the first set any elements also found in the second set, and the member operator tells whether a given element is contained in a set.

> {a, b, c, d} union {d, e, f};

$$\{a, d, b, c, f, e\}$$

> {1, 2, 3, 4, 5} intersect {2, 4, 6, 8, 10};

$$\{2, 4\}$$

> {x1, x2, x3} minus {x1, y1};

$$\{x2, x3\}$$

> member(n, {W, e, d, n, e, s, d, a, y});

$$true$$

## Lists

Though similar in syntax, lists and sets have significant differences. Both sets and lists are created from expression sequences, but lists are enclosed with right and left brackets, [], as opposed to braces. As well, lists are *well-ordered* objects, meaning that when you specify a list in Maple the ordering that you indicated is preserved. A third fundamental difference is that duplicate elements are valid within a list. Following are some examples.

> [1, 2, 3, 4, 5, 4, 3, 2, 1];

$$[1, 2, 3, 4, 5, 4, 3, 2, 1]$$

> [a, d, c, b, e];

$$[a, d, c, b, e]$$

> [{c,a,t}, {d,o,g}, {m,o,u,s,e}];

$$[\{a, t, c\}, \{d, g, o\}, \{s, e, u, m, o\}]$$

In the last example, the three sets are elements of the list enclosing them. While the ordering of the elements within the sets may vary, the ordering of the three sets themselves remains constant.

While the union, intersect, and minus operators do not work on lists, the commands op and nops may be used to access and manipulate elements of a list. (There is more explanation of these commands in the *Data Manipulation* section.)

# Calling Maple Commands

## Command Names

Learning about Maple expressions and other data types, such as lists and sets, has lead to their most common use—as parameters in commands. As mentioned earlier, Maple has a wealth of built-in commands stored in the Maple library. Each command is called in the following manner:

commandname(parameter$_1$, parameter$_2$, ..., parameter$_n$);

Command names have been chosen to best represent the functionality of a command and, at the same time, require the least amount of typing possible. For example, the command for integration by parts is called intparts and the command for changing variables is called changevar. Some command names are as short as one character long (e.g., D), while others are over ten characters long (e.g., completesquare).

The large majority of Maple commands are written entirely with lowercase letters. One notable exception to this rule is *inert* functions. Inert functions are "placeholders" for their active counterparts, and they are *usually* spelled the same except for having the initial letter capitalized. For example, Int is the inert form of the int command. When Int is called, no actual calculations take place, and the input is echoed back to you. Most of these inert functions appear in the areas of calculus and polynomials and they are further explained in the sections on those discipline. [Note: not *all* commands that start with a capital letter are inert (e.g., D, C, GAMMA).]

Again, remember that Maple is *case sensitive*, which means that diff, Diff, and DiFf are all different in Maple; whether they are used as command names or variable names, Maple considers them three separate entities.

While blank spaces can be inserted before a command name or between a command name and its parameters, they cannot appear in the middle of a command name.

## Composition of Commands

By using the @ operator, you can compose two or more commands. For example, (cos@sin)(x) is equivalent to cos(sin(x)), and (exp@@4)(x) is equivalent to exp(exp(exp(exp(x)))). Create a few examples yourself to understand better the workings of @.

## Parameter Sequences

Each Maple command takes a parameter sequence as input. This sequence may contain several numbers, expressions, sets, or lists, or it may contain no parameters at all. Regardless of how many parameters are specified, always make sure to enclose the sequence in parentheses, (). Other types of brackets cause the input to be interpreted not as parameters, but as something very different.

Any type of Maple element that we have learned about (and a few that we haven't) can be used as a parameter. Entire command calls can be used as input as well; such commands are evaluated and their results are inserted in the parameter sequence. Some commands have restrictions on what type of elements they accept as input and with most commands the ordering of parameters is also important. As well, all commands have a minimum number of parameters with which they can be called. (For example, int must have at least two parameters, an expression and a variable of integration). Many commands, though, can handle more than their minimum number of parameters. These "extra" parameters can represent many things, including additional data and options controlling the functioning of the command.

The following are example of calls to Maple commands.

```
> isprime(10889);
```

$$true$$

```
> diff(3*x^2+2*x-6, x);
```

$$6x + 2$$

```
> diff(3*x^2+2*x-6, x, x);
```

$$6$$

```
> int(int(x^2*y^3, x), y);
```

$$\frac{1}{12} x^3 y^4$$

## Where's That Command?

It is not always sufficient simply to know the name of the command that you want to enter—sometimes you must explicitly load the command from some part of the Maple library before you can execute it. Because Maple is very forgiving in its nature, it lets you issue a command that has not been loaded or does not yet exist and simply echoes the input back at you as if to say, "OK, I'll let you use that command name, even though it doesn't mean anything to me right now." Following are a few examples of such behavior.

```
> INT(x^2, x);
```

$$INT(x^2, x)$$

```
> realroot(x^3+37*x-21,1/100);
```

$$realroot\left(x^3 + 37x - 21, \frac{1}{100}\right)$$

```
> mean(1, 2, 3, 4, 5, 6);
```

$$mean(1, 2, 3, 4, 5, 6)$$

When this happens, check to make sure that you have spelled the command correctly (including proper lowercase and uppercase letters) and loaded the command into Maple's memory.

## Automatically Loaded and readlib Defined Functions

When Maple starts up, it does not have *any* commands entirely loaded into memory. There are, however, many standard commands that have pointers to their locations loaded, so that when you call them for the first time, Maple automatically knows where to go to load them. Some other functions that reside in the library are not automatically loaded, but must be explicitly loaded with the Maple command readlib (read from the library). If you try entering a command that is in the standard library but it does not seem to work, try doing a readlib and calling it again.

The following are some examples of both automatically loaded and readlib defined functions.

```
> expand((x-2)*(x+5));
```

$$x^2 + 3x - 10$$

```
> realroot(x^3+37*x-21, 1/100);
```

$$realroot\left(x^3 + 37x - 21, \frac{1}{100}\right)$$

```
> readlib(realroot);
> realroot(x^3+37*x-21, 1/100);
```

$$\left[\left[\frac{9}{16}, \frac{73}{128}\right]\right]$$

Once a command has been loaded into memory, it does not need to be reloaded during the current Maple session.

## Functions in Packages

Maple contains over a dozen specialized sets of commands called *packages* (e.g., linalg, liesymm, etc.). The routines in these packages are not automatically loaded, nor can they be accessed with the readlib command. The first method to access these commands is to use the with command, which loads in pointers to all the commands in a particular package. Then, when any command in that

Getting Started    27

package is called, it is automatically loaded into memory. Another way is to call the command with its package name prepended to it. A few examples will illustrate these methods.

> `with(combinat);`

[*Chi, bell, binomial, cartprod, character, choose, composition,*
  *conjpart, decodepart, encodepart, fibonacci, firstpart,*
  *graycode, inttovec, lastpart, multinomial, nextpart,*
  *numbcomb, numbcomp, numbpart, numbperm, partition,*
  *permute, powerset, prevpart, randcomb, randpart,*
  *randperm, stirling1, stirling2, subsets, vectoint*]

> `numbperm([1,2,3,4]);`

$$24$$

> `numtheory[euler](6,x);`

$$x^6 - 3x^5 + 5x^3 - 3x$$

The commands in four packages of Maple code cannot be accessed with the long form of their names. Before the commands in these packages are used, a with(packagename) or with(packagename, commandname) command must be entered. The packages are:

- geometry
- geom3d
- NPspinor
- projgeom

## Assignments and Equations

This section explains the difference between the assignment operator, := (the colon character immediately followed by the equal sign), and the equation operator, =. It is very important to understand the distinction between the two.

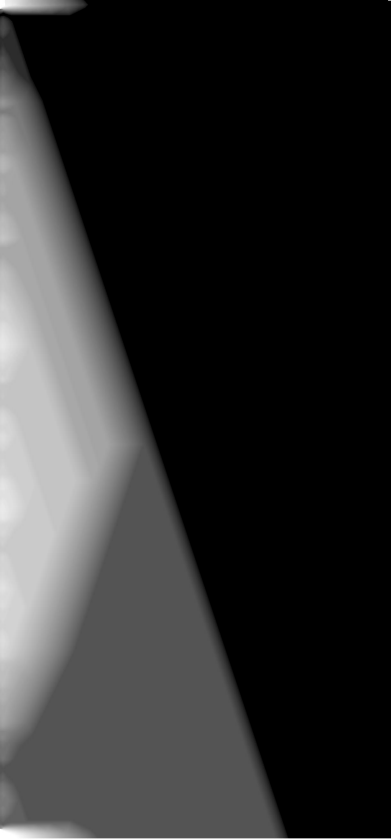

# Equations

The most important thing to realize about equations is that they are not the same as assignments. Equations are simply mathematical expressions that show relationships between certain variables and values; they do not infer any *explicit* values on the variables they contain. For example:

```
> x = y + 3;
```

$$x = y + 3$$

```
> x;
```

$$x$$

```
> y;
```

$$y$$

As you can see, the variables x and y are still unassigned.

The = operator is most frequently seen in either input to a Maple command or output from a Maple command. One common family of commands that makes extensive use of the = operator is the solve family (i.e., solve, rsolve, dsolve, etc.). These commands take equations of various forms and try to find a solution for a given set of variables.

For example, solve takes a set of linear or nonlinear equations and tries to find a closed-form solution.

```
> sols := solve({x + y = 3,x - y = 1}, {x, y});
```

$$sols := \{x = 2, y = 1\}$$

```
> x, y;
```

$$x, y$$

As you can see, the solution that you get is a set of equations for the specified variables. If there are multiple solutions, they are all presented. Be aware that x and y have *not* been assigned to the values 2 and 1, respectively. If you wish to make such an assignment, use the assign command, which takes an equation or set of equations and changes each equation to an assignment.

## 30    Getting Started

```
> assign(sols);
> x, y;
```

$$2, 1$$

Another common use of the equation operator is in boolean (true or false) statements. When you want to make a decision how to proceed dependant on the relationship between the values of two variables, the = operator can be used to construct boolean statements. Other such relational operators include: <, <=, <>, and, or, and not. The Maple command evalb can then be used to evaluate a boolean statement. Some examples follow.

```
> evalb(3! = 4!/2^2);
```

$$true$$

```
> evalb(isprime(5) and isprime(541));
```

$$true$$

## The Use of Quotes in Maple

There are three types of quotes used in Maple. Each has a separate meaning, and it is very important that you understand how to use each of them correctly.

### Double Quote

The double quote operator, ", is perhaps the easiest quote to remember. Double quote recalls previous output in a Maple session. One double quote recalls the most recent output, two (" ") recall the second most previous result, and three (" " ") recall the third most previous output. You cannot go further back than three outputs.

Using double quote remains very straightforward when in command line mode. Even if you use the colon terminator to suppress the display of output from a particular command, the double quote operator can be used to display that previously suppressed output.

Another way of looking at the double quote operator is as a short-term replacement for assignment. In most cases, it is better to immediately assign output to some variable name (which allows you to refer to it *any* time later in that session).

```
> expand((x-2)^3*(x-1));
```

$$x^4 - 7x^3 + 18x^2 - 20x + 8$$

```
> factor(");
```

$$(x-2)^3 (x-1)$$

```
> ""/(x^2+3*x-2);
```

$$\frac{x^4 - 7x^3 + 18x^2 - 20x + 8}{x^2 + 3x - 2}$$

## Backward Quote

The backward quote operator, ', is used (as seen in a previous section) to enclose Maple strings. While only strings with special characters (e.g., /, *, !) need backward quotes around them, it is recommended that you get in the habit of using them on all your strings. For more information on strings, see the section *Numbers and Constants, Strings and Names*.

Two backward quotes appearing in succession after the beginning of a string parse as a single backward quote. This allows you to include that character as part of a string. The following examples display this feature as well as two of the commands available for string manipulation.

```
> `backward``quote`;
```

$$backward`quote$$

```
> length(`This is a long string, don't you think?`);
```

$$39$$

```
> substring(`abcdefghijklmnopqrstuvwxyz`, 15..20);
```

$$opqrst$$

## Forward Quote

Perhaps the most difficult quote to use effectively, the forward quote (') can both eliminate ambiguities and cause confusion. Put simply, enclosing an expression in forward quotes delays evaluation of that expression for one trip through Maple's parser. Another way of thinking of this is that each time the parser encounters an expression enclosed by forward quotes, the *evaluation* that is performed consists of stripping away one layer of these quotes. Therefore, if you want to delay evaluation of the expression for two trips through the parser, use two layers of forward quotes.

> ''factor(x^2-x-2)'';

$$'factor(\,x^2 - x - 2\,)'$$

> ";

$$factor(\,x^2 - x - 2\,)$$

> ";

$$(\,x+1\,)(\,x-2\,)$$

The trick to using forward quotes is understanding when to apply them and what exactly their effect is on an *entire* calculation. Two common examples are illustrated here.

Forward quotes can be used to unassign a variable that was previously assigned to a value.

> x := 3;

$$x := 3$$

> x := 'x';

$$x := x$$

As well, forward quotes can be used for clarification within commands that use indices (e.g., sum, product).

```
> i := 3:
> sum(i^2, i=1..6);
Error, (in sum) summation variable previously assigned,
second argument evaluates to, 3 = 1 .. 6
> sum('i^2', 'i'=1..6);
```

# Calculus

## Introduction

Calculus is a broad discipline and Maple contains commands that cover most of the types of calculations that might interest you.

The input to a standard calculus command typically consists of an expression in one or more unknowns and a specification of the variable upon which you wish to calculate (e.g., a variable of integration). The following is a basic call to the diff command that differentiates the expression $x^4$ with respect to $x$.

> diff(x^4, x);

$$4x^3$$

As you can see, an expression is typically the result from such a command. If you enter a calculus command and the result displayed is not an expression but an echoing of the initial command, this means one of the following things: Maple is unable to find the answer to your question, Maple has determined that there is no possible solution for that question, or you have not provided enough information to produce an answer that differs from the input. (In most cases, it is the first reason.) You may have to rely on your mathematical skills and common sense to determine which is the case.

> int(exp(x^3), x=1..2);

$$\int_1^2 e^{(x^3)}\, dx$$

While it may not be immediately obvious that this output represents exactly the question you entered, after a little examination you see that it is simply a "prettified" version of the input. (As noted before, the quality with which these output expressions are displayed depends upon the capabilities of your monitor.)

At this point there are still many options open to you for getting a solution to your problem. One choice is numerical evaluation, for which Maple has many built-in algorithms. The previous output can be integrated numerically by performing an evalf on it.

```
> evalf(");
```

$$275.5109838$$

Numerical evaluation can be performed without first attempting symbolical evaluation. In most cases, this can be achieved by calling evalf on the *inert* form of the particular command. As explained in the *Getting Started with Maple* section, inert commands are placeholders for their active partners, and are usually spelled with a leading capital letter (e.g., Diff vs. diff). If a numerical approximation is all you require, this method can greatly speed up your overall calculations.

```
> evalf(Int(exp(x^3), x=0..1));
```

$$1.341904418$$

Another, albeit more vague, way of continuing on with a problem that Maple initially cannot solve is to *reconstruct* the problem in a way that lends itself to solution methods available in Maple, methods that are tried internally, but unsuccessfully, with the original formation of the question. This may involve expressing the input expression in a ''simplified'' form using several of Maple's data manipulation commands (e.g., normal, simplify, collect, etc.) or rethinking the problem and splitting it into simpler components. Mastering the use of these manipulation techniques is an art unto itself, and is discussed further in the *Data Manipulation* section.

Other methods for solving problems that Maple may not be able to handle in their initial form include integration by parts and change of variables. Again, both of these methods are automatically applied to problems, but Maple may not have chosen the particular elements in such a way to make the problem solvable.

Apart from integrals and derivatives, Maple also has commands for calculating and representing limits (limit and Limit), summations (sum and Sum), and products (product and Product). While all these commands share common calling structures and syntax, there are some basic differences of which you must be aware.

Calculus    37

For example, while the commands for integration, summation, and finding products can all contain specified *ranges*, these ranges have inherently different meanings. In integration, a range indicates the interval over which definite integration is to be done. Endpoints for the interval can be symbolic or numeric, whole numbers, fractions, or constants and Maple treats them accordingly. The ranges for taking both sums and products must be integer values or they have no intrinsic meaning—you cannot define step size in the sum or product commands, so taking a sum from 1 to 3.8 is meaningless. As you can see, taking such a sum only results in Maple truncating the non-integer value.

```
> sum(i, i=1..3.8);
```

$$6$$

Another difference in the calling structures occurs when performing multiple ''layers'' of a specific calculation (e.g., taking multiple integrals). Of all the standard calculus commands, only diff allows you to perform multiple iterations by simply adding extra variables to the parameter sequence. For other commands, you must embed one call as input within another call.

```
> diff(x^5*y^2, x, y);
```

$$10\,x^4\,y$$

```
> int(int(x^2*y^3, x), y);
```

$$\frac{1}{12}\,x^3\,y^4$$

```
> sum(sum(x^5*y^2, x=1..5), y=1..5);
```

$$243375$$

[Note: Because of the large and complicated *intermediate expressions* that can result when embedding command calls, performing multiple integration, summation, etc. often fails unless steps are performed to simplify these intermediate expressions.]

In addition to the calculus routines found in the standard Maple library, there are three packages that contain supplementary commands. The student package contains utility routines that allow you to carry out calculations step by step, picking and choosing which methods are used.

```
> with(student):
> Int((cos(x)+1)^3*sin(x), x);
```

$$\int (cos(x)+1)^3 \, sin(x) \, dx$$

```
> changevar(cos(x)+1=u, Int((cos(x)+1)^3*sin(x), x), u);
```

$$\int -u^3 \, du$$

```
> value(");
```

$$-\frac{1}{4} u^4$$

```
> subs(u=cos(x)+1, ");
```

$$-\frac{1}{4}(cos(x)+1)^4$$

```
> diff(", x);
```

$$(cos(x)+1)^3 \, sin(x)$$

## Parameter Types Specific to This Chapter

Fnc             an unevaluated function
pseries         a formal power series

## Command Listing

### asympt(expr, var)

Computes an asymptotic expansion of expr as variable var approaches infinity. The series is calculated up to the order specified by the global variable Order.
**Output:** Unlike other series commands which return a series data type, asympt returns a sum of products.
**Argument options:** (expr, var, posint) to compute the asymptotic expansion to degree posint.
**See also:** Order, op(series), op(polynom), series, taylor
**FL = 46** *LA = 14, 24−25* **LI = 14**

## chebyshev(expr, var=a..b)

Computes the Chebyshev series expansion of expr with respect to var on the interval [a..b].

**Output:** An expression (sum) containing unevaluated Chebyshev polynomials (expressed by calls to T) is returned.

**Argument options:** (expr, var) to compute the series expansion on the default interval -1..1. ♣ (expr, var=a..b, expr$_{error}$) to compute the series expansion to a tolerance specified by error expression expr$_{error}$. The default is 10^(-Digits).

**Additional information:** expr must be analytic over the defined interval. ♣ To convert from unevaluated T format to actual polynomials, it is only necessary to load the orthopoly[T] command using with(orthopoly, T).

**See also:** series, taylor, *orthopoly(T)*

FL = 221−223  *LI = 17*


## coeftayl(expr, var=a, n)

Computes the coefficient of (var - a)$^n$ in the Taylor series expansion of expr with respect to var and about a.

**Output:** An expression is returned.

**Argument options:** (expr, [var$_1$, ..., var$_m$] = [a$_1$, ..., a$_m$], [n$_1$, ..., n$_m$]) in the multivariate case to compute the coefficient of the (var$_1$-a$_1$)$^{n_1}$ * $\cdots$ * (var$_m$-a$_m$)$^{n_m}$ term.

**Additional information:** This command needs to be defined by readlib(coeftayl) before being invoked. ♣ Instead of computing the entire Taylor series, only the single coefficient is calculated.

**See also:** taylor, *coeff*

*LI = 270*


## combine(expr)

Combines terms in expr, an expression containing multiple calls to one of Int, Diff, Limit, Sum, or Product, as well as exponential, trigonometric, and other special functions.

**Output:** A simplified expression is returned.

**Argument options:** (expr, option) where option is one of *exp*, *ln*, *power*, *Psi*, or *trig* to force specific transformations. ♣ ([expr$_1$, ..., expr$_n$]) or ({expr$_1$, ..., expr$_n$}) to apply combine to all components of a list or set of expressions.

**Additional information:** In most cases, combine does not need the optional parameter, as it can decide for itself which transformation(s) are applicable.

**See also:** *collect, factor, expand*

*LI = 332*

**convert(expr, *hypergeom*)**

Converts calls to sum or Sum in expr into hypergeometric functions, if possible.
**Output:** An expression in hypergeom is returned.
**Additional information:** The resulting hypergeometric functions are not checked for convergence or termination.
See also: *hypergeom*, Sum, sum
*LI = 41*


**convert(s, *polynom*)**

Converts series s to a polynomial. The order term is eliminated.
**Output:** If s is a Taylor series, then a polynomial data type is returned; if not, a sum (of products) structure is returned.
**Additional information:** Before a series can be passed to many commands (e.g., coeff), it must first be converted to a polynomial. The reason is that, while series and polynomial data types may look very similar, their internal representation is very different.
See also: convert(*ratpoly*), series, taylor, chebyshev, op(series), op(polynom)
**FL = 46, 219**  *LI = 48*


**convert(s, *ratpoly*)**

Converts series s to a rational polynomial expression.
**Output:** If s is a Taylor or Laurent series, then a rational expression which is a Padé approximation is returned. If s is a Chebyshev series, then a rational expression which is a Chebyshev-Padé approximation is returned.
**Argument options:** (s, *ratpoly*, $int_n$, $int_d$) to produce a rational polynomial expression whose numerator and denominator are of degree $int_n$ and $int_d$, respectively.
See also: convert(*polynom*), series, taylor, chebyshev, op(series)
**FL = 220** *LA = 15 LI = 51*


**D(fnc)**

Represents the differential operator, where fnc is typically a function (e.g., tan, exp) or an expression consisting of functions.
**Output:** A function or functional expression that represents the derivative of fnc is returned.
**Argument options:** (var -> expr) for the partial derivative of a functional operator. ♣ [n](fnc) for the partial derivative of fnc with respect to its $n^{th}$ argument. ♣ [$n_1$, $n_2$](fnc) for the partial derivative of fnc with respect to its $n_1^{th}$ and then its $n_2^{th}$ argument, in that order.

**Additional information:** Constants are mapped to zero. ♣ If fnc contains unknowns, they are mapped to the appropriate unknown functional operators.
**See also:** diff, Diff
**FL = 137** *LA = 9–10, 134–135* **LI = 57**

### diff(expr, var)

Differentiates expr with respect to var.
**Output:** In almost all cases, an expression in var is returned. If expr is an unevaluated function call, then an unevaluated derivative is returned.
**Argument options:** (expr, var$n) for $n^{th}$-order derivative of expr with respect to var. ♣ (expr, $var_1$, $var_2$, ..., $var_n$) for partial differentiation of expr with respect to $var_1$ through $var_n$, in sequence.
**See also:** Diff, D, *dsolve*, int
**FL = 41** *LI = 63*

### Diff(expr, var)

Represents the inert derivative of expr with respect to var.
**Output:** An unevaluated Diff command, that can be evaluated as a diff call with value, is returned.
**Argument options:** (expr, var$n) for $n^{th}$-order derivative of expr with respect to var. ♣ (expr, $var_1$, $var_2$, ..., $var_n$) for partial differentiation of expr with respect to $var_1$ through $var_n$, in sequence.
**Additional information:** Unevaluated Diff calls can be combined with the combine command.
**See also:** diff, D, op(*Diff*), Int, combine
*LI = 63*

### difforms package

Provides commands for creating, manipulating, and calculating with differential forms.
**Additional information:** To access the command fnc from the difforms package, use the long form of the name difforms[fnc], or with(difforms, fnc) to load in a pointer for fnc only, or with(difforms) to load in pointers to all the commands. ♣ Before calculations with differential forms can be performed, formal definitions of the forms must be made with the difforms[defform] command.

### difforms[&^]($expr_1$, ..., $expr_n$)

Represents the wedge product of expressions in differential forms $expr_1$ through $expr_n$.

**Output:** An expression in wedge products is returned.
**Argument options:** expr$_1$ &^ expr$_2$ &^ ... &^ expr$_n$ to produce the same results.
**Additional information:** Elementary simplifications are always done before a result is returned. Examples of such simplifications are distribution of &^ over + and transforming &^(expr, expr) to 0 if expr is of odd degree. ♣ Results returned by difforms[&^] are not guaranteed to be in simplest form. To insure simplest form, apply expand and difforms[simpform] to the result.
**See also:** difforms[defform], difforms[simpform], *expand*


### difforms[d](expr)

Computes the exterior derivative of expr, an expression in differential forms.
**Output:** An expression in differential forms is returned.
**Argument options:** ([expr$_1$, ..., expr$_n$]) to return the exterior derivative of expr$_1$ through expr$_n$ in a list of length n. ♣ command (expr, [name$_1$, ..., name$_n$]), where name$_1$ through name$_n$ are forms of degree 1, to expand any scalars in expr in terms of these 1-forms.
**See also:** difforms[defform], difforms[&^], difforms[mixpar], D, diff


### difforms[defform](name$_1$ = expr$_1$, ..., name$_n$ = expr$_n$)

Defines the basic variables name$_1$ through name$_n$ used in differential form calculations.
**Output:** A NULL expression is returned.
**Additional information:** If expr$_i$ is one of the special values *const* (constant degree), *scalar*, *form*, *odd* (odd degree), *even* (even degree), *-1* (constant degree), *0* (even constant degree), or *1* (odd constant degree), then name$_i$ has a special definition. Otherwise, the meaning is that the degree of the form name$_i$ is set to expr$_i$. ♣ The exterior derivative of an expression can be defined with difforms[d]. Such definitions are remembered for the rest of your session.
**See also:** type(*form*), type(*const*), type(*scalar*)


### difforms[formpart](expr)

Extracts the *form* part of expr, an expression in differential forms.
**Output:** An expression is returned.
**Additional information:** Because it is most effective on products, difforms[formpart] tries to convert expr to that form. Scalars and

constants are removed from the converted expression. ♣ Use difforms[scalarpart] to extract the *scalar* part of an expression.
**See also:** difforms[defform], difforms[&^], difforms[scalarpart], type(*form*), type(*const*), type(*scalar*)

### difforms[mixpar](expr)

Sorts the sequence of differentiations in expr which are created by nested calls to the difforms[d] command.
**Output:** An expression is returned. Typically, the value is 0.
**Additional information:** The sorting is done in lexicographical order on the variables of differentiation. ♣ The purpose of this command is to make sure that mixed partial derivatives are considered equal. If the result is not 0, then there is something wrong.
♣ See the on-line help page for examples.
**See also:** difforms[defform], difforms[d]

### difforms[parity](expr)

Calculates the parity of expr, an expression in differential forms.
**Output:** An expression is returned.
**Additional information:** The difforms[parity] command is used to determine whether a form is even or odd. ♣ Undefined variables used as exponents are considered to be integers. If expr contains other undefined variables, they are returned in the result.
**See also:** difforms[defform], difforms[wdegree]

### difforms[scalarpart](expr)

Extracts the *scalar* part of expr, an expression containing differential forms.
**Output:** An expression is returned.
**Additional information:** Because it is most effective on products, difforms[scalarpart] tries to convert expr to that form. Scalars and constants are extracted from the converted expression. ♣ Use difforms[formpart] to extract the *form* part of an expression.
**See also:** difforms[defform], difforms[&^], difforms[formpart], type(*form*), type(*const*), type(*scalar*)

### difforms[simpform](expr)

Simplifies expr, an expression in differential forms.
**Output:** An expression in differential forms is returned.
**Additional information:** Operations done to simplify differential forms include collecting like terms, simplifying wedge products,

and extracting scalar factors. ♣ The simplify command has no effect on expressions of differential forms.
**See also:** difforms[defform], difforms[d], difforms[&^]

### difforms[wdegree](expr)

Determines the degree of expr, an expression in differential forms.
**Output:** An expression is returned.
**Additional information:** A value of nonhmg is returned if expr is a sum of differential forms whose individual degrees differ. ♣ All forms of type scalar or const have degree of $0$.
**See also:** difforms[defform], difforms[parity], type(*form*), type(*const*), type(*scalar*)

### eulermac(expr, var, n)

Calculates the Euler-Maclaurin summation formula of degree n for expr, an expression in var.
**Output:** A sum of terms in var is returned.
**Argument options:** (expr, var) to use the default degree corresponding to the value of global variable Order less 1.
**Additional information:** This command needs to be defined by readlib(eulermac) before being invoked. ♣ The result contains n terms of the expansion plus an order (O) term if applicable.
**See also:** sum, series
*LI = 281*

### evalf(*int*(expr, var=#..#))

Converts result of successful symbolic integration of expr to numeric values. Attempts numeric algorithms on unsuccessful symbolic integration of expr, initially trying the Chebyshev series method, then trying Newton-Côtes rule.
**Output:** In most cases, a numeric expression is returned.
**Additional information:** To skip the symbolic integration attempt, use evalf(*Int*). ♣ To skip both the symbolic integration attempt and the Chebyshev series method, use 'evalf/int'.
**See also:** int, Int, evalf, evalf(*Int*), 'evalf/int'
**FL = 42−43, 217−218** *LA = 11−12*

### evalf(*Int*(expr, var=#..#))

Attempts numeric algorithms on inert integral of expr. Initially tries Chebyshev series method, then Newton-Côtes rule.
**Output:** In most cases, a numeric expression is returned.
**Additional information:** To skip Chebyshev series method, use

'evalf/int'.
**See also:** int, Int, evalf, evalf(*int*), 'evalf/int'

### 'evalf/int'(expr, var=#..#)

Automatically tries numeric integration using Newton-Côtes rule.
**Output:** In most cases, a numeric expression is returned.
**Additional information:** This command needs to be defined by readlib('evalf/int') before being invoked. ✦ To try symbolic integration first, use evalf(*int*). ✦ To try Chebyshev series method, use evalf(*int*).
**See also:** int, Int, evalf, evalf(*int*), evalf(*Int*)

### extrema(expr, eqn, var)

Calculates the relative extreme value points or candidates for extrema of expr with respect to var and under the constraint eqn.
**Output:** If successful, a set containing the extrema is returned.
**Argument options:** (expr, eqn) to find the extrema with respect to all the unassigned variables in both expr and eqn. ✦ (expr, {}, var) to find extrema of expr with respect to var under no constraints. ✦ (expr, {eqn$_1$, ..., eqn$_n$}, {var$_1$, ..., var$_n$}) to find extrema of expr under constraints eqn$_1$ through eqn$_n$ with respect to variables var$_1$ through var$_n$. ✦ Use (expr, eqn, var, name) when no assured extrema can be found. A set of sets of possible candidates is returned in name.
**Additional information:** This command needs to be defined by readlib(extrema) before being invoked.
**See also:** maximize, minimize, diff
*LI = 283*

### int(expr, var)

Calculates the indefinite integral of expr with respect to var.
**Output:** If successful, an expression in var is returned.
**Argument options:** (expr, var=a..b) for symbolic definite integration. ✦ (expr, var=#..#) for numeric definite integration. ✦ (expr, var=a..b, continuous) to turn off automatic checking for continuity on the given interval.
**Additional information:** If int cannot find a closed form for the integral, an unevaluated Int command is returned which can be evaluated approximately by evalf, series, etc. ✦ Operation of int can be controlled by using evalf(*int*) or 'evalf/int'.
**See also:** Int, evalf, 'evalf/int', evalf(*int*), series, diff
**FL = 42** *LA = 10−12* **LI = 110**

## Int(expr, var)

Represents the inert indefinite integral of expr with respect to var.
**Output:** An unevaluated Int command is returned that can be evaluated approximately by evalf, series, etc. or as an int call by value.
**Argument options:** See command entry for int for other valid parameter options.
**Additional information:** Unevaluated Int calls can be combined, expanded, and simplified with the combine, expand, and simplify commands, respectively.
**See also:** int, evalf, 'evalf/int', evalf(*Int*), op(*Int*), series, value, diff, combine
*LI = 110, 334*


## iscont(expr, var=a..b)

Determines whether expr is continuous for var in the range [a, b].
**Output:** If continuity can be established, a boolean value of either true or false is returned. Otherwise, FAIL is returned.
**Argument options:** (expr, var=a..b, *open*) to specify that a..b is an open range. The default is a closed range.
**Additional information:** This command needs to be defined by readlib(iscont) before being invoked. ♣ All variables in expr are assumed to be real. ♣ a and b must be real numeric values, infinity, or -infinity.
**See also:** int, limit
*LI = 297*


## limit(expr, var=a)

Computes the limiting value of expr as var approaches a. Unless otherwise specified, the limit is considered *real* and *bidirectional*.
**Output:** If successful, an expression is typically returned. Occasionally, a range is returned indicating a bounded limit.
**Argument options:** (expr, var=a, option), where option is either *left* or *right*, to compute a directional limit. ♣ (expr, var=a, *complex*) to compute a complex limiting value. ♣ (expr, {$var_1=a_1$, ..., $var_n=a_n$}) to compute the limiting value of expr in n-dimensional space.
**Additional information:** If limit cannot find a closed form, an unevaluated Limit command is returned. ♣ The values infinity and -infinity can be used as input to limit or seen as output from limit. A result of infinity when limit is called with the *complex* option denotes both positive and negative infinity. ♣ Since most limits

are calculated using series, increasing the value of global variable Order may increase the capabilities of limit.
**See also:** Limit, Order
**FL = 47–49** *LA = 13* **LI = 127–131**

### Limit(expr, var=a)

Represents the inert limiting value of expr as var approaches a.
**Output:** An unevaluated Limit command that can be evaluated as a limit call with value is returned.
**Argument options:** See command entry for limit for other valid parameter options.
**Additional information:** Unevaluated Limit calls can be combined, expanded, and simplified with the combine, expand, and simplify commands, respectively.
**See also:** limit, op(*Limit*), value, combine
*LI = 128*

### maximize(expr)

Calculates the maximum value of the expression expr. All unassigned variables are taken into account.
**Output:** A value representing the maximum is returned.
**Argument options:** (expr, {$var_1$, ..., $var_n$}) to find the maximum value of expr with respect to variables $var_1$ through $var_n$.
**Additional information:** The maxima found are unconstrained and over the real numbers, $\infty$, and $-\infty$.
**See also:** minimize, extrema
*LI = 301*

### minimize(expr)

Calculates the minimum value of the expression expr. All unassigned variables are taken into account.
**Output:** A value representing the minimum is returned.
**Argument options:** (expr, {$var_1$, ..., $var_n$}) to find the minimum value of expr with respect to variables $var_1$ through $var_n$. ♣ (expr, {$var_1$, ..., $var_n$}, {$a_1..b_1$, ..., $a_n..b_n$}) to limit the ranges of variables $var_1$ through $var_n$ to the ranges $a_1..b_1$ through $a_n..b_n$, respectively. By default, all ranges are infinite.
**Additional information:** The minima found are unconstrained and over the real numbers, $\infty$, and $-\infty$.
**See also:** maximize, extrema
*LI = 301*

## mtaylor(expr, [var$_1$, ..., var$_n$])

Computes the multivariate Taylor series expansion of expr with respect to variables var$_1$ through var$_n$. The series is calculated up to the order specified by the global variable Order.

**Output:** A multivariate polynomial is returned. This is *not* a series data structure.

**Argument options:** (expr, {var$_1$, ..., var$_n$}) to produce the same result. ♣ (expr, [var$_1$, ..., var$_n$], n) to compute the series expansion to order n. ♣ (expr, [var$_1$, ..., var$_n$], n, [posint$_1$, ..., posint$_n$]) to specify variable weights for var$_1$ through var$_n$ of posint$_1$ through posint$_n$, respectively. By default, all variables have a weight of 1.

**Additional information:** This command needs to be defined by readlib(mtaylor) before being invoked. ♣ While the order of the result can be specified, no order term is present in the result. ♣ The returned polynomials have strictly non-negative powers in the variables.

**See also:** poisson, taylor, series, op(polynom)

*LI = 303*


## op(*Diff*(expr, var))

Displays the internal structure of an unevaluated Diff command and allows you access to its operands.

**Output:** An expression sequence of two operands is returned.

**Argument options:** (*1, Diff*(expr, var)) to extract the first operand, expr. ♣ (*2, Diff*(expr, var)) to extract the second operand, var. ♣ (*Diff*(expr, var$_1$, var$_2$, ..., var$_n$)) for multiple differentiation. Operands 2 through $n + 1$ are variables of differentiation var$_1$ through var$_n$.

**Additional information:** Unless a call to diff returns unevaluated, op(*diff*(expr, var)) returns the operands of the result, not the diff call itself. ♣ To substitute for an operand in a Diff structure, use the subsop command.

**See also:** diff, Diff, *op, subsop*


## op(*Int*(expr, var))

Displays the internal structure of an unevaluated Int command and allows you access to its operands.

**Output:** An expression sequence of two operands is returned.

**Argument options:** (*1, Int*(expr, var)) to extract the integrand, expr. ♣ (*2, Int*(expr, var)) to extract the variable of integration, var. ♣ For definite integration, (*Int*(expr, var=a..b)), the second operand is the equation var=a..b. ♣ When the *continuous* option

Calculus    49

is supplied, it is stored as the third operand.
**Additional information:** Unless a call to int returns unevaluated, op(*int*(expr, var)) returns the operands of the result, not the Int call itself. ♣ To substitute for an operand in an Int structure, use the subsop command.
**See also:** int, Int, student[integrand], *op*, *subsop*

### op(*Limit*(expr, var=a))

Displays the internal structure of an unevaluated Limit command and allows you access to its operands.
**Output:** An expression sequence containing two operands is returned.
**Argument options:** (*1, Limit*(expr, var=a)) to extract the first operand, expr. ♣ (*2, Limit*(expr, var=a)) to extract the second operand, var=a. ♣ When the *right*, *left*, *real*, or *complex* option is supplied, it is stored as a third operand. ♣ When computing multi-dimensional limits with (*Limit*(expr, {var$_1$=a$_1$, ..., var$_n$=a$_n$})), the expression sequence of equations var$_i$=a$_i$ is stored as the second operand.
**Additional information:** Unless a call to limit returns unevaluated, op(*limit*(expr, var)) returns the operands of the result, not the limit call itself. ♣ To substitute for an operand in a Limit structure, use the subsop command.
**See also:** limit, Limit, *op*, *subsop*

### op(*Product*('expr', 'var'=n$_1$..n$_2$))

Displays the internal structure of an unevaluated Product command and allows you access to its operands.
**Output:** An expression sequence containing two operands is returned.
**Argument options:** (*1, Product*('expr', 'var'=n$_1$..n$_2$)) to extract the first operand, expr. ♣ (*2, Product*('expr', 'var'=n$_1$..n$_2$)) to extract the second operand, var=n$_1$..n$_2$. ♣ For indefinite products, (*Product*('expr', 'var')), the second operand is the name var.
**Additional information:** Unless a call to product returns unevaluated, op(*product*('expr', 'var'=n$_1$..n$_2$)) returns the operands of the result, not the product call itself. ♣ To substitute for an operand in a Product structure, use the subsop command.
**See also:** product, Product, *op*, *subsop*

### op(pseries)

Displays the internal structure of pseries, a power series data structure, and allows you access to its operands.

**Output:** A procedure of the form proc(powparm) ... end is returned.

**Argument options:** (4, op(pseries)) to extract the table that defines all coefficients of pseries.

**Additional information:** To substitute for an operand in pseries, use the subsop command.

**See also:** powseries[powcreate], powseries[powpoly], *op*, *subsop*

### op(series)

Displays the internal structure of series, a series data structure, and allows you access to its operands.

**Output:** An expression sequence with $2n$ operands, where $n$ is the number of terms in the series, including the order term, is returned. For $i$ from 1 to $n$, the $(2i-1)^{th}$ operand is the $i^{th}$ term coefficient and the $(2i)^{th}$ operand is the $i^{th}$ term exponent.

**Argument options:** If the series was calculated with respect to var about the point a, then (0, series) extracts the expression var=a.

**Additional information:** Zero coefficient terms are eliminated and are not represented in the data structure. ♣ Terms are sorted from lowest exponent to highest exponent. ♣ If there is an order term, the coefficient operand is O(1) and the exponent is the order of truncation for the series. ♣ To substitute for an operand in a series structure, use the subsop command.

**See also:** series, Order, taylor, *op*, *subsop*

### op(Sum('expr', 'var'=$n_1$..$n_2$))

Displays the internal structure of an unevaluated Sum command and allows you access to its operands.

**Output:** An expression sequence containing two operands is returned.

**Argument options:** (1, Sum('expr', 'var'=$n_1$..$n_2$)) to extract the first operand, expr. ♣ (2, Sum('expr', 'var'=$n_1$..$n_2$)) to extract the second operand, var=$n_1$..$n_2$. ♣ (Sum('expr', 'var')) for indefinite summation. The second operand is the name var.

**Additional information:** Unless a call to sum returns unevaluated, op(sum('expr', 'var'=$n_1$..$n_2$)) returns the operands of the result, not the sum call itself. ♣ To substitute for an operand in a Sum structure, use the subsop command.

**See also:** sum, Sum, *op*, *subsop*

## order(s)

Determines the truncation order of series s.
**Output:** If s contains an order term, an integer value is returned. Otherwise, $\infty$ is returned.
**Additional information:** s must be of type series.
**See also:** Order, series, type(*series*)
*LI = 148*

## Order := n

Sets the global variable Order, which specifies what order is used in series calculations, to a value of n.
**Output:** The input is echoed.
**Argument options:** Order to return the current value of the global variable.
**Additional information:** The default value of Order is 6. ♣ In a series, the O term, which is always the last term, represents the remainder of the series expansion beyond the precision of Order.
**See also:** order, series, op(series), convert(polynom)
*FL = 45, 141*

## poisson(expr, [$var_1$, ..., $var_n$])

Computes a multivariate Taylor series expansion in Poisson form of expr with respect to variables $var_1$ through $var_n$. The series is calculated up to the order specified by global variable Order.
**Output:** A multivariate polynomial is returned. The result is *not* a series data structure.
**Argument options:** (expr, {$var_1$, ..., $var_n$}) to produce the same result. ♣ (expr, [$var_1$, ..., $var_n$], n) to compute the series expansion to order n. ♣ (expr, [$var_1$, ..., $var_n$], n, [$posint_1$, ..., $posint_n$]) to assign variable weights of $posint_1$ through $posint_n$ to $var_1$ through $var_n$, respectively. By default, all variables have a weight of 1.
**Additional information:** This command needs to be defined by readlib(poisson) before being invoked. ♣ While the order of the result can be specified, no order term is present in the result. ♣ The returned polynomials have strictly *non-negative* powers in the variables. ♣ The only difference between poisson and mtaylor is that in poisson the sines and cosines in coefficients of the resulting polynomial are combined in *Fourier canonical form*.
**See also:** mtaylor, taylor, series, op(polynom)
*LI = 309*

## powseries package

Provides commands for creating, manipulating, and calculating with formal power series in general form.

**Additional information:** To access the command fnc from the powseries package, use the long form of the name powseries[fnc], or with(powseries, fnc) to load in a pointer for fnc only, or with(powseries) to load in pointers to all the commands. ♣ All power series are represented by Maple procedures. ♣ Two commands for creating power series in proper form are powseries[powcreate] and powseries[powpoly]. ♣ To convert a power series structure to a standard series data structure, use powseries[tpsform].

## powseries[add](pseries$_1$, ..., pseries$_n$)

Adds the power series pseries$_1$ through pseries$_n$.
**Output:** A power series is returned.
**Additional information:** Each coefficient of the result is the sum of the corresponding coefficients from pseries$_1$ through pseries$_n$.
**See also:** powseries[powcreate], powseries[powpoly], powseries[subtract], powseries[multiply], powseries[quotient], powseries[negative], powseries[tpsform]

## powseries[compose](pseries$_1$, pseries$_2$)

Calculates the composition of power series pseries$_1$ with power series pseries$_2$.
**Output:** A power series is returned.
**Additional information:** The composition a(b) is defined as a(0) + a(1)*b + a(2)*b^2 + .... ♣ To insure that the resulting power series is well-defined, pseries$_2$ must have a non-zero first term.
**See also:** powseries[powcreate], powseries[powpoly], powseries[reversion], powseries[tpsform]

## powseries[evalpow](expr$_{ps}$)

Evaluates expr$_{ps}$, an arithmetic expression containing formal power series structures and special power series operators.
**Output:** A power series is returned.
**Additional information:** Any of the mathematical operators +, -, *, /, or ^ can be used in expr$_{ps}$. Special operators powexp, powinv, powlog, powneg, powrev, powdiff, powint, and powquo can only be used within powseries[evalpow]. They correspond di-

rectly to commands in the powseries package.
**See also:** powseries[powcreate], powseries[powpoly], powseries[powexp], powseries[inverse], powseries[powlog], powseries[negative], powseries[reversion], powseries[powdiff], powseries[powint], powseries[quotient], powseries[subtract], powseries[tpsform]

### powseries[inverse](pseries)

Computes the multiplicative inverse of power series pseries.
**Output:** A power series structure is returned.
**See also:** powseries[powcreate], powseries[powpoly], powseries[multiply], powseries[tpsform]

### powseries[multconst](pseries, expr)

Multiplies each coefficient of power series pseries by the expression expr.
**Output:** A power series is returned.
**Additional information:** Typically, expr is a scalar constant.
**See also:** powseries[powcreate], powseries[powpoly], powseries[multiply], powseries[tpsform]

### powseries[multiply](pseries$_1$, pseries$_2$)

Multiplies two power series, pseries$_1$ and pseries$_2$, together.
**Output:** A power series is returned.
**See also:** powseries[powcreate], powseries[powpoly], powseries[add], powseries[subtract], powseries[quotient], powseries[multconst], powseries[inverse], powseries[tpsform]

### powseries[negative](pseries)

Computes the additive inverse of the power series pseries.
**Output:** A power series is returned.
**See also:** powseries[powcreate], powseries[powpoly], powseries[multiply], powseries[tpsform]

### powseries[powcreate](eqn$_c$, eqn$_1$, eqn$_2$, ..., eqn$_n$)

Creates a formal power series, where eqn$_c$ is an equation that defines the coefficients and whose initial conditions are defined by the equations eqn$_1$ through eqn$_n$.
**Output:** A NULL expression is returned.
**Additional information:** The equation eqn$_c$ must be of the form name(var) = expr$_{var}$, where expr$_{var}$ is an expression in var. ♣

The initial condition equations must be in the form name(var) = expr. They define specific values of coefficients for the power series. ♣ If there are any conflicts, then the initial conditions equations take precedence.

**See also:** op(pseries), powseries[powpoly], powseries[tpsform]

### powseries[powdiff](pseries)

Calculates the power series that is the derivative of the given power series pseries with respect to its variable.
**Output:** A power series is returned.
**See also:** powseries[powcreate], powseries[powpoly], powseries[powint], powseries[tpsform]

### powseries[powexp](pseries)

Calculates the power series that is the exponential of the given power series pseries with respect to its variable.
**Output:** A power series is returned.
**See also:** powseries[powcreate], powseries[powpoly], powseries[powlog], powseries[tpsform]

### powseries[powint](pseries)

Calculates the power series that is the indefinite integral of the given power series pseries with respect to its variable.
**Output:** A power series is returned.
**See also:** powseries[powcreate], powseries[powpoly], powseries[powdiff], powseries[tpsform]

### powseries[powlog](pseries)

Calculates the power series that is the natural logarithm of the given power series pseries with respect to its variable.
**Output:** A power series is returned.
**Additional information:** To insure that the resulting power series is well-defined, pseries must have a nonzero first term.
**See also:** powseries[powcreate], powseries[powpoly], powseries[powlog], powseries[tpsform]

### powseries[powpoly](poly, var)

Creates a formal power series equivalent to poly, a polynomial in var.
**Output:** A power series is returned.
**Additional information:** The coefficients of poly are echoed in

the power series.
**See also:** op(pseries), powseries[powcreate], powseries[tpsform], *coeff*

### powseries[quotient](pseries$_1$, pseries$_2$)

Divides power series pseries$_1$ by power series pseries$_2$.
**Output:** A power series is returned.
**Additional information:** Each coefficient of the result is the difference of the corresponding coefficients from pseries$_1$ and pseries$_2$.
**See also:** powseries[powcreate], powseries[powpoly], powseries[add], powseries[subtract], powseries[quotient], powseries[tpsform]

### powseries[reversion](pseries$_1$, pseries$_2$)

Calculates the reversion of power series pseries$_1$ with respect to power series pseries$_2$.
**Output:** A power series is returned.
**Argument options:** (pseries) to calculate the reversion of pseries with respect to the power series pseries$_2$, which has one non-zero coefficient, pseries$_2$(1) = 1.
**Additional information:** Reversion of power series is the inverse operation of composition. ♣ To insure that the resulting power series is well-defined, pseries$_1$ and pseries$_2$ must have a nonzero first term and pseries$_1$ must have the term pseries$_1$(1) = 1.
**See also:** powseries[powcreate], powseries[powpoly], powseries[compose], powseries[tpsform]

### powseries[subtract](pseries$_1$, pseries$_2$)

Subtracts power series pseries$_2$ from power series pseries$_1$.
**Output:** A power series is returned.
**Additional information:** Each coefficient of the result is the difference of corresponding coefficients from pseries$_1$ and pseries$_2$.
**See also:** powseries[powcreate], powseries[powpoly], powseries[add], powseries[multiply], powseries[quotient], powseries[tpsform]

### powseries[tpsform](pseries, var, posint)

Transforms pseries, a formal power series structure, into a standard Maple series structure with variable var and order posint.
**Output:** A series is returned.

## product('expr', 'var'=a..b)

Computes the definite product of expr while var ranges from a to b. Use forward quote characters around expr and var to prevent premature evaluation of these elements.

**Output:** If successful, an expression is returned.

**Argument options:** ('expr', 'var') to compute the indefinite product of expr with respect to var. ♣ ('expr', 'var'=$RootOf$(expr$_2$)) to compute the definite product of expr over the roots of expr$_2$. ♣ ('expr', 'var'=#..#) to compute the definite product of expr over numerical bounds.

**Additional information:** Typically, a and b are integer values; but they can be symbolic or floating-point values as well. ♣ If product cannot find a closed form for the summation, an unevaluated Product command is returned. ♣ If the bounds have a small difference, product directly multiplies the elements. Otherwise, product finds a closed form and evaluates at the bounds. ♣ In many cases, using seq or a for loop produces quicker results.

**See also:** Product, sum, *RootOf*, *seq*

*LI = 169*


## Product('expr', 'var'=a..b)

Represents the inert definite product of expr while var ranges from a to b.

**Output:** An unevaluated Product command that can be evaluated as a product call with value is returned.

**Argument options:** ('expr', 'var') to compute the indefinite product of expr with respect to var. ♣ ('expr', 'var'=$RootOf$(expr$_2$)) to compute the definite product of expr over roots of expr$_2$. ♣ ('expr', 'var'=#..#) to compute the definite product of expr over numerical bounds.

**Additional information:** Unevaluated Product calls can be combined, expanded, and simplified with the combine, expand, and simplify commands, respectively.

**See also:** product, op(*Product*), Sum, *RootOf*, value, combine

*LI = 169*

### residue(expr, var=a)

Computes the algebraic residue of expr, an algebraic expression in var, about the point a.
**Output:** An expression is returned.
**Additional information:** The algebraic residue is defined to be the coefficient of (var-a)^(-1) in the Laurent series expansion of expr.
**See also:** series, singular
*LI = 316*

### series(expr, var=a)

Computes the truncated, generalized series expansion of expr with respect to var about a. The series is calculated up to the order specified by the global variable Order.
**Output:** Typically, a series ending in an order term is returned.
**Argument options:** (expr, var) to compute series expansion with respect to var about 0 ♣ (*Int*(expr, var), var=a) to compute series expansion of an unevaluated integral. ♣ (*leadterm*(expr), var=a) to compute only the lowest degree term in the series expansion. ♣ (expr, var=a, n) to compute series expansion to order n.
**Additional information:** The generalized series expansion could be any of a Taylor series, Laurent series, or more general series. ♣ Zero coefficients terms are not displayed. ♣ Before you can use a series as input to many commands, you must convert it to a polynomial with convert(*polynom*). ♣ If fractional exponents are necessary to express a series, then a sum (+) data structure is returned rather than a series data structure.
**See also:** Order, order, convert(polynom), op(series), op(polynom), Int, asympt, taylor, chebyshev, powseries
**FL = 45−47, 218−219** *LA = 14−15* ***LI = 185−188***

### singular(expr)

Computes singularities of expr, an algebraic expression.
**Output:** An expression sequence of sets each containing one equation, representing a singularity, is returned.
**Argument options:** (expr, var) to treat every variable in expr, except var, as a constant. ♣ (expr, {$var_1$, ..., $var_n$}) to treat every variable in expr, except $var_1$ through $var_n$, as a constant.
**Additional information:** This command needs to be defined by readlib(singular) before being invoked. ♣ Both removable and non-removable singularities are found. ♣ If there are infinitely many singularities, an expression in _N or _NN, where these variables represent any natural number, is returned.
**See also:** solve, roots, residue
*LI = 319*

## student package

Provides commands designed to allow students to perform calculus operations in a step-by-step manner and help them examine these operations more closely.

**Additional information:** To access the command fnc from the student package, use the long form of the name student[fnc], or with(student, fnc) to load in a pointer for fnc only, or with(student) to load in pointers to all the commands. ♣ There are several commands in the student package that duplicate similarly named commands in the main library. These include student[combine], student[D], student[extrema], student[Int], student[isolate], student[Limit], student[maximize], student[minimize], student[Sum], and student[value]. To learn more about these commands, see their main library counterparts combine, D, extrema, etc.

## student[changevar](expr$_1$ = expr$_2$, Fnc, var)

Performs a change of variables on unevaluated command Fnc (one of Int, Limit, or Sum). The new variable is defined in terms of the old one in expr$_1$ := expr$_2$, where expr$_1$ and expr$_2$ are expressions in the variable in Fnc and the new variable var, respectively.

**Output:** A unevaluated command (in terms of var) is returned.

**Argument options:** ({eqn$_1$, eqn$_2$}, Doubleint, [var$_1$, var$_2$]) when working with double integrals. ♣ ({eqn$_1$, eqn$_2$, eqn$_3$}, Tripleint, [var$_1$, var$_2$, var$_3$]) when working with triple integrals.

**Additional information:** student[changevar] is used on explicitly unevaluated commands or failed attempts at corresponding active commands, and works equally well with indefinite and definite integrations and summations. ♣The results of student[changevar] can be evaluated using value.

**See also:** Int, Sum, Limit, student[Doubleint], student[Tripleint], student[intparts], *student[powsubs]*, *DEtools[Dchangevar]*

*LI = 330*

## student[completesquare](expr)

Rewrites the expression expr as a perfect square plus a remainder.

**Output:** An expression is returned.

**Argument options:** (expr, var) to complete the square with respect to the variable var only.

**See also:** Int

*LI = 331*

## student[distance](pt$_1$, pt$_2$)

Computes the distance between two points, pt$_1$ and pt$_2$, in two or more dimensions.

**Output:** An expression is returned.

**Argument options:** (expr$_1$, expr$_2$) to find the distance, or absolute difference, of two expressions in one dimension. ♣ (pt, name) or (name, pt) to find the distance between explicit point pt and undefined point name.

**See also:** student[midpoint], student[intercept], student[showtangent], student[slope]

*LI = 333*


## student[Doubleint](expr, var$_1$, var$_2$, name)

Represents the inert indefinite double integral of expr with respect to var$_1$ and var$_2$. In the display of the double integral, the domain of integration is called name.

**Output:** Two nested Int commands are returned.

**Argument options:** (expr, var$_1$=a$_1$..b$_1$, var$_2$=a$_2$..b$_2$) to represent the inert definite integration.

**See also:** int, Int, value, student[Tripleint]


## student[equate](list$_1$, list$_2$)

Creates a set of equations of the form {list$_1$[1]=list$_2$[1], ..., list$_1$[n]=list$_2$[n]} from lists list$_1$ and list$_2$.

**Output:** A set of equations is returned.

**Argument options:** (list) to convert list to a set of expressions {list[1]=0, ..., list[n]=0]}. ♣ (array$_1$, array$_2$) to convert two arrays to a set of equations. ♣ (table$_1$, table$_2$) to convert two tables to a set of equations. ♣ (expr$_1$, expr$_2$) to convert two expressions to a set containing one equation, expr$_1$=expr$_2$.

**Additional information:** All non-list data types are converted to lists before the equations are created, so lists, arrays (of any dimension), and tables can be mixed and matched as long as they contain an equal number of elements. Thus a $3 \times 4$ matrix can be matched with a 12-element list.

**See also:** *subs*


## student[integrand](expr)

Extracts from expr, an expression containing unevaluated integrals, all the integrands.

**Output:** If expr contains one unevaluated integral, it is returned. If expr contains more than one unevaluated integral, they are returned in a set. If expr contains no unevaluated integrals, the empty

set, {}, is returned.
**Additional information:** Unevaluated integrals can be created with Int, student[Doubleint], or student[Tripleint].
**See also:** Int, student[Doubleint], student[Tripleint]

### student[intercept](eqn$_1$, eqn$_2$)

Computes the points of intersection of bivariate curves eqn$_1$ and eqn$_2$.
**Output:** A set containing coordinates is returned. If more than one intercept exists, the sets are returned in an expression sequence.
**Argument options:** (eqn$_1$, eqn$_2$, {var$_1$, var$_2$}), where var$_1$ and var$_2$ represent the coordinate variables, if any ambiguity is present.
✦ (eqn) to find the $y$-intercept of eqn.
**See also:** student[distance], student[midpoint], student[showtangent], student[slope]
*LI = 334*

### student[intparts](*Int*(expr$_1$, var), expr$_2$)

Performs integration by parts on unevaluated integral *Int*(expr$_1$, var). expr$_2$ is the factor of expr$_1$ to be differentiated and to become part of the new integrand.
**Output:** A new expression containing an unevaluated integral is returned.
**Additional information:** student[intparts] can be used on explicitly unevaluated Int commands or on failed attempts at int.
✦ student[intparts] works equally well with indefinite and definite integrations. ✦ The results of student[intparts] can be evaluated using value.
**See also:** Int, student[changevar], value, *student[powsubs]*
*LI = 335*

### student[leftbox](expr, var=a..b)

Creates a graphical representation of the numerical approximation of an integral performed by a similar call to student[leftsum].
**Output:** A two-dimensional plot structure is returned.
**Argument options:** (expr, var=a..b, n) to indicate that n rectangles be used in the approximation. ✦ (expr, var=a..b, options) to set options that could normally be set in the plot command.
**See also:** *plot*, int, student[middlebox], student[rightbox], student[leftsum]

### student[leftsum](expr, var=a..b)

Performs numerical approximation, using rectangles, of the definite integral of expr as var ranges from a to b. The height of each rectangle is the value of expr at the left side of the subinterval. By default, four equal-sized intervals are used.
**Output:** An unevaluated Sum command that represents the approximation is returned.
**Argument options:** (expr, var=a..b, n) to indicate that n rectangles should be used in the approximation.
**Additional information:** To evaluate symbolically the Sum command returned as output, use the value command. ♣ To evaluate numerically the Sum command returned as output, use the evalf command. ♣ A two-dimensional graph representing the approximation can be created with student[leftbox].
**See also:** evalf, Sum, student[middlesum], student[rightsum], student[trapezoid], student[simpson], student[leftbox], value

### student[makeproc](expr, var)

Converts expr, an expression in var, into functional operator notation.
**Output:** A functional operator that takes one parameter is returned.
**Argument options:** ([exprx$_1$, expry$_1$], [exprx$_2$, expry$_2$]) to create a functional operator that defines the line passing through the two points [exprx$_1$, expry$_1$] and [exprx$_2$, expry$_2$]. ♣ ([exprx, expry], *slope*=expr) or (*slope*=expr, [exprx, expry]) to create a functional operator that defines the line passing through the the point [exprx, expry] and with a slope of expr.
**See also:** *unapply*

### student[middlebox](expr, var=a..b)

Creates a graphical representation of the numerical approximation of an integral performed by a similar call to student[middlesum].
**Output:** A two-dimensional plot structure is returned.
**Argument options:** (expr, var=a..b, n) to indicate that n rectangles be used in the approximation. ♣ (expr, var=a..b, options) to set options that could normally be set in the plot command.
**See also:** *plot*, int, student[leftbox], student[rightbox], student[middlesum]

### student[middlesum](expr, var=a..b)

Performs numerical approximation, using rectangles, of the definite integral of expr as var ranges from a to b. The height of each rectangle is the value of expr at the middle of the subinterval. By default, four equal-sized intervals are used.

**Output:** An unevaluated Sum command that represents the approximation is returned.

**Argument options:** (expr, var=a..b, n) to indicate that n rectangles should be used in the approximation.

**Additional information:** To evaluate symbolically the Sum command returned as output, use the value command. ♣ To evaluate numerically the Sum command returned as output, use the evalf command. ♣ A two-dimensional graph representing the approximation can be created with student[middlebox].

**See also:** evalf, Sum, student[leftsum], student[rightsum], student[trapezoid], student[simpson], student[middlebox], value


### student[midpoint]($pt_1$, $pt_2$)

Computes the midpoint of a line segment connecting two points, $pt_1$ and $pt_2$.

**Output:** A point is returned.

**Argument options:** ($expr_1$, $expr_2$) to find the average of two expressions. ♣ (pt, name) or (name, pt) to find the midpoint of the line segment between pt and undefined point name.

**See also:** student[distance], student[intercept], student[showtangent], student[slope]


### student[powsubs]($expr_1$ = $expr_2$, expr)

Substitute all occurrences of $expr_1$ in expr with $expr_2$.

**Output:** An expression is returned.

**Argument options:** (\{$eqn_1$, ..., $eqn_n$\}, expr) or ([$eqn_1$, ..., $eqn_n$], expr) to perform the n substitutions corresponding to $eqn_1$ through $eqn_n$ in the order they are provided.

**Additional information:** student[powsubs] differs from subs in that instead of examining the underlying structure of expr directly, algebraic factors are considered. Because of this, student[powsubs] cannot find expressions that occur as *part* of a sum. ♣ See the on-line help pages for subs and student[powsubs] for examples.

**See also:** *subs*


### student[rightbox](expr, var=a..b)

Creates a graphical representation of the numerical approximation of an integral performed by a similar call to student[rightsum].

**Output:** A two-dimensional plot structure is returned.

**Argument options:** (expr, var=a..b, n) to indicate that n rectangles be used in the approximation. ✦ (expr, var=a..b, options) to set options that could normally be set in the plot command.
**See also:** *plot*, int, student[leftbox], student[middlebox], student[rightsum]

### student[rightsum](expr, var=a..b)

Performs numerical approximation, using rectangles, of the definite integral of expr as var ranges from a to b. The height of each rectangle is the value of expr at the right side of the subinterval. By default, four equal-sized intervals are used.
**Output:** An unevaluated Sum command which represents the approximation is returned.
**Argument options:** (expr, var=a..b, n) to indicate that n rectangles should be used in the approximation.
**Additional information:** To evaluate symbolically the Sum command returned as output, use the value command. ✦ To evaluate numerically the Sum command returned as output, use the evalf command. ✦ A two-dimensional graph representing the approximation can be created with student[rightbox].
**See also:** evalf, Sum, student[leftsum], student[middlesum], student[trapezoid], student[simpson], student[rightbox], value

### student[showtangent](expr, var=a)

Plots expr, an expression in var, and its tangent line at the point var=a on one set of two-dimensional axes.
**Output:** A two-dimensional plot structure is returned.
**Additional information:** You cannot alter any default option settings for plots from within the student[showtangent] command.
**See also:** student[distance], student[midpoint], student[showtangent], student[slope], *plot*

### student[simpson](expr, var=a..b)

Performs numerical approximation, using Simpson's rule, of the definite integral of expr as var ranges from a to b. By default, four equal-sized intervals are used.
**Output:** An unevaluated Sum command that represents the approximation is returned.
**Argument options:** (expr, var=a..b, n) to indicate that n intervals should be used in the approximation.
**Additional information:** To evaluate symbolically the Sum command returned as output, use the value command. ✦ To evaluate

numerically the Sum command returned as output, use the evalf command.
**See also:** int, evalf, Sum, student[leftsum], student[middlesum], student[rightsum], student[trapezoid], value

### student[slope](eqn)

Computes the slope of eqn, which is of the form y=f(x). y is the dependent variable and x is the independent variable.
**Output:** A expression is returned.
**Argument options:** (eqn, y, x) or (eqn, y(x)), when any ambiguity is present. y then becomes the dependent variable and x becomes the independent variable. ♣ (eqn, $pt_1$, $pt_2$) to compute the slope between points $pt_1$ and $pt_2$.
**See also:** student[distance], student[midpoint], student[intercept], student[showtangent]

### student[trapezoid](expr, var=a..b)

Performs numerical approximation, using the Trapezoidal rule, of the definite integral of expr as var ranges from a to b. By default, four equal-sized intervals are used.
**Output:** An unevaluated Sum command that represents the approximation is returned.
**Argument options:** (expr, var=a..b, n) to indicate that n trapezoids should be used in the approximation.
**Additional information:** To evaluate symbolically the Sum command returned as output, use the value command. ♣ To evaluate numerically the Sum command returned as output, use the evalf command.
**See also:** int, evalf, Sum, student[leftsum], student[middlesum], student[rightsum], student[simpson], value

### student[Tripleint](expr, $var_1$, $var_2$, $var_3$, name)

Represents the inert indefinite triple integral of expr with respect to $var_1$, $var_2$, and $var_3$. In the display of the triple integral, the domain of integration is called name.
**Output:** Three nested Int commands are returned.
**Argument options:** (expr, $var_1$=$a_1$..$b_1$, $var_2$=$a_2$..$b_2$, $var_3$=$a_3$..$b_3$) to represent the inert definite integration.
**See also:** int, Int, value, student[Doubleint]

## sum('expr', 'var'=a..b)

Computes the definite summation of expr while var ranges from a to b. Use forward quote characters around expr and var to prevent premature evaluation of these elements.

**Output:** If successful, an expression is returned.

**Argument options:** ('expr', 'var') to compute indefinite summation of expr with respect to var. ♣ ('expr', 'var'=*Rootof*(expr$_2$)) to compute definite summation of expr over roots of expr$_2$. ♣ ('expr', 'var'=#..#) to compute definite summation of expr over numerical bounds.

**Additional information:** Typically, a and b are integer values; but they can be symbolic or floating-point values as well. ♣ If sum cannot find a closed form for the summation, an unevaluated Sum command is returned. ♣ If the bounds have a small difference, sum directly adds the elements. Otherwise, sum finds a closed form and evaluates at the bounds. ♣ In many cases, using the seq command or a for loop produces quicker results.

**See also:** Sum, *seq*, product

**FL = 49−50, 68, 70** *LA = 12−13* **LI = 209**

## Sum('expr', 'var'=a..b)

Represents the inert definite summation of expr while var ranges from a to b.

**Output:** An unevaluated Sum command which can be evaluated as a sum call with value is returned.

**Argument options:** ('expr', 'var') to compute indefinite summation of expr with respect to var. ♣ ('expr', 'var'=*Rootof*(expr$_2$)) to perform definite summation of expr over roots of expr$_2$. ♣ ('expr', 'var'=#..#) to compute the definite summation of expr over numerical bounds.

**Additional information:** Unevaluated Sum calls can be combined, expanded, and simplified with the combine, expand, and simplify commands, respectively.

**See also:** sum, op(*Sum*), value, combine

*LI = 209*

## taylor(expr, var=a)

Computes a truncated Taylor series expansion of expr with respect to var about a. The series is calculated up to the order specified by the global variable Order.

**Output:** Typically, a series is returned. If no Taylor series expansion exists, you are asked to use the series command.

**Argument options:** (expr, var) to compute the series with respect to var about zero. ♣ (*Int*, var=a) to compute Taylor series expansion of an unevaluated integral. ♣ (*leadterm*(expr), var=a) to compute only the lowest degree term in Taylor series expansion. ♣ (expr, var=a, n) to compute Taylor series expansion to order n.

**Additional information:** Zero coefficient terms are not displayed. ♣ Before you can use a series as input to many commands, you must convert it to a polynomial with convert(*polynom*).

**See also:** mtaylor, series, asympt, chebyshev, Order, op(series), op(polynom), convert(*polynom*), Int

*LA = 12* **LI = 213**

### type(expr, *const*)

Determines whether expr is a differential form of constant type.
**Output:** A boolean value of either true or false is returned.
**Additional information:** expr is of type *const* if it consists of a single constant declared through difforms[defform], a sum or product strictly consisting of constant variables, or a constant variable to a power that is not a form or scalar. ♣ If *any* of the variables in expr are not variables defined by difforms[defform], then false is automatically returned. ♣ No expression can cause more than one of type(*form*), type(*scalar*), or type(*const*) to be true.
**See also:** difforms[defform], type(*form*), type(*scalar*)

### type(expr, *form*)

Determines whether expr is a differential form of type *form*.
**Output:** A boolean value of either true or false is returned.
**Argument options:** (expr, *form*, posint) to determine if expr is a form with degree equal to posint.
**Additional information:** expr is of type *form* if it consists of a single form declared through difforms[defform], a sum or product containing at least one such form, or such a form to a power. ♣ If any of the variables in expr are not variables defined by difforms[defform], then false is automatically returned. ♣ No expression can cause more than one of type(*form*), type(*scalar*), or type(*const*) to be true.
**See also:** difforms[defform], type(*const*), type(*scalar*)

### type(expr, *laurent*)

Determines whether expr is a Laurent series.
**Output:** A boolean value of either true or false is returned.
**See also:** op(series), type(*series*), type(*taylor*)

*LI = 230*

### type(expr, *scalar*)

Determines whether expr is a differential form of scalar type.
**Output:** A boolean value of either true or false is returned.
**Additional information:** expr is of type *const* if it consists of a single scalar declared through difforms[defform], a sum or product strictly consisting of one or more scalar variables and any number of constant variables, or a scalar variable to a power that is not a form. ♣ If any of the variables in expr are not variables defined by difforms[defform], then false is automatically returned. ♣ No expression can cause more than one of type(*form*), type(*scalar*), or type(*const*) to be true.
**See also:** difforms[defform], type(*form*), type(*const*)

### type(expr, *series*)

Determines whether expr is a series data type.
**Output:** A boolean value of either true or false is returned.
**Additional information:** For more information on what makes a series structure special, see the entry for op(series).
**See also:** op(series), type(*taylor*), type(*laurent*)
*LA = 72−73* **LI = 247**

### type(expr, *taylor*)

Determines whether expr is a Taylor series.
**Output:** A boolean value of either true or false is returned.
**See also:** op(series), type(*series*), type(*laurent*)
***LI = 254***

### value(expr)

Evaluates expr, an expression containing one or more unevaluated calls to Int, Diff, Product, Sum, or Limit.
**Output:** An expression is returned.
**Additional information:** The value command works by converting the inert uppercase commands to their corresponding active lowercase equivalents and evaluating. ♣ If the active commands are not solvable, then the result still contains unevaluated command calls. ♣ The inert commands that work with value are different than those that work with mod, modp1, evalgf, or evala.
**See also:** Int, Diff, Sum, Product, Limit, int, diff, sum, product, limit, *mod*, *modp1*, *evala*, *evalgf*

# Linear Algebra

## Introduction

Linear algebra is one of Maple's most well-researched disciplines. Most of the commands for linear algebra are contained in the linalg package. If you are going to be doing extensive linear algebra calculations, enter the with(linalg) command before starting. This loads pointers to the over 100 commands in linalg.

Before you can begin using the individual commands, you need an understanding of the data structures they use and create.

## Arrays

An array is a generic high-level Maple structure (of one or more dimensions) containing a well-ordered collection of individual elements. Each dimension of an array has predefined ranges. Think of one-dimensional arrays as special lists, two-dimensional arrays as lists of lists, and so on.

Multiple-dimension arrays are always *rectangular* in nature. That is, you cannot create a two-dimensional array with three elements in the first row, two in the second row, and four in the third row.

Arrays can be created with the array command. The individual elements of an array can either be specified beforehand or left as unknowns. The following is an example of a $3 \times 3$ array that contains predefined elements.

```
> array(1..3, 1..3, [[a,b,c],[d,e,f],[g,h,i]]);
```

$$\begin{bmatrix} a & b & c \\ d & e & f \\ g & h & i \end{bmatrix}$$

When you don't want all or some of the elements predefined, simply leave out all or part of the last parameter and Maple fills in the missing elements with default variables.

As you can see, Maple prints out the array in a special matrix display form. This special form is only presented when the array is one- or two-dimensional and *all* the indices begin with one. If there is any deviation from this, the array is printed out in a form more representative of its internal structure (i.e., as the bounding function(s) and a table representing the elements).

```
> array(3..6, [a,b,c,d]);
```

$$array(3..6, [\\ (3) = a \\ (4) = b \\ (5) = c \\ (6) = d \\ ])$$

## Vectors and Matrices

In the linalg package there are commands that allow you to create a *vector* or a *matrix* data structure (the vector and matrix commands, respectively). These commands are simply special cases of the array command.

A vector is equivalent to a one-dimensional array with range starting at one. Instead of having a range as the first parameter, vector uses an integer that represents the size. When doing matrix calculations, one-dimensional arrays are thought of as *column vectors*. If you want a *row vector*, use the matrix command with 1 as the first parameter.

```
> linalg[vector](3, [x1,y1,z1]);
```

$$[x1\ y1\ z1]$$

```
> array(1..3, [x1,y1,z1]);
```

$$[x1\ y1\ z1]$$

Linear Algebra    71

A matrix is equivalent to a two-dimensional array with both indices starting at one; and matrix also takes integer dimensions.

> linalg[matrix](2, 2, [1,2,3,4]);

$$\begin{bmatrix} 1 & 2 \\ 3 & 4 \end{bmatrix}$$

> array(1..2, 1..2, [[1,2], [3,4]]);

$$\begin{bmatrix} 1 & 2 \\ 3 & 4 \end{bmatrix}$$

While vector and matrix may help you avoid confusion in defining arrays, it can also be claimed that exclusively using the array command may help keep things clear. Which commands you use is really a matter of personal choice and style.

## Special Types of Arrays

Normally, when defining an array, you need to enter every element in the array for it to be defined fully. However, there are a few special types of arrays (or matrices) that can be automatically generated in Maple. Symmetric, antisymmetric, sparse, diagonal, and identity matrices are all supported. To create such matrices, define the necessary elements and add the matrix type as an additional option to the array command.

> array(1..2, 1..2, identity);

$$\begin{bmatrix} 1 & 0 \\ 0 & 1 \end{bmatrix}$$

> array(1..2, 1..6, [(1,3)=p, (2,4)=r], sparse);

$$\begin{bmatrix} 0 & 0 & p & 0 & 0 & 0 \\ 0 & 0 & 0 & r & 0 & 0 \end{bmatrix}$$

## Manipulating Elements of Arrays

You have already learned a little bit about how Maple evaluates different types of objects. There is one major difference in the evaluation rules for arrays. If you assign an array to a variable name and

## 72  Linear Algebra

then type in that name, the result displayed is not the array itself but the variable name to which it is assigned. This is called *last name evaluation*. Two commands that force full evaluation to the elements of the array are eval and op.

> M := array(1..2, 1..2, [[3,4],[6,5]]);

$$M := \begin{bmatrix} 3 & 4 \\ 6 & 5 \end{bmatrix}$$

> M;

$$M$$

> eval(M);

$$\begin{bmatrix} 3 & 4 \\ 6 & 5 \end{bmatrix}$$

> op(M);

$$\begin{bmatrix} 3 & 4 \\ 6 & 5 \end{bmatrix}$$

Last name evaluation also holds for Maple tables and procedures.

Individual array elements can be referenced by specifying the element's row and column. Note, though, that when referencing an array element, you do *not* need to worry about last name evaluation. As well, elements of an array can be redefined (after the array is created) by assigning new values to their referenced names.

> M := array(1..2, 1..3, [[2,8,32], [45,-1,0]]);

$$M := \begin{bmatrix} 2 & 8 & 32 \\ 45 & -1 & 0 \end{bmatrix}$$

> M[1,3];

$$32$$

> M[2,3] := x - 3;

$$M_{2,3} := x - 3$$

```
> op(M);
```

$$\begin{bmatrix} 2 & 8 & 32 \\ 45 & -1 & x-3 \end{bmatrix}$$

# Command Listing

### array(*1*..n, list)

Creates a one-dimensional array data structure with n elements whose values are defined in list.
**Output:** An array is returned and displayed as a vector.
**Argument options:** (1..n) to create a one-dimensional array with n undefined elements. ♣ (n..m, list) to create a one-dimensional array with m−n+1 elements defined by list. If n ≠ 1 this array is displayed in internal form. ♣ (list) to create a one-dimensional array with number of elements equal to the size of list and index starting at one. ♣ (1..n, 1..m, [list$_1$, ..., list$_n$]) to create a two-dimensional array of n rows and m columns, where list$_1$ through list$_n$, each of m elements, contain the row elements. ♣ (1..n, *option*) or (1..n, 1..m, *option*) to create a special type of array, where option can be one of *symmetric*, *antisymmetric*, *diagonal*, *identity*, or *sparse*.
**Additional information:** Arrays have a special evaluation rule called *last name evaluation*. See eval(array) for more details.
♣ Use map to apply a function or command to each element of an array. ♣ You can program your own special matrix types to supplement the array command. Just place the appropriate Maple command in the file '*index*/mycommand', where mycommand is the option name to be invoked inside array.
**See also:** eval(array), op(*eval*(array)), linalg[vector], linalg[matrix], table, *map*
**FL = 87**−**89** *LA = 18, 95*−*112* ***LI* = *9***

### convert([list$_1$, ..., list$_n$], *matrix*)

Converts a list of lists to a matrix with the same elements.
**Output:** A matrix is returned.
**Argument options:** (A, *matrix*) to convert two-dimensional array A to a matrix with the same elements. This just changes the index bounds to start with 1.

74    Linear Algebra

**Additional information:** The lists $list_1$ through $list_n$ must be of the proper number of elements to create a rectangular matrix.
**See also:** convert(*vector*), type(*matrix*)
**FL = 207−208  *LI = 43***

### convert(list, *vector*)

Converts list to a vector with the same elements.
**Output:** A vector is returned.
**Argument options:** (A, *vector*) to convert a one-dimensional array A to a vector with the same elements. This just changes the index bound to start with 1.
**Additional information:** If a list of lists is passed to convert(*vector*), convert(*matrix*) is called. ♣ If a two-dimensional array is passed to convert(*vector*), the array is "flattened" into a vector.
**See also:** convert(*matrix*), type(*vector*)
*LI = 55*

### copy(A)

Creates a copy of array A (or a vector or a matrix).
**Output:** An array is returned.
**Argument options:** (T) to copy table T. ♣ (expr) to copy another type of expression. This simply returns expr.
**Additional information:** This command is necessary because of the *last name evaluation* rule for tables and arrays. ♣ For more information, see the on-line help page.
**See also:** array, table, matrix, vector
**FL = 87  *LI = 56***

### Det(M)

Represents the inert determinant command for matrix M.
**Output:** An unevaluated Det call is returned.
**Argument options:** Det(M) *mod* posint to evaluate the inert command call modulo posint.
**Additional information:** More information on other parameter sequences for Det can be found in the listing for linalg[det].
**See also:** linalg[det], modp1(*Det*)
*LI = 63*

### Eigenvals(M)

Computes eigenvalues of M, a square matrix with all numeric elements. Eigenvals is inert and is evaluated with evalf.

**Output:** An array data structure containing the eigenvalues of M is returned.

**Argument options:** ($M_1$, $M_2$) to solve the *generalized eigenvalue problem*, that is, find the roots of linalg[det](lambda &* $M_2$ - $M_1$) for scalar variable lambda. ♣ (M, name) to assign the eigenvectors of M to the variable name. A two-dimensional array is returned where, if all the eigenvalues are real, the $i^{th}$ column represents the eigenvector of the $i^{th}$ eigenvalue of M. If the $i^{th}$ eigenvalue is complex, then the $i^{th}$ and the $(i+1)^{th}$ columns represent the real and imaginary parts of the corresponding eigenvector.

**Additional information:** Eigenvalues are computed by solving the *characteristic polynomial* defined by linalg[det](lambda &* I - M), where lambda is a scalar variable and I is the identity matrix. ♣ If M is a symmetric matrix, then a faster algorithm is used to compute the eigenvalues.

**See also:** linalg[eigenvals], linalg[eigenvects], linalg[charpoly], linalg[det], linalg[matrix], *evalf*

*LI = 71*

### entries(A)

Displays the entries in array A (or a vector or a matrix).

**Output:** An expression sequence of lists is returned.

**Argument options:** (T) to display the entries of table T.

**Additional information:** To display the indices, use the indices command. ♣ While the order in which the entries are displayed may appear arbitrary, there is a direct correspondence between the results of entries and indices.

**See also:** indices, array, table, matrix, vector

**FL = 85** *LA = 104* **LI = 110**

### eval(array)

Displays the elements of an array data structure.

**Output:** If array's indices all start with one, a two-dimensional representation of the array is displayed. If not, array is displayed in internal form, which includes indexing function, indices, and a table of values.

**Additional information:** eval is needed to access the elements because of the *last name evaluation* rule for arrays. ♣ To access the *operands* of array, use op(*eval*(array)).

**See also:** array, op(*eval*(array)), table, linalg[vector], linalg[matrix]

### eval(table)

Displays the structure of an table data structure.
**Output:** A table data structure is displayed.
**Additional information:** The eval command is needed to access the elements because of the *last name evaluation* rules for tables.
♣ To access the *operands* of table, use the op(*eval*(table)) command.
**See also:** table, op(*eval*(table)), array

### evalm(expr$_M$)

Evaluates expr$_M$, an expression involving matrices and the operators &*, +, -, and ^.
**Output:** A matrix is returned.
**Additional information:** The evalm command is an alternative to the commands linalg[multiply], linalg[add], and linalg[scalarmul].
♣ Because matrix multiplication is *non-commutative*, the &* operator must be used instead of the * operator. ♣ Use 0 to represent the matrix or scalar zero and &*() to represent the identity matrix.
♣ Keep in mind that evalm(M^(-1)) is another way of finding the inverse of M. ♣ Since any purely scalar value in expr$_M$ is seen to represent an appropriately sized matrix with all elements equal to that scalar value, matrix polynomials can be entered quite intuitively. ♣ Matrix expressions are left unevaluated unless used as a parameter to evalm, with the exception that M^0 is always simplified to scalar value 1.
**See also:** matrix, linalg[add], linalg[multiply], linalg[scalarmul], linalg[inverse]
**FL** = 94−95, 101, 210  *LI = 86*

### indices(A)

Displays the indices of array A (or a vector or a matrix).
**Output:** An expression sequence of lists is returned.
**Argument options:** (T) to display the indices of table T.
**Additional information:** To display the entries, use the entries command. ♣ While the order in which the indices are displayed may appear arbitrary, there is a direct correspondence between the results of indices and entries.
**See also:** entries, array, table, matrix, vector
**FL** = 85 *LA = 104 LI = 110*

### lattice([list$_1$, ..., list$_n$])

Computes a reduced basis for the lattice defined by lists list$_1$ through list$_n$.

**Output:** A list of lists representing the reduced basis vectors is returned.

**Argument options:** ($[V_1, ..., V_n]$) to use a list of vectors. The result is still a list of lists. ♣ ($[list_1, ..., list_n]$, *integer*), when the lattice is defined solely by integer elements, to use more efficient integer arithmetic.

**Additional information:** This command needs to be defined by readlib(lattice) before being invoked. ♣ This command only works for vectors with *rational* elements; unknowns are not allowed. ♣ For more information on the algorithm used, see the on-line help page.

**See also:** *minpoly*, linalg[basis], linalg[intbasis], linalg[sumbasis]

*LI = 300*


### linalg package

Provides commands for creating, manipulating, and calculating with *linear algebra* structures, such as vectors and matrices. As well, several types of predefined special matrices are available.

**Additional information:** To access the command fnc from the linalg package, use the long form of the name linalg[fnc], or with(linalg, fnc) to load in a pointer for fnc only, or with(linalg) to load in pointers to all the commands.


### linalg[add]($M_1$, $M_2$)

Adds matrix $M_1$ to $M_2$. $M_1$ and $M_2$ must be of equal dimensions.

**Output:** A matrix is returned.

**Argument options:** ($V_1$, $V_2$) to add two vectors of equal dimension. ♣ ($M_1$, $M_2$, $c_1$, $c_2$), where $c_1$ and $c_2$ are scalars, to compute the matrix $c_1 * M_1 + c_2 * M_2$.

**Additional information:** The evalm command can be used with the + operator to perform the same calculation.

**See also:** evalm, linalg[matrix], linalg[vector], linalg[multiply], linalg[scalarmul], linalg[exponential]

**FL = 105**


### linalg[addcol](M, $j_1$, $j_2$, expr)

Creates a new matrix by replacing column $j_2$ of matrix M with the column formed by expr $*$ column $j_1$ + column $j_2$.

**Output:** A matrix is returned.

**Additional information:** linalg[addcol] has no effect upon M.

**See also:** linalg[addrow], linalg[mulcol], linalg[col], linalg[swapcol], linalg[pivot], linalg[matrix]

### linalg[addrow](M, $i_1$, $i_2$, expr)

Creates a new matrix by replacing row $i_2$ of matrix M with the row formed by expr $*$ row $i_1$ + row $i_2$.
**Output:** A matrix is returned.
**Additional information:** linalg[addrow] has no effect upon M.
**See also:** linalg[addcol], linalg[mulrow], linalg[row], linalg[swaprow], linalg[pivot], linalg[matrix]
**FL = 102**

### linalg[adj](M)

This command is identical to linalg[adjoint].
**See also:** linalg[adjoint]

### linalg[adjoint](M)

Computes the adjoint of a square matrix M.
**Output:** A matrix of the same dimensions as M is returned.
**Additional information:** The adjoint of M is defined as the matrix $M_1$ such that M times $M_1$ is equal to the determinant of M times the identity matrix of appropriate dimensions.
**See also:** linalg[det], linalg[minor]

### linalg[angle]($V_1$, $V_2$)

Computes the angle between $V_1$ and $V_2$, two vectors with equal numbers of elements.
**Output:** An expression representing the angle in radians is returned.
**See also:** linalg[norm], linalg[dotprod], linalg[innerprod], linalg[crossprod], linalg[vector]
*LA = 19*

### linalg[augment]($M_1$, $M_2$)

Creates a new matrix by prepending matrix $M_1$ directly to the left of matrix $M_2$. $M_1$ and $M_2$ must have the same number of rows.
**Output:** A matrix is returned.
**Additional information:** linalg[augment] has no effect upon $M_1$ or $M_2$.
**See also:** linalg[stack], linalg[extend], linalg[col], linalg[delcols], linalg[matrix]
**FL = 103**

# Linear Algebra 79

## linalg[backsub](M)

Performs back substitution on matrix M, which must be of *upper-triangular* form.

**Output:** If there is a unique solution, a vector is returned containing back-substituted values for each row. The dimension of this vector equal the number of nonzero rows in M. If the solution is not unique, then a parameterized solution in terms of t1, t2, etc. is returned. If there is no solution, an error message is returned.

**Additional information:** Normally, upper-triangular matrix M is generated with linalg[gausselim] or linalg[gaussjord].

**See also:** linalg[gausselim], linalg[gaussjord], linalg[linsolve], linalg[pivot]

## linalg[band](V, n)

Creates a banded square matrix of dimension n (which must be an odd integer) with diagonals initialized by the elements of vector V.

**Output:** A matrix is returned.

**Additional information:** The diagonals are initialized as follows. The main diagonal is initialized with the middle element of V. The first subdiagonal is initialized with the element to the left of the middle element of V, the second subdiagonal is initialized with the element two to the left of the middle element of V, and so on. The first superdiagonal is initialized with the element to the right of the middle element of V, the second superdiagonal is initialized with the element two to the right of the middle element of V, and so on. The remaining diagonals are initialized to zero. ✦ Of course, V must also have an odd number of elements.

**See also:** linalg[diag]

## linalg[basis]({$V_1$, ..., $V_n$})

Computes a basis for the vector space spanned by $V_1$ through $V_n$, vectors that all have the same number of elements.

**Output:** A set of vectors is returned. The number of vectors in this basis is equal to or less than n.

**Argument options:** ([$V_1$, ..., $V_n$]) to return a *list* of vectors representing a basis for the vector space.

**Additional information:** Calling linalg[basis] with a single vector returns that same vector in a single element set.

**See also:** linalg[intbasis], linalg[sumbasis], linalg[rowspace], linalg[colspace], linalg[rowspan], linalg[colspan], linalg[kernel]

### linalg[bezout](poly₁, poly₂, var)

Creates the Bezout matrix of the polynomials poly₁ and poly₂ in the variable var.
**Output:** A matrix is returned.
**Additional information:** The result is a square matrix of dimension max(m,n), where m and n are the degrees in var of poly₁ and poly₂. ♣ Both poly₁ and poly₂ must be in expanded form for linalg[bezout] to work properly. ♣ The determinant of the result is always equal to resultant(poly₁, poly₂, var).
**See also:** linalg[sylvester], linalg[matrix], linalg[det]

### linalg[BlockDiagonal](M₁, ..., Mₙ)

This command is identical to the linalg[diag] command.
**See also:** linalg[diag], linalg[JordanBlock], linalg[companion]

### linalg[blockmatrix](m, n, M₁, M₂, ..., M_{m*n})

Creates a square matrix constructed out of the m*n square matrices $M_1$ through $M_{m*n}$.
**Output:** A square matrix is returned.
**Additional information:** The matrices must all have the same number of rows. ♣ The resulting matrix is filled in *row major order*, that is, left to right, top to bottom.
**See also:** linalg[JordanBlock] linalg[matrix]

### linalg[charmat](M, name)

Creates the characteristic matrix of square matrix M, defined by name * I − M, where I is the appropriately sized identity matrix.
**Output:** A matrix is returned.
**Argument options:** (M, expr) to compute the characteristic matrix of M, defined by expr * I − M.
**Additional information:** If a name data type is supplied, the determinant of the result equals the characteristic polynomial of M.
**See also:** linalg[charpoly], linalg[eigenvals], linalg[nullspace]
*LA = 20*

### linalg[charpoly](M, name)

Creates the characteristic polynomial of square matrix M, defined by the determinant of the result of name * I − M, where I is the appropriately sized identity matrix.
**Output:** A matrix is returned.
**Argument options:** (M, expr) to compute the characteristic polynomial of M, defined by the determinant of the result of expr * I

– M. If expr is a numeric value, then a numeric value is returned. Otherwise an expanded polynomial is returned.
**See also:** linalg[charmat], linalg[minpoly], linalg[eigenvals], linalg[nullspace]
**FL = 98**

### linalg[col](M, j)

Extracts the $j^{th}$ column of matrix M.
**Output:** A vector is returned.
**Argument options:** (M, $j_a..j_b$) to extract columns $j_a$ through $j_b$ of M. A sequence of vectors is returned.
**Additional information:** linalg[col] has no effect upon M.
**See also:** linalg[row], linalg[swapcol], linalg[delcols], linalg[augment], linalg[submatrix], linalg[matrix], linalg[vector]

### linalg[coldim](M)

Determines the number of columns in matrix M.
**Output:** An integer is returned.
**See also:** linalg[rowdim], linalg[col], linalg[vectdim], linalg[matrix]

### linalg[colspace](M)

Computes a basis for the columns of matrix M.
**Output:** A set of vectors in canonical form is returned.
**Argument options:** (M, name) to return the rank or dimension of the column space of M in name.
**See also:** linalg[range], linalg[rowspace], linalg[colspan], linalg[rank], linalg[basis], linalg[kernel], linalg[intbasis], linalg[sumbasis], linalg[matrix]

### linalg[colspan](M)

Computes a spanning set of vectors for the column space of matrix M, whose elements are multivariate polynomials over the rational numbers.
**Output:** A set of vectors of polynomials is returned.
**Argument options:** (M, name) to return the rank or dimension of the column space of M in name.
**Additional information:** linalg[colspan] uses *fraction-free Gaussian elimination* to triangularize M. ♣ Because no fractions are

used, there is no possibility of encountering a division by zero error when values are substituted into the result.
**See also:** linalg[rowspan], linalg[colspace], linalg[rank], linalg[basis], linalg[intbasis], linalg[sumbasis], linalg[matrix]

### linalg[companion](poly, var)

Creates the companion matrix associated with the univariate polynomial poly in the variable var.
**Output:** A square matrix is returned.
**Additional information:** The elements of companion matrix M for poly of degree n are defined by: M[i, n] = -coeff(poly, var, i-1), where i=1..n; M[i, i-1] = 1, where i=2..n; and all other elements being zero.
**See also:** linalg[diag], linalg[JordanBlock], linalg[matrix]

### linalg[concat]($M_1$, $M_2$)

This command is identical to linalg[augment].
**See also:** linalg[augment]

### linalg[cond](M)

Computes the standard matrix condition number of square matrix M, which is the product of the infinity norm of M and the infinity norm of the inverse of M.
**Output:** An expression is returned.
**Argument options:** (M, option) to compute other types of condition numbers of matrix M. Available options are *1*, *2*, *'infinity'*, and *'frobenius'*. They correspond to using these options in the calls to linalg[norm] when calculating the condition numbers.
**Additional information:** For more information on the options for linalg[cond] see the on-line help page for linalg[norm].
**See also:** linalg[norm], linalg[inverse], linalg[innerprod], linalg[dotprod], linalg[crossprod], linalg[angle], linalg[vector]

### linalg[copyinto]($M_1$, $M_2$, i, j)

Updates matrix $M_2$ by copying the elements of matrix $M_1$ into $M_2$ beginning at the $(i, j)^{th}$ entry of $M_2$. The corresponding elements of $M_2$ are overwritten by the new elements from $M_1$.
**Output:** A matrix is returned. Note that $M_2$ is assigned to this new matrix as well.
**Additional information:** $M_1$ and $M_2$ must be of adequate dimensions so that $M_1$ can fit entirely into $M_2$ at the desired position.

Linear Algebra 83

♣ This command can be used in conjunction with linalg[extend].
**See also:** linalg[extend], linalg[col], linalg[row], linalg[submatrix], linalg[matrix]

### linalg[crossprod]($V_1$, $V_2$)

Calculates the cross product of $V_1$ and $V_2$, each vectors of three elements.
**Output:** A vector with three elements is returned.
**Argument options:** ($list_1$, $list_2$) to return a vector representing the cross product of the three-element lists $list_1$ and $list_2$.
**Additional information:** The elements of the cross product are defined to equal $V_1[2] * V_2[3] - V_1[3] * V_2[2]$, $V_1[3] * V_2[1] - V_1[1] * V_2[3]$, and $V_1[1] * V_2[2] - V_1[2] * V_2[1]$, respectively.
**See also:** linalg[innerprod], linalg[dotprod], linalg[angle], linalg[norm], linalg[vector]

### linalg[curl]([$expr_1$, $expr_2$, $expr_3$], [$var_1$, $var_2$, $var_3$])

Computes the curl of expressions $expr_1$, $expr_2$, and $expr_3$ with respect to the variables $var_1$, $var_2$, and $var_3$.
**Output:** A vector containing three expressions is returned.
**Argument options:** ($V_1$, $V_2$), where $V_1$ is a vector of three expressions and $V_2$ is a vector of three variables, to produce the same result. ♣ ([$expr_1$, $expr_2$, $expr_3$], [$var_1$, $var_2$, $var_3$], *coords*=option), where option is one of *cartesian*, *spherical*, and *cylindrical*, to operate in one of three coordinate systems. See the help page for information on the computations performed in the different coordinate systems. *cartesian* is the default.
**Additional information:** The elements of the curl are defined by diff($expr_3$, $var_2$) − diff($expr_2$, $var_3$), diff($expr_1$, $var_3$) − diff($expr_3$, $var_1$), and diff($expr_2$, $var_1$) − diff($expr_1$, $var_2$), respectively.
**See also:** linalg[vecpotent], linalg[grad], linalg[diverge], linalg[crossprod]
**FL = 100**

### linalg[definite](M, option)

Determines if square matrix M has the property option, where option is '*positive_def*' (for positive definite), '*positive_semidef*' (for positive semidefinite), '*negative_def*', or '*negative_semidef*'.
**Output:** If all the elements of M are numerical, then a boolean value of true or false is returned. Otherwise, a conjunction of boolean expressions that must hold if the answer is to be true are given.
**Additional information:** For M to be any of the types described

above, it must be a symmetric matrix. ♣ To see what general conditions must hold for each case, call linalg[definite] with a matrix with no defined elements.
**See also:** linalg[det], linalg[matrix]

### linalg[delcols](M, $j_1..j_2$)

Creates a matrix by removing columns $j_1$ through $j_2$ of matrix M.
**Output:** A matrix is returned.
**Additional information:** linalg[delcols] has no effect upon M.
**See also:** linalg[delrows], linalg[minor], linalg[col], linalg[swapcol], linalg[augment], linalg[extend], linalg[submatrix], linalg[matrix]

### linalg[delrows](M, $i_1..i_2$)

Creates a matrix by removing rows $i_1$ through $i_2$ of matrix M.
**Output:** A matrix is returned.
**Additional information:** linalg[delrows] has no effect upon M.
**See also:** linalg[delcols], linalg[minor], linalg[row], linalg[swaprow], linalg[stack], linalg[extend], linalg[submatrix], linalg[matrix]

### linalg[det](M)

Computes the determinant of square matrix M.
**Output:** An expression is returned.
**Argument options:** (M, *sparse*), when M is a sparse matrix, to use the method of minor expansion when calculating the determinant.
**Additional information:** If the *sparse* option is not supplied, linalg[det] decides itself whether the use the method of minor expansion or Gaussian elimination or both.
**See also:** linalg[gausselim], linalg[trace], linalg[minor], linalg[permanent], linalg[adjoint], linalg[matrix]
**FL = 98** *LA = 20*

### linalg[diag]($M_1$, ..., $M_n$)

Creates a square matrix whose diagonals (starting from the top left) are square matrices $M_1$ through $M_n$. All other elements are 0.
**Output:** A matrix is returned.
**Additional information:** Scalar values substituted for any of $M_1$ through $M_n$ are treated as $1 \times 1$ matrices. ♣ The dimension of the resulting matrix equals the sum of the dimensions of all the matrices making up its diagonal. ♣ In combination with

Linear Algebra 85

linalg[JordanBlock], linalg[diag] can create matrices in Jordan form.
**See also:** linalg[BlockDiagonal], linalg[JordanBlock], linalg[band], linalg[companion]

### linalg[diverge]([expr$_1$, ..., expr$_n$], [var$_1$, ..., var$_n$])

Computes the divergence of the list of expressions expr$_1$ through expr$_n$ with respect to the list of variables var$_1$ through var$_n$.
**Output:** An expression is returned.
**Argument options:** (V$_1$, V$_2$), where V$_1$ is a vector of n expressions and V$_2$ is a vector of n variables, to compute the divergence of V$_1$. ♣ ([expr$_1$, ..., expr$_n$], [var$_1$, ..., var$_n$], *coords*=option), where option is one of *cartesian*, *spherical*, and *cylindrical*, to operate in one of three coordinate systems. See the help page for information on the computations performed in the different coordinate systems. *cartesian* is the default.
**Additional information:** Vectors and lists may be used interchangeably for either parameter of linalg[diverge]. ♣ The divergence equals the sum of diff(expr$_1$, var$_1$) through diff(expr$_n$, var$_n$).
**See also:** linalg[curl], linalg[grad]
**FL = 101**

### linalg[dotprod](V$_1$, V$_2$)

Computes the dot product of V$_1$ and V$_2$, two vectors of identical numbers of elements.
**Output:** An expression is returned.
**Argument options:** (list$_1$, list$_2$) to compute the dot product of list$_1$ and list$_2$, two lists of identical numbers of elements. ♣ (V$_1$, V$_2$, '*orthogonal*') to assume an orthogonal space and calculate the dot product by summing V$_1$[i] ∗ V$_2$[i].
**See also:** linalg[innerprod], linalg[crossprod], linalg[angle], linalg[norm], linalg[vector], *conjugate*

### linalg[eigenvals](M)

Computes the eigenvalues of square matrix M.
**Output:** An expression sequence is returned. In some cases, two or more of the eigenvalues may be expressed in RootOf notation.
**Argument options:** (M, '*implicit*') to force expression of the eigenvalues in RootOf notation. ♣ (M, '*radical*') to force expression of the eigenvalues in terms of radicals if possible. ♣ (M$_1$, M$_2$) to solve the "generalized eigenvalue problem," that is, find the roots of linalg[det](lambda &∗ M$_2$ - M$_1$) for scalar variable lambda.
**Additional information:** Eigenvalues are computed by solving the

*characteristic polynomial.* ♣ If M contains only floating-point numbers or complex numbers with floating-point elements, then a numerical method at Digits precision is used.
**See also:** Eigenvals, linalg[eigenvects], linalg[singularvals], linalg[charpoly], linalg[det], *Digits*
**FL = 98, 209** *LA = 20*

### linalg[eigenvects](M)

Computes the eigenvalues and eigenvectors of square matrix M.
**Output:** An expression sequence of lists is returned. Each list consists of an eigenvalue (or eigenvalues expressed in RootOf notation) of M, that eigenvalue's algebraic multiplicity, and a set of basis vectors for the eigenspace of the eigenvalue(s).
**Argument options:** (M, *'implicit'*) to force expression of the eigenvectors in RootOf notation. This is the default. ♣ (M, *'radical'*) to force expression of the eigenvalues in terms of radicals if possible.
**Additional information:** If M contains only floating-point numbers or complex numbers with floating-point elements, then a numerical method at Digits precision is used. ♣ For more information on linalg[eigenvects], see the on-line help page.
**See also:** linalg[eigenvals], Eigenvals, linalg[charmat]

### linalg[entermatrix](M)

Provides a basic interface to entering elements of matrix M. You are prompted to enter elements followed by semicolons.
**Output:** A matrix with the newly entered elements is returned.
**Additional information:** Normally, M is set up as a matrix with unassigned elements before linalg[entermatrix] is called. ♣ Any previously existing elements of M are overwritten with the new expressions. ♣ If a null expression (just a semicolon) is entered for a previously defined element of M, the original value is retained. Otherwise, null elements are left unassigned.
**See also:** linalg[matrix]

### linalg[equal]($M_1$, $M_2$)

Determines if matrices $M_1$ and $M_2$ are equal (i.e., have the same dimensions and equal elements at corresponding positions).
**Output:** A boolean value of either true or false is returned.
**Argument options:** ($V_1$, $V_2$) to determine if vectors $V_1$ and $V_2$ are equal.
**See also:** linalg[iszero]

### linalg[exponential](M, var)

Computes the matrix exponential of the square matrix M with respect to the parameter var.
**Output:** A square matrix of the same dimension as M is returned.
**Argument options:** (M) to compute the matrix exponential of M with respect to the first unknown variable found in M.
**Additional information:** Calculation of the matrix exponential depends on finding eigenvalues for M. A symbolic answer can be achieved only if symbolic eigenvalues can be found. To force numeric solution, include at least one floating-point element in M.
**See also:** linalg[add], linalg[multiply], linalg[eigenvals], Eigenvals
**FL = 99** *LA = 21*


### linalg[extend](M, m, n)

Creates an extended matrix by adding m rows to the right of and n columns to the bottom of matrix M.
**Output:** A matrix is returned. The new elements are unassigned.
**Argument options:** (M, m, n, expr) to initialize the new elements to the value of expr.
**Additional information:** Letting either m or n be zero allows you to add just columns or rows, respectively.
**See also:** linalg[copyinto], linalg[delrows], linalg[delcols], linalg[col], linalg[row], linalg[augment], linalg[stack], linalg[submatrix], linalg[matrix]


### linalg[ffgausselim](M)

Performs fraction-free Gaussian elimination on matrix M, which must consist of multivariate polynomials over the rationals.
**Output:** A matrix is returned. The result has the same dimension as M, is in upper-triangular form, and consists of multivariate polynomials over the rationals.
**Argument options:** (M, j), where j is a non-negative integer, to perform the elimination up to and including the $j^{th}$ column only.
♣ (M, name) to assign the rank of M to name. Because of the reduced form, the rank is equal to the number of nonzero rows in the result. ♣ (M, name$_1$, name$_2$), to assign the rank r of M to name$_1$ and the determinant of the matrix defined by linalg[submatrix](M, 1..r, 1..r) to name$_2$.
**See also:** linalg[gausselim], linalg[gaussjord], linalg[hermite], linalg[backsub], linalg[pivot]

## linalg[fibonacci](n)

Creates the $n^{th}$ Fibonacci matrix, $F(n)$.
**Output:** A square matrix is returned.
**Additional information:** The dimension of $F(n)$ is equal to the sum of the dimensions of $F(n-1)$ and $F(n-2)$. The formulas to define the entries of a Fibonacci matrix are detailed in the on-line help page, but be wary of the misleading definition for $F(n)$.
**See also:** *combinat[fibonacci]*, linalg[matrix]

## linalg[frobenius](M)

Computes the Frobenius form (rational canonical form) of square matrix M.
**Output:** A matrix with the same dimensions as M is returned.
**Argument options:** (M, name) to assign the transformation matrix to name such that linalg[frobenius](M) = evalm(name &* M).
**Additional information:** For further description of the Frobenius form, see the on-line help page.
**See also:** linalg[ratform], linalg[jordan]

## linalg[gausselim](M)

Performs Gaussian elimination on matrix M.
**Output:** A matrix is returned. The result has the same dimensions as M and is in upper-triangular form.
**Argument options:** (M, j), where j is a non-negative integer, to perform the elimination up to and including the $j^{th}$ column only. ♣ (M, name), to assign the rank of M to name. Because of the reduced form, the rank is equal to the number of nonzero rows in the result. ♣ (M, $name_1$, $name_2$), to assign the rank r of M to $name_1$ and the determinant of the matrix defined by linalg[submatrix](M, 1..r, 1..r) to $name_2$.
**Additional information:** Ordinary Gaussian elimination is used unless M contains solely floating-point numbers, in which case Gaussian elimination with partial pivoting is used at a precision specified by the global variable Digits.
**See also:** linalg[gaussjord], linalg[hermite], linalg[backsub], linalg[pivot], *Digits*

## linalg[gaussjord](M)

Performs elementary row operations to reduce matrix M to Gauss-Jordan (row reduced echelon) form.
**Output:** A matrix is returned. The result has the same dimensions

Linear Algebra    89

as M and is in upper-triangular form with all leading nonzero entries equal to one.

**Argument options:** (M, j), where j is a non-negative integer, to perform the reduction up to and including the $j^{th}$ column only. ♣ (M, name), to assign the rank of M to name. Because of the reduced form, the rank is equal to the number of nonzero rows in the result. ♣ (M, $name_1$, $name_2$) to assign the rank r of M to $name_1$ and the determinant of the matrix defined by linalg[submatrix](M, 1..r, 1..r) to $name_2$.

**Additional information:** Ordinary Gaussian elimination is used unless M contains only floating-point numbers. In that case, Gaussian elimination with partial pivoting is used at a precision specified by the global variable Digits.

**See also:** linalg[gausselim], linalg[hermite], linalg[backsub], linalg[pivot], *Digits*

**FL = 103**


### linalg[genmatrix]([$eqn_1$, ..., $eqn_m$], [$var_1$, ..., $var_n$])

Generates the coefficient matrix corresponding to linear equations $eqn_1$ through $eqn_m$ with respect to variables $var_1$ through $var_n$.

**Output:** A matrix is returned.

**Argument options:** ({$eqn_1$ ..., $eqn_m$}, {$var_1$, ..., $var_n$}) to generate the coefficient matrix from a set of equations and a set of variables. ♣ ([$eqn_1$, ..., $eqn_m$], [$var_1$, ..., $var_n$], *flag*) to include the right hand sides of $eqn_1$ through $eqn_m$ as the last column.

**Additional information:** linalg[genmatrix] can be used in conjunction with linalg[linsolve] to solve a system of linear equations.

**See also:** linalg[linsolve]


### linalg[grad](expr, [$var_1$, ..., $var_n$])

Computes the gradient of expression expr with respect to variables $var_1$ through $var_n$.

**Output:** A vector containing n expressions is returned.

**Argument options:** (expr, V), where V is a vector of n variables, to compute the gradient of expr. ♣ (expr, [$var_1$, ..., $var_n$], *coords*=option), where option is one of *cartesian*, *spherical*, and *cylindrical*, to operate in one of three coordinate systems. See the help page for information on the computations performed in the different coordinate systems. *cartesian* is the default.

**Additional information:** The elements of the gradient of expr are

equal to the n partial derivatives diff(expr, $var_1$) through diff(expr, $var_n$), respectively.
**See also:** linalg[potential], linalg[curl], linalg[diverge], linalg[dotprod], linalg[hessian], linalg[jacobian], linalg[laplacian]
**FL = 100**

### linalg[GramSchmidt]({$V_1$, ..., $V_n$})

Computes a set of orthogonal vectors for the vector space spanned by $V_1$ through $V_n$, vectors all with the same number of elements.
**Output:** A set of orthogonal vectors is returned. The number of vectors returned is equal to or less than n.
**Argument options:** ([$V_1$, ..., $V_n$]) to return a list of orthogonal vectors for the vector space spanned by the vectors.
**Additional information:** Calling linalg[GramSchmidt] with a single vector returns that same vector in a single element set. ♣ The vectors returned are *not* normalized.
**See also:** linalg[basis], linalg[intbasis], linalg[normalize]

### linalg[hadamard](M)

Computes the bound on the maxnorm of the determinant of M, a square matrix in the domain $Z$, $Z[x]$, $Q$, $Q[x]$, $R$, or $R[x]$.
**Output:** An expression is returned.
**See also:** linalg[det], linalg[norm], linalg[matrix]

### linalg[hermite](M, var)

Computes the Hermite Normal Form (reduced row echelon form) of matrix M. M must have elements that are univariate polynomials in var over the rational numbers.
**Output:** A matrix with the same dimensions as M is returned.
**Argument options:** (M, var, 'M1') to assign a transformation matrix to M1, such that linalg[hermite](M, var) = evalm(M1 &* M).
**Additional information:** The Hermite Normal Form is obtained through a series of row operations on M. ♣ Since linalg[hermite] uses the normal command extensively, it is only as effective as that command. ♣ To compute the reduced column echelon form of M use linalg[transpose] on the result of linalg[ihermite].
**See also:** linalg[smith], linalg[ihermite], linalg[gaussjord], linalg[gausselim], *Hermite*

### linalg[hessian](expr, [var$_1$, ..., var$_n$])

Computes the Hessian matrix of expression expr with respect to variables var$_1$ through var$_n$.
**Output:** A square matrix of dimension n is returned.
**Argument options:** (expr, V) to create the Hessian matrix of expr with respect to the elements in the vector V.
**Additional information:** The element $[i, j]$ of a Hessian matrix is calculated by the formula diff(expr, var$_i$, var$_j$). Because of this, the resulting matrix is always symmetric about the main diagonal.
**See also:** linalg[grad], linalg[jacobian], linalg[matrix], *diff*


### linalg[hilbert](n)

Creates the n × n generalized Hilbert matrix.
**Output:** A square matrix is returned.
**Argument options:** (n, expr) to create the Hilbert matrix where expr is subtracted from the denominator of each element.
**Additional information:** The formula to determine the $[i, j]^{th}$ entry of a Hilbert matrix is $1/(i+j-\text{expr})$. ✦ If expr is not explicitly supplied, it is set to 1.
**See also:** linalg[matrix]
*LA = 21*


### linalg[htranspose](M)

Computes the Hermitian transpose of matrix M.
**Output:** A matrix is returned. If M is an $n \times m$ matrix, then the result is an $m \times n$ matrix.
**Argument options:** (V) to find the transpose of vector V. V is treated as a column vector, so the result is a row vector.
**Additional information:** The $(i, j)^{th}$ element of a Hermitian transpose is defined as the conjugate of the $(j, i)^{th}$ element of M. ✦ If M has special indexing, then the Hermitian transpose of M also has that indexing.
**See also:** linalg[transpose], linalg[hermite], linalg[matrix], linalg[vector]


### linalg[ihermite](M, var)

Computes the Hermite Normal Form (reduced row echelon form) of matrix M. M must be a square matrix with all integer elements.
**Output:** A diagonal matrix is returned. The result is a square matrix of the same dimension as M.
**Argument options:** (M, var, 'M1') to assign the transformation

matrix to M1, such that linalg[ihermite](M, var) = evalm(M1 &* M).

**Additional information:** The integer Hermite Normal Form is obtained through performing a series of row and column operations on M. ♣ To compute the reduced *column* echelon form of M use linalg[transpose] on the result of linalg[ihermite].

**See also:** linalg[hermite], linalg[ismith]

### linalg[indexfunc](A)

Displays the special indexing function (if any) of array A.

**Output:** A name representing the indexing function of A is returned. If A has no special indexing function, then NULL is returned.

**See also:** linalg[vector], linalg[matrix]

### linalg[innerprod]($V_1$, M, $V_2$)

Calculates the inner product of vectors $V_1$ and $V_2$ and matrix M.

**Output:** An expression is returned.

**Argument options:** ($V_1$, $M_1$, $M_2$, ..., $M_n$, $V_2$) to compute the inner product of vectors $V_1$ and $V_2$ and matrices $M_1$ through $M_n$.

**Additional information:** Of course, all parameters must have dimensions consistent with what is needed for matrix/vector multiplication.

**See also:** linalg[dotprod], linalg[crossprod], linalg[angle], linalg[norm], linalg[vector]

### linalg[intbasis]({$Vs_1$}, {$Vs_2$}, ..., {$Vs_n$})

Computes a basis for the intersection of the vector spaces spanned by the sets of vectors {$Vs_1$} through {$Vs_n$}. Every vector must have an equal number of elements, but the individual sets can have different numbers of vectors in them.

**Output:** A set of vectors is returned. The number of vectors in this basis is equal to or less than the number of vectors in the smallest of the {$Vs_i$}s.

**Argument options:** ([$Vs_1$], [$Vs_2$], ..., [$Vs_n$]) to return a list of vectors representing a basis of the intersection of the vector spaces spanned by the lists of vectors [$Vs_1$] through [$Vs_n$].

**Additional information:** Calling linalg[basis] with a single set or list of vectors is equivalent to calling linalg[basis]. ♣ A basis for the zero-dimensional space, when the intersection is null, is the empty set, {}, or the empty list, [].

**See also:** linalg[basis], linalg[sumbasis], linalg[rowspace], linalg[colspace], linalg[rowspan], linalg[colspan]

Linear Algebra 93

### linalg[inverse](M)

Computes the inverse of square matrix M.
**Output:** A square matrix is returned.
**Additional information:** If M is a singular matrix, then an error occurs. ♣ For square matrices of dimension four or less (or with sparse indexing), Cramer's rule is used to calculate the inverse. Otherwise, Gauss-Jordan reduction is applied. ♣ Keep in mind that evalm(M^(-1)) is another way of finding the inverse of M.
**See also:** linalg[gaussjord], linalg[matrix], evalm
*LA = 20−21*

### linalg[ismith](M, var)

Computes the integer Smith Normal Form of matrix M. M must be a square matrix with all integer elements.
**Output:** A diagonal matrix is returned. The result is a square matrix of the same dimensions as M.
**Argument options:** (M, var, $name_1$, $name_2$) to assign the transformation matrices to $name_1$ and $name_2$, such that linalg[ismith] (M, var) = evalm($name_1$ &* M &* $name_2$).
**Additional information:** The integer Smith Normal Form is obtained through a series of row and column operations on M.
**See also:** linalg[smith], linalg[ihermite]

### linalg[iszero](M)

Determines if matrix M is composed *solely* of zero entries.
**Output:** A boolean value of either true or false is returned.
**Argument options:** (V) to determine if vector V is composed *solely* of zero entries.
**See also:** linalg[equal]

### linalg[jacobian]([$expr_1$, ..., $expr_m$], [$var_1$, ..., $var_n$])

Computes the Jacobian matrix of expressions $expr_1$ through $expr_m$ with respect to variables $var_1$ through $var_n$.
**Output:** A m × n matrix is returned.
**Argument options:** ($V_1$, $V_2$), where $V_1$ is a vector of expressions and $V_2$ is a vector of variables, to produce the same result.
**Additional information:** The $[i, j]^{th}$ element of a Jacobian matrix is calculated by the formula diff($expr_i$, $var_j$).
**See also:** linalg[grad], linalg[hessian], *diff*
**FL = 101** *LA = 25*

94    Linear Algebra

### linalg[jordan](M)

Computes the Jordan form of square matrix M.
**Output:** A square matrix is returned.
**Argument options:** (M, name) to assign the transformation matrix to name such that linalg[jordan](M) = evalm(name &* M).
**Additional information:** The Jordan form has along its diagonal Jordan block matrices whose diagonal entries are eigenvalues of M. All other elements are zero. ♣ If the eigenvalues cannot be solved exactly, then floating-point approximations are substituted.
**See also:** linalg[eigenvals], Eigenvals, linalg[frobenius], linalg[JordanBlock]

### linalg[JordanBlock](expr, n)

Creates a square matrix of dimension n in Jordan block form.
**Output:** A square matrix is returned.
**Additional information:** The elements of the main diagonal of a matrix in Jordan block form are equal to the expression expr, the elements on the first super-diagonal are equal to one, and all other elements are equal to zero.
**See also:** linalg[diag], linalg[companion], linalg[jordan], linalg[blockmatrix], linalg[matrix]

### linalg[kernel](M)

Computes a basis for the kernel (null space) of the linear transformation defined by M.
**Output:** A set of vectors is returned. This set may be empty, in which case the null space is simply the origin.
**Argument options:** (M, name) to assign a non-negative integer representing the dimension of the kernel the variable to name. This value should always equal the number of vectors in the basis set.
**Additional information:** If vector V is in the null vector space of M, then linalg[multiply](M, V) = [0, 0, 0].
**See also:** linalg[rowspace], linalg[colspace], linalg[basis]

### linalg[laplacian](expr, [$var_1$, ..., $var_n$])

Computes the laplacian of expression expr with respect to variables $var_1$ through $var_n$.
**Output:** An expression is returned.
**Argument options:** (expr, V), where V is a vector of n variables, to produce the same result. ♣(expr, [$var_1$, ..., $var_n$], *coords*=option), where option is one of *cartesian*, *spherical*, and *cylindrical*, to operate in one of three coordinate systems. See the help page for

Linear Algebra    95

information on the computations performed in the different coordinate systems. *cartesian* is the default.
**Additional information:** The laplacian of expr is equal to the sum of second derivatives diff(expr, var$_1$ \$2) through diff(expr, var$_n$ \$2).
**See also:** linalg[grad]
**FL = 101**


### linalg[leastsqrs](M, V)

Computes a vector (call it $V_x$) that satisfies the equation A &* $V_x$ = V optimally in the *least-squares* sense.
**Output:** A vector is returned.
**Argument options:** ({eqn$_1$, ..., eqn$_m$}, {var$_1$, ..., var$_n$}) to compute the values of var$_1$ through var$_n$ that optimally satisfy equations eqn$_1$ through eqn$_m$ in the least-squares sense. The result is a set of equations for var$_1$ through var$_n$.
**Additional information:** You can also think of $V_x$ as the vector that minimizes the *2-norm* of M &* $V_x$ - V.
**See also:** linalg[norm], linalg[linsolve]


### linalg[linsolve](M, V)

Computes a vector (call it $V_x$) satisfying the equation A &* $V_x$ = V. The number of elements in V must equal the number of rows in M.
**Output:** A vector with a number of elements equal to linalg[coldim(M)] is returned. If no solution can be found, NULL is returned.
**Argument options:** ($M_1$, $M_2$) to compute a matrix (call it $M_x$), such that each column of $M_x$ satisfies the equation evalm($M_1$ &* linalg[col]($M_x$)) = linalg[col]($M_2$). The dimensions of $M_1$, $M_2$, and $M_x$ are all identical. ♣  (M, V, name, var) to assign the rank of M to name. If any global names need to be created to express a solution, they are named var[1], var[2], etc. If this fourth parameter is not supplied, the default names are _t[1], _t[2], etc.
**See also:** linalg[backsub], linalg[leastsqrs], linalg[vectdim], linalg[coldim], linalg[rowdim], linalg[col]
**FL = 98**


### linalg[matrix](m, n, list)

Creates a two-dimensional array with m rows and n columns, with indices starting at one and elements initialized to those in list.
**Output:** An array is returned and displayed as a matrix.

**Argument options:** (m, n) to create a matrix of m rows and n columns with undefined elements. ♣ ([list$_1$ ..., list$_m$]) to create a matrix of m rows and a number of columns equal to the number of elements in any of the lists. ♣ (m, n, option) to create a special type of matrix, where option can be *symmetric*, *antisymmetric*, *diagonal*, *identity*, or *sparse*. ♣ (m, n, fnc) to create a matrix, where the elements are defined by the function fnc acting on the points [i, j], where i=1..m and j=1..n.

**Additional information:** Arrays have a special evaluation rule called *last name evaluation*. See eval(array) for more details. ♣ The evalm command is used to perform many operations on matrices. ♣ Use the map command to apply a function or command to each element of a matrix.

**See also:** evalm, array, eval(array), op(eval(array)), type(matrix), linalg[vector], *map*

**FL = 102** *LA = 19–21*


### linalg[minor](M, i, j)

Creates the $(i, j)^{th}$ minor of matrix M by removing the $i^{th}$ row and $j^{th}$ column of M.

**Output:** A matrix is returned. The result has one less row and one less column than M.

**Additional information:** Minors are used in many linear algebra calculations including those for determinants and adjoints.

**See also:** linalg[delrows], linalg[delcols], linalg[submatrix], linalg[det], linalg[permanent], linalg[adjoint], linalg[matrix]


### linalg[minpoly](M, var)

Computes the minimum polynomial of matrix M with respect to variable var.

**Output:** A polynomial in var is returned.

**Additional information:** The minimum polynomial of M is the polynomial of smallest degree that "annihilates" M. A polynomial annihilates a matrix if a zero matrix is returned when the matrix is substituted in for the polynomial's variable. ♣ The minimum polynomial divides the characteristic polynomial exactly.

**See also:** linalg[charpoly], *minpoly*


### linalg[mulcol](M, j, expr)

Creates a new matrix by multiplying the $j^{th}$ column of matrix M by expr.

**Output:** A matrix is returned.
**Additional information:** linalg[mulcol] has no effect upon M.
**See also:** linalg[mulrow], linalg[addcol], linalg[col], linalg[swapcol], linalg[pivot], linalg[matrix]

### linalg[mulrow](M, i, expr)

Creates a new matrix by multiplying the $i^{th}$ row of matrix M by expr.
**Output:** A matrix is returned.
**Additional information:** linalg[mulrow] has no effect upon M.
**See also:** linalg[mulcol], linalg[addrow], linalg[row], linalg[swaprow], linalg[pivot], linalg[matrix]
**FL = 102**

### linalg[multiply]($M_1$, $M_2$)

Multiplies matrices $M_1$ and $M_2$ together. $M_1$ and $M_2$ must have dimensions appropriate for matrix multiplication.
**Output:** A matrix is returned.
**Argument options:** (M, V) to compute the matrix/vector product of matrix M and vector V. A vector is returned. ♣ ($M_1$, ..., $M_n$) to compute the product of n matrices.
**Additional information:** The evalm command and the &* operator can be used to perform the same calculation as linalg[multiply].
**See also:** evalm, linalg[matrix], linalg[vector], linalg[add], linalg[scalarmul], linalg[exponential]

### linalg[norm](M)

Computes the infinity norm of matrix M, which is defined as the maximum of the sums of the magnitudes of the elements in each row of M.
**Output:** An expression is returned.
**Argument options:** (M, option) to compute other matrix norms. Available options are *1*, *2*, *'infinity'*, and *'frobenius'*. ♣ (V) to compute the infinity norm of vector V, which is defined as the maximum magnitude of all the elements of V. ♣ (V, option) to compute other norms of vector V. Available options are *'infinity'*, *'frobenius'*, and any positive integer.
**Additional information:** For more information on the options for linalg[norm] see the on-line help page.
**See also:** Norm, linalg[hadamard], linalg[cond], linalg[innerprod], linalg[dotprod], linalg[angle], linalg[vector]

## linalg[normalize](V)

Normalizes vector V with the 2-norm of V.
**Output:** A vector representing the normalized vector is returned.
**Argument options:** (list) to normalize list with its 2-norm.
**Additional information:** Each element of V is divided by the 2-norm of V. A normalized vector always has a 2-norm of one.
**See also:** Norm, linalg[norm], linalg[vector]

## linalg[nullspace](M)

This command is identical to linalg[kernel].
**See also:** linalg[kernel]
**FL = 209**

## linalg[orthog](M)

Determines whether the matrix M is orthogonal or not.
**Output:** A boolean value of true or false is returned.
**Additional information:** M is orthogonal if the inner product of any column of M with itself is equal to 1 and the inner product of any two columns of M is equal to 0. ♣ The command testeq is used to test for equality to 0 or 1.
**See also:** linalg[innerprod], *testeq*

## linalg[permanent](M)

Computes the permanent of square matrix M.
**Output:** An expression is returned.
**Additional information:** The permanent of M is computed using minor expansion, exactly like computing the determinant of M except that while expanding along a row or column of M, alternating signs are not used. For example, the permanent of $2 \times 2$ matrix linalg[matrix](2, 2, [a,b,c,d]), is $ad + bc$.
**See also:** linalg[det], linalg[minor], linalg[matrix]

## linalg[pivot](M, i, j)

Pivots the matrix M about the element M[i, j].
**Output:** A matrix is returned.
**Argument options:** (M, i, j, $i_a..i_b$) to perform the pivot transformation only on rows $i_a$ through $i_b$ of M.
**Additional information:** This operation adds suitable multiples of the $i^{th}$ row of M to every other row in M, such that the $j^{th}$ entry in each of these rows is zero. ♣ linalg[pivot] has no effect upon M.
**See also:** linalg[gaussjord], linalg[gausselim], linalg[addrow], linalg[addcol], linalg[mulrow], linalg[mulcol], linalg[matrix]

## linalg[potential]([expr$_1$, ..., expr$_n$], [var$_1$, ..., var$_m$], name)

Determines if the vector field defined by expr$_1$ through expr$_n$ has a *scalar potential* on variables var$_1$ through var$_m$ (where m $\geq$ n).
**Output:** A boolean value of either true or false is returned. If true, then a scalar expression such that the gradient of name is equal to [expr$_1$, ..., expr$_n$] is assigned to the third parameter, name.
**Additional information:** Basically, linalg[potential] computes the inverse of linalg[grad], if such an inverse exists. ♣ Besides a list, a vector may be used for the first parameter.
**See also:** linalg[grad], linalg[vecpotent], linalg[vector]

## linalg[randmatrix](m, n)

Creates a m × n matrix with random entries.
**Output:** A matrix is returned.
**Argument options:** (m, n, option) to create a special type of random matrix, where option can be *symmetric*, *antisymmetric*, *diagonal*, *identity*, or *sparse*. ♣ (m, n, *entries*=fnc) to create the m × n matrix with random entries defined by repeated calls to the command fnc, which usually contains some type of random generator itself.
**Additional information:** By default, the random generator used is rand(-99..99), which creates random integers between -99 and 99. ♣ The randmatrix command is particularly useful for creating example matrices for testing and debugging purposes.
**See also:** linalg[randvector], linalg[matrix], rand
**FL = 105**

## linalg[randvector](n)

Creates a vector with n random entries.
**Output:** A vector is returned.
**Argument options:** (m, n, *entries*=fnc) to create a vector with n random entries defined by repeated calls to the command fnc, which usually contain some type of random generator itself.
**Additional information:** By default, the random generator used is rand(-99..99), which creates random integers between -99 and 99. ♣ The randvector command is particularly useful for creating example vectors for testing and debugging purposes.
**See also:** linalg[randmatrix], linalg[vector], rand

## linalg[range](M)

This command is identical to the linalg[colspace] command.
**See also:** linalg[colspace], linalg[colspan], linalg[matrix]

### linalg[rank](M)

Computes the rank of matrix M.
**Output:** A non-negative integer is returned.
**Additional information:** The rank of M is equal to the number of nonzero rows in M after Gaussian elimination is performed on it.
**See also:** linalg[rowspace], linalg[rowspan], linalg[colspace], linalg[colspan], linalg[matrix]

### linalg[ratform](M)

This command is identical to linalg[frobenius].
**See also:** linalg[frobenius]

### linalg[row](M, i)

Extracts the $i^{th}$ row of matrix M.
**Output:** A vector is returned.
**Argument options:** (M, $i_1..i_2$) to extract rows $i_1$ through $i_2$ of M. A sequence of vectors is returned.
**Additional information:** linalg[row] has no effect upon M.
**See also:** linalg[col], linalg[swaprow], linalg[delrows], linalg[stack], linalg[submatrix], linalg[matrix], linalg[vector]
**FL = 102**

### linalg[rowdim](M)

Determines the number of rows in matrix M.
**Output:** An integer is returned.
**See also:** linalg[coldim], linalg[row], linalg[vectdim], linalg[matrix]

### linalg[rowspace](M)

Computes a basis for the rows of matrix M.
**Output:** A set of vectors in canonical form is returned.
**Argument options:** (M, name) to return the rank or dimension of the row space of M in name.
**See also:** linalg[colspace], linalg[rowspan], linalg[rank], linalg[basis], linalg[kernel], linalg[intbasis], linalg[sumbasis], linalg[matrix]

### linalg[rowspan](M)

Computes a spanning set of vectors for the row space of matrix M, whose elements are multivariate polynomials over the rational numbers.

**Output:** A set of vectors of polynomials is returned.
**Argument options:** (M, name) to return the rank of M in name.
**Additional information:** linalg[rowspan] uses *fraction-free Gaussian elimination* to triangularize M. ♣ Because no fractions are used, there is no possibility of encountering a division by zero error when values are substituted into the result.
**See also:** linalg[colspan], linalg[rowspace], linalg[rank], linalg[basis], linalg[intbasis], linalg[sumbasis], linalg[matrix]

### linalg[rref](M)

This command is identical to linalg[gaussjord]. rref is an acronym for *row reduced echelon form*.
**See also:** linalg[gaussjord]
**FL = 105**

### linalg[scalarmul](M, expr)

Multiplies matrix M by the scalar expression expr.
**Output:** A matrix is returned.
**Additional information:** Every element of M is multiplied by expr.
**See also:** evalm, linalg[multiply], linalg[add], linalg[addrow], linalg[addcol]

### linalg[singularvals](M)

Computes the singular values of square matrix M.
**Output:** A list of values is returned.
**Additional information:** The singular values equal the square roots of the eigenvalues of the matrix obtained from the transpose of M times M itself. ♣ Warning: linalg[singularvals] works for matrices with elements containing unknown quantities, but the calculations can take an enormous amount of time and memory.
**See also:** linalg[eigenvals], linalg[transpose], linalg[matrix], Svd

### linalg[smith](M, var)

Computes the Smith Normal Form of matrix M. M must be a square matrix with elements that are univariate polynomials in var.
**Output:** A diagonal matrix is returned. The result is a square matrix of the same dimension as M.
**Argument options:** (M, var, name$_1$, name$_2$) to assign transformation matrices to name$_1$ and name$_2$, such that linalg[smith](M, var) = evalm(name$_1$ &* M &* name$_2$).
**Additional information:** The Smith Normal Form is obtained through a series of row and column operations on M. ♣ Since

linalg[smith] uses normal extensively, it is only as effective as that command.
**See also:** linalg[ismith], *Smith*, linalg[hermite]
*LA = 21* **LI = 105**

### linalg[stack]($M_1$, $M_2$)

Creates a new matrix by stacking matrix $M_1$ directly on top of matrix $M_2$. $M_1$ and $M_2$ must have the same number of columns.
**Output:** A matrix is returned.
**Additional information:** linalg[stack] has no effect upon $M_1$ or $M_2$.
**See also:** linalg[augment], linalg[extend], linalg[row], linalg[delrows], linalg[matrix]
**FL = 102**

### linalg[submatrix](M, $i_1..i_2$, $j_1..j_2$)

Extracts the submatrix of matrix M defined by the elements contained in both rows $i_1$ through $i_2$ and columns $j_1$ through $j_2$.
**Output:** A matrix is returned.
**Argument options:** (M, list$_r$, list$_c$) to extract the submatrix of M, where element $[i, j]$ is set to M[list$_r$[i], list$_c$[j]] for each element i in list$_r$ and each element j in list$_c$. The resulting matrix has a number of rows and columns equal to the number of elements in list$_r$ and list$_c$, respectively.
**Additional information:** The ranges $i_1..i_2$ and $j_1..j_2$ must fall completely within the bounds of M and, therefore, the resulting matrix is always of smaller or equal dimensions to M. ♣ The lists listrows and listcols may have multiple occurrences of rows or columns within the bounds of M and, therefore, it is possible to create a ''submatrix'' of M of greater dimensions than M.
**See also:** linalg[subvector], linalg[row], linalg[col], linalg[delrows], linalg[delcols], linalg[minor], linalg[copyinto], linalg[matrix], linalg[vector]
**FL = 102**

### linalg[subvector](M, i, $j_1..j_2$)

Extracts the subvector of matrix M defined by the elements in row i between columns $j_1$ and $j_2$ inclusively.
**Output:** A vector is returned.
**Argument options:** (M, $i_1..i_2$, j) to extract the subvector defined by elements in column j between rows $i_1$ and $i_2$ inclusively.

**Additional information:** Both calling sequences return a vector that is interpreted as a *row* vector.
**See also:** linalg[submatrix], linalg[row], linalg[col], linalg[matrix], linalg[vector]

## linalg[sumbasis]($\{Vs_1\}$, ..., $\{Vs_n\}$)

Computes a basis for the sum of the vector spaces spanned by the sets of vectors $\{Vs_1\}$ through $\{Vs_n\}$. Every vector must have an equal number of elements, but the individual sets can have differing numbers of vectors in them.
**Output:** A set of vectors is returned. The number of vectors is equal to or less than the number of vectors in the largest of the $Vs_i$s.
**Argument options:** ($[Vs_1]$, ..., $[Vs_n]$) to return a list of vectors representing a basis of the sum of the vector spaces spanned by the lists of vectors $[Vs_1]$ through $[Vs_n]$.
**Additional information:** Calling linalg[sumbasis] with a single set or list of vectors is equivalent to calling linalg[basis]. ♣ A basis for the zero-dimensional space, when the intersection is null, is the empty set, {}, or the empty list, [].
**See also:** linalg[basis], linalg[intbasis], linalg[rowspace], linalg[colspace], linalg[rowspan], linalg[colspan]

## linalg[swapcol](M, $j_1$, $j_2$)

Exchanges column $j_1$ and column $j_2$ of matrix M.
**Output:** A new matrix is returned.
**Additional information:** linalg[swapcol] has no effect upon M.
**See also:** linalg[swaprow], linalg[col], linalg[delcols], linalg[augment], linalg[matrix]

## linalg[swaprow](M, $i_1$, $i_2$)

Exchanges row $i_1$ and row $i_2$ of matrix M.
**Output:** A new matrix is returned.
**Additional information:** linalg[swaprow] has no effect upon M.
**See also:** linalg[swapcol], linalg[row], linalg[delrows], linalg[stack], linalg[matrix]
**FL = 102**

## linalg[sylvester]($poly_1$, $poly_2$, var)

Creates the Sylvester matrix of polynomials $poly_1$ and $poly_2$ in variable var.

**Output:** A matrix is returned. The result is a square matrix of dimension $m+n$, where $m$ and $n$ are the degrees in var of poly$_1$ and poly$_2$.
**Additional information:** poly$_1$ and poly$_2$ must be in expanded form. ♣ The determinant of the resulting matrix always equals resultant(poly$_1$, poly$_2$, var).
**See also:** linalg[bezout], linalg[matrix], linalg[det]

### linalg[toeplitz]([expr$_1$, ..., expr$_n$])

Creates the symmetric Toeplitz matrix formed by the elements expr$_1$ through expr$_n$.
**Output:** A square matrix of dimension n is returned.
**Additional information:** A Toeplitz matrix has expr$_1$ all along the diagonal and subsequent expressions along progressive sub- and super-diagonals.
**See also:** linalg[matrix]

### linalg[trace](M)

Computes the trace of square matrix M.
**Output:** An expression is returned.
**Additional information:** The trace of M is defined to be the sum of the elements on the main diagonal of M.
**See also:** linalg[gausselim], linalg[det], linalg[matrix]

### linalg[transpose](M)

Computes the transpose of matrix M.
**Output:** A matrix is returned. If M was an $n \times m$ matrix, then the result is an $m \times n$ matrix.
**Argument options:** (V) to find the transpose of vector V. V is treated as a column vector, so the result is a row vector.
**Additional information:** If M has special indexing, then the transpose of M also has that indexing.
**See also:** linalg[htranspose], linalg[matrix], linalg[vector]

### linalg[vandermonde]([expr$_1$, ..., expr$_n$])

Creates the Vandermonde matrix formed by the expressions expr$_1$ through expr$_n$.
**Output:** A square matrix of dimension n is returned.
**Additional information:** The $[i,j]^{th}$ element of a Vandermonde matrix is calculated by the formula expr$_i^{j-1}$.
**See also:** linalg[matrix]
*LA = 21*

### linalg[vecpotent]([expr$_1$, expr$_2$, expr$_3$], [var$_1$, var$_2$, var$_3$], name)

Determines if the vector field defined by expressions expr$_1$, expr$_2$, and expr$_3$ has a vector potential on variables var$_1$, var$_2$, and var$_3$.
**Output:** A boolean value of either true or false is returned. If true, a vector with three elements representing the curl of [expr$_1$, expr$_2$, expr$_3$] is assigned to the third parameter, name.
**Additional information:** Basically, linalg[vecpotent] computes the inverse of linalg[curl], if such an inverse exists. ♣ A vector potential exists if and only if the divergence of [expr$_1$, expr$_2$, expr$_3$] equals 0. ♣ Besides a list, a vector may be used for the first parameter.
**See also:** linalg[curl], linalg[potential], linalg[vector]

### linalg[vectdim](V)

Determines the dimension of vector V.
**Output:** An integer is returned.
**Argument options:** (list) to return the number of elements in list.
**See also:** linalg[coldim], linalg[rowdim], linalg[vector]

### linalg[vector](m, list)

Creates a one-dimensional array with m elements and index starting at 1.
**Output:** An array is returned and displayed as a *row* vector.
**Argument options:** (m) to create a vector of m elements with no predefined elements. ♣ (list) to create a vector with a number of elements equal to the number of elements in list. ♣ (m, fnc) to create a vector with m elements, where the elements are defined by the function fnc acting on 1 through m.
**Additional information:** Arrays have a special evaluation rule called *last name evaluation*. See eval(array) for more details. ♣ Use map to apply a function or command to each element of a vector.
**See also:** array, eval(array), op(*eval*(array)), type(vector), linalg[matrix], *map*
**FL = 103** *LA = 19*

### linalg[Wronskian](V, var)

Creates the Wronskian matrix formed by the elements of vector V with respect to var.
**Output:** A matrix is returned. The result is a square matrix of dimension $n$, where $n$ is the number of elements in V.

**Argument options:** (list, var) to create the Wronskian matrix of the elements of list.
**Additional information:** The $[i, j]^{th}$ element of a Wronskian matrix is calculated by the formula diff(f[j], v$(j-1)). ✿ The determinant of this matrix is called the *Wronskian determinant*.
**See also:** linalg[matrix]

### Nullspace(M)

Represents the inert linalg[nullspace] command for matrix M.
**Output:** An unevaluated Nullspace call is returned.
**Argument options:** Nullspace(M) *mod* posint to evaluate the inert command call modulo posint.
**Additional information:** For more information on valid parameter sequences, see the command listing for linalg[kernel].
**See also:** linalg[kernal]
*LI = 145*

### op(*eval*(array))

Displays the internal structure of array and allows you access to its operands.
**Output:** An expression sequence is returned.
**Argument options:** (*1*, *eval*(array)) to extract the indexing function of array. ✿ (*2*, *eval*(array)) to extract an expression sequence of the specified ranges or indices of array. ✿ (*3*, *eval*(array)) to extract a table data structure representing the values of the individual elements of array.
**Additional information:** The eval command is needed to access elements because of the *last name evaluation* rules for arrays.
**See also:** array, eval(array), linalg[vector], linalg[matrix]

### op(*eval*(table))

Displays the internal structure of table and allows you access to its operands.
**Output:** An expression sequence is returned.
**Argument options:** (*1*, *eval*(table)) to extract the indexing function of table. ✿ (*2*, *eval*(array)) to extract a list of equations whose lefthand and righthand sides represent the indices and entries of table, respectively.
**Additional information:** The eval command is needed to access elements because of the *last name evaluation* rules for tables.
**See also:** table, eval(table)
*LA = 103−104*

### Svd(M)

Computes singular values of M, a matrix with all numeric elements.
**Output:** An unevaluated Svd command is returned.
**Argument options:** (M, name, option), where option is *left* or *right*, to assign the *left* or *right* singular vectors of M to name.
* (M, name$_1$, name$_2$) to assign the *left* and *right* singular vectors of M to name$_1$ and name$_2$, respectively.
**Additional information:** In order to evaluate an unevaluated Svd command, apply evalf to it.
**See also:** *evalf*, linalg[singularvals]
*LI = 210*

### table([(expri$_1$) = expre$_1$, ..., (expri$_n$) = expre$_n$])

Creates a table with initial indices expri$_1$ through expri$_n$ and corresponding entries expre$_1$ through expre$_n$.
**Output:** A table is returned and displayed.
**Argument options:** ([expr$_1$, ..., expr$_n$]) to create a table whose entries corresponding to indices 1 through $n$ are expr$_1$ through expr$_n$.
* (option, [(expri$_1$) = expre$_1$, ..., (expri$_n$) = expre$_n$]), where option can be one of the indexing functions *symmetric*, *antisymmetric*, *diagonal*, *identity*, or *sparse* to create a specialized table.
**Additional information:** Tables have a special evaluation rule called *last name evaluation*. See eval(table) for more details.
* Unlike arrays, the indices for tables can be *any* expression.
* A new entry with index expr$_1$ and entry expr$_n$ can be added to table T by entering assignment T[expr$_1$] := expr$_2$.
* Use map to apply a function or command to each entry of a table. The indices are left alone.
**See also:** eval(table), op(*eval*(table)), array, *indices*, *entries*, *map*
**FL = 83–87, 89–90** *LA = 18–19, 95–112* ***LI = 212***

### type(expr, *array*)

Determines whether expr is an array data type.
**Output:** A boolean value of either true or false is returned.
**Argument options:** (expr, 'array'(type)) to determine if expr is an array with all elements in domain type.
**Additional information:** If any entry of expr is undefined, then its type is not checked. That is, it is assumed to be of *all* possible

types. ♣ The forward quote characters around *array* are necessary to stop the invocation of the array command.
**See also:** array, type(vector), type(matrix), type(table), op(eval(array))
*LA = 76 LI = 224*


### type(expr, *indexed*)

Determines whether expr is an indexed value.
**Output:** A boolean value of either true or false is returned.
**Additional information:** An indexed type is a subscripted table or array entry whose value is *not* assigned (i.e., is equal to its own name).
**See also:** table, array
**FL =** *LA = LI =*


### type(expr, *matrix*)

Determines whether expr is a matrix data type (that is, a two-dimensional array data type with both indices starting at 1).
**Output:** A boolean value of either true or false is returned.
**Argument options:** (expr, *'matrix'*(type)) to determine if expr is a matrix with all elements in domain type. ♣ (expr, *'matrix'*(type, *square*)) to determine if expr is a *square* matrix with all elements in domain type.
**Additional information:** The forward quote characters around *matrix* are necessary to stop the invocation of the linalg[matrix] command.
**See also:** evalm, array, linalg[matrix], type(array), type(vector), type(table), op(eval(array))
*LI = 233*


### type(expr, *scalar*)

Determines whether expr is a scalar in the matrix sense.
**Output:** A boolean value of either true or false is returned.
**Additional information:** A scalar is basically any expression that is not an array or table.
**See also:** array, type(array), type(vector), type(matrix), type(table), linalg[scalarmul]

**type(expr, *table*)**

Determines whether expr is a table data type.
**Output:** A boolean value of either true or false is returned.
**See also:** array, type(array), type(vector), type(matrix), op(eval(table))
*LA = 76*

**type(expr, *vector*)**

Determines whether expr is a vector data type (i.e., a one-dimensional array data type with index starting at 1).
**Output:** A boolean value of either true or false is returned.
**Argument options:** (expr, 'vector'(type)) to determine if expr is a vector with all elements in domain type.
**Additional information:** The forward quote characters around *vector* are necessary to stop the invocation of the linalg[vector] command.
**See also:** evalm, array, linalg[vector], type(array), type(matrix), type(table), op(eval(array))
*LI = 256*

# Solving Equations

## Introduction

One area where Maple excels at saving you time and frustration is in solving equations and systems of equations. Everyone who has tried to solve even a simple system of equations by hand has experienced the havoc wreaked by making a simple mistake in arithmetic. Maple doesn't make arithmetic mistakes or transcription errors. (But this doesn't mean you should trust every Maple answer without question—checking the feasibility of your results is necessary whether you are using pencil and paper or a computer!)

## Defining Equations

As noted earlier in this book, *equations* use the = character as opposed to the := characters, which are used for assignment. For a more thorough explanation of the differences, see the sections *Assignments* and *Equations* in the chapter *Getting Started With Maple*.

Standard linear and nonlinear equations can usually be defined with as much information as desired on either side of the = character. If a Maple command needs equations in a specific form, transforms them itself before calculations start. The few exceptions to this rule are outlined in the individual command entries.

```
> a + b = 3;
```

$$a + b = 3$$

```
> 4*x^2 + y^3 = exp(z^4)/10;
```

$$4x^2 + y^3 = \frac{1}{10} e^{(z^4)}$$

```
> (x+2)^3 = expand((x+2)^3);
```

$$(x+2)^3 = x^3 + 6x^2 + 12x + 8$$

The preceding examples show you how to define one equation at a time. In many cases, you want to solve a *system* of equations. These systems are normally enclosed in set brackets ({}). It is a good idea to assign the set of equations to a variable name. This makes the completed commands much easier to read.

## solve, isolve, msolve, and fsolve

The four commands solve, isolve, msolve, and fsolve all act on the same type of standard linear or non-linear equations; but they all produce different types of results.

With some of these commands, you need to indicate exactly which variables you are solving for. These variables follow the equation(s) that you have specified. If you are solving for more than one variable, enclose the names in a set.

solve is Maple's general purpose symbolic equation solver. Following are a few examples of how it works.

```
> solve(x^2 + 3*x = 6, x);
```

$$-\frac{3}{2} + \frac{1}{2}\sqrt{33}, -\frac{3}{2} - \frac{1}{2}\sqrt{33}$$

```
> solve({x^2*y^2=0, x+y = 1}, {x,y});
```

$$\{y = 0, x = 1\}, \{x = 0, y = 1\}$$

```
> eqns := {x + y + z = 6, x - y - z = 1, x + y - z = 4};
```

$$eqns := \{x + y - z = 4, x - y - z = 1, x + y + z = 6\}$$

```
> solve(eqns, {x,y,z});
```

$$\left\{ x = \frac{7}{2}, y = \frac{3}{2}, z = 1 \right\}$$

As you can see, the result of a solve command is an expression sequence of solutions. (This also hold true for the other solving commands.) If there is more than one variable being solved for, each set of equations represents one possible solution to the system. With this format, it is easy to check your answers using the assign and subs commands. Just remember to *unassign* the variables after you are finished!

```
> assign(");
> x, y, z;
```

$$\frac{7}{2}, \frac{3}{2}, 1$$

```
> eqns;
```

$$\{1 = 1, 4 = 4, 6 = 6\}$$

```
> x := 'x': y := 'y': z := 'z':
> x, y, z;
```

$$x, y, z$$

isolve and msolve are specialty solvers. Integer solutions are found with isolve and msolve solves over the integers modulo a certain value. Both commands automatically solve with respect to *all* the unknown variables they are given.

```
> isolve(2*x + 3*y = 7);
```

$$\{x = 3\_N1 - 7, y = -2\_N1 + 7\}$$

```
> msolve({2*x - 3*y = 5, x + y = -3}, 6);
```

$$\{y = 5, x = 4\}$$

The _N1 in the solution to isolve(2*x + 3*y = 7) signifies that there are infinitely many integer solutions, which can be obtained by substituting any integer for _N1.

fsolve returns floating-point solutions to equations or systems of equations. Keep in mind that this command does not just use solve and then evaluate the answers in floating-point form. fsolve has its own special routines for finding approximate answers to systems of equations that the other solvers may be unable to handle.

```
> poly := 45*x^6 - 2*x^5 - 45*x^4 + 10*x^2 - 12*x + 1;
```

$$poly := 45\,x^6 - 2\,x^5 - 45\,x^4 + 10\,x^2 - 12\,x + 1$$

```
> solve(poly, x);
```

$$RootOf(\,45\,\_Z^6 - 2\,\_Z^5 - 45\,\_Z^4 + 10\,\_Z^2 - 12\,\_Z + 1\,)$$

```
> fsolve(poly, x);
```

$$.08981219600, 1.029868758$$

```
> fsolve(poly, x, 1..2);
```

$$1.029868758$$

```
> fsolve(poly, x, complex);
```

$$-.8341755041 - .3242897760\,I, -.8341755041 + .3242897760\,I,$$
$$.08981219600, .2965572492 - .4604252441\,I,$$
$$.2965572492 + .4604252441\,I, 1.029868758$$

As the last two examples illustrate, fsolve takes extra parameters that allow you to specify a solution range or that *all* solutions, including complex ones, should be found.

## Differential Equations

dsolve is Maple's command for solving differential equations and systems of differential equations. Before we get into examples of how dsolve works, we need to take a quick look at how to create differential equations and initial conditions.

Unevaluated derivatives are created with either the diff or the D command. We use diff in our examples, as does most of the existing Maple literature. When defining an unevaluated derivative, the standard notation is diff(fnc(var), var), where fnc is any unassigned function name and var is an unassigned variable name.

```
> diff(y(x), x);
```

$$\frac{\partial}{\partial x}\,y(\,x\,)$$

```
> diff(myfunc(t), t$2);
```

$$\frac{\partial^2}{\partial t^2} myfunc(t)$$

Using similar syntax, differential *equations* can be created.

```
> diff(y(x), x) = 2*y(x)/(x^2 + 1);
```

$$\frac{\partial}{\partial x} y(x) = 2 \frac{y(x)}{x^2 + 1}$$

```
> deqn := 2*diff(y(x),x$2) + 3*diff(y(x),x) - 5*y(x) = 0;
```

$$deqn := 2\left(\frac{\partial^2}{\partial x^2} y(x)\right) + 3\left(\frac{\partial}{\partial x} y(x)\right) - 5 y(x) = 0$$

Once an equation or system of equations has been defined, calling dsolve is relatively simple.

```
> dsolve(deqn, y(x));
```

$$y(x) = \_C1\, e^x + \_C2\, e^{(-5/2\,x)}$$

As you can see, until some initial conditions are specified for the differential equation, the answer is going to contain some arbitrary constants. To define initial conditions, use the following form.

```
> initcons := y(0) = 0, y(1) = 1;
```

$$initcons := y(0) = 0, y(1) = 1$$

```
> dsolve({deqn, initcons}, y(x));
```

$$y(x) = \frac{e^x}{e - e^{(-5/2)}} - \frac{e^{(-5/2\,x)}}{e - e^{(-5/2)}}$$

```
> simplify(");
```

$$y(x) = \frac{e^x - e^{(-5/2\,x)}}{e - e^{(-5/2)}}$$

Again, you can use assign to check the answer.

```
> assign(");
> deqn;
```

$$2\,\frac{e^x - \dfrac{25}{4}\,e^{(-5/2\,x)}}{e - e^{(-5/2)}} + 3\,\frac{e^x + \dfrac{5}{2}\,e^{(-5/2\,x)}}{e - e^{(-5/2)}} - 5\,\frac{e^x - e^{(-5/2\,x)}}{e - e^{(-5/2)}} = 0$$

```
> simplify(");
```

$$0 = 0$$

```
> y(x) := 'y(x)':
```

If the initial conditions for a derivative of a function need to be specified, use the D operator.

```
> deqn := 2/3*diff(y(x), x$2) + 4/5*y(x) = 0;
```

$$deqn := \frac{2}{3}\left(\frac{\partial^2}{\partial x^2}\,y(x)\right) + \frac{4}{5}\,y(x) = 0$$

```
> initcons := y(0) = 1/3, D(y)(0) = 0:
> dsolve({deqn, initcons}, y(x));
```

$$y(x) = \frac{1}{3}\,cos\left(\frac{1}{5}\,\sqrt{30}\,x\right)$$

When an *explicit* or *implicit* solution cannot be found with the above methods, dsolve provides three alternate methods—a series method, a laplace transforms method, and a numerical method. The following examples take an identical set containing a differential equation and some initial conditions and show the result from each method.

```
> deqn := 4*diff(y(x),x$2) - diff(y(x),x) + 3*y(x) = 0:
> initcons := y(0) = 0, D(y)(0) = 1:
> dsolve({deqn, initcons}, y(x), series);
```

$$y(x) = x + \frac{1}{8}\,x^2 - \frac{11}{96}\,x^3 - \frac{23}{1536}\,x^4 + \frac{109}{30720}\,x^5 + O(x^6)$$

```
> dsolve({deqn, initcons}, y(x), laplace);
```

$$y(x) = \frac{8}{47}\,e^{(1/8\,x)}\,sin\left(\frac{1}{8}\,\sqrt{47}\,x\right)\sqrt{47}$$

```
> F := dsolve({deqn, initcons}, y(x), numeric);
F := proc(rkf45_x) ... end
```

The result of dsolve with the *numeric* option is a Maple procedure. This procedure can then be evaluated at particular values of x or it can be plotted over a given range.

```
> F(0), F(1), F(Pi), F(10);
```

$$\left[x = 0, y(x) = 0, \frac{\partial}{\partial x} y(x) = 1.\right],$$

$$\left[x = 1, y(x) = .9994599306107751,\right.$$
$$\left.\frac{\partial}{\partial x} y(x) = .8668511568713577\right],$$

$$\left[x = \pi, y(x) = .7507380807587012,\right.$$
$$\left.\frac{\partial}{\partial x} y(x) = -1.240091954640905\right],$$

$$\left[x = 10, y(x) = 3.073891559319550,\right.$$
$$\left.\frac{\partial}{\partial x} y(x) = -1.905637782125056\right]$$

```
> plots[odeplot](F, [x, y(x)], 0..20, labels = [x, y]);
```

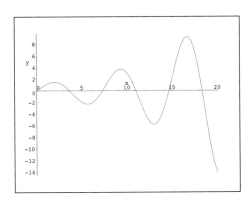

You may have noticed that evaluating F at specific values produced a list with equations for *three* values, $x$, $y(x)$, and $\frac{d}{dx} y(x)$ — even though you never asked for the value of $\frac{d}{dx} y(x)$. This is a bug and will be fixed in future releases of Maple.

## Recurrence Equations

rsolve is Maple's command for solving recurrence relations. The equations must be stated in terms of an unevaluated function of some variable (typically f(n), though any name can be used). As with differential equations, initial conditions help simplify the results.

```
> rsolve({f(n) = -3*f(n-1) -2*f(n-2), f(0)=1, f(1)=1},
> f(n));
```

$$3(-1)^n - 2(-2)^n$$

```
> rsolve(f(a*n) = b*f(n) + c, f(n));
```

$$f(1)\, n^{\left(\frac{ln(b)}{ln(a)}\right)} + n^{\left(\frac{ln(b)}{ln(a)}\right)} \left( -\frac{c \left(\frac{1}{b}\right)^{\left(\frac{ln(n)}{ln(a)}+1\right)} b}{-1+b} + \frac{c}{-1+b} \right)$$

For more examples, see the on-line help page for rsolve.

## Parameter Types Specific to This Chapter

| | |
|---|---|
| deqn | a differential equation |
| pde | a partial differential equation |

## Command Listing

**assign({var$_1$ = expr$_1$, ..., var$_n$ = expr$_n$})**

Assigns expressions expr$_1$ through expr$_n$ to variables var$_1$ through var$_n$, respectively.
**Output:** A NULL statement is returned.
**Argument options:** ([var$_1$ = expr$_1$, ..., var$_n$ = expr$_n$]) to produce the same result. ♣ (name = expr) or (name, expr) to assign expr to the variable name.
**Additional information:** A set or list of equations is typically sent to assign to validate the results of one of the solving commands. Be sure to unassign those variables after the verification. ♣ For

examples of using this command, see the introduction to this chapter.
**See also:** solve, isolve, msolve, rsolve, dsolve, fsolve, unassign
*FL = 17* **LI = 12**

### classi()

Determines the *Bianchi type* of a three-dimensional Lie algebra.
**Output:** Strings reporting the Bianchi type and other information are returned.
**Additional information:** This command needs to be defined by readlib(bianchi) before being invoked. ♣ For information on how to define the Lie algebra, see the on-line help page.
**See also:** liesymm package
*LI = 264*

### convert(expr, D)

Converts all the derivatives written with diff in expr into derivatives written with D.
**Output:** An expression is returned.
**See also:** convert(*diff*), D, diff
*LI = 34*

### convert(expr, *diff*)

Converts all the derivatives written with D in expr into derivatives written with diff.
**Output:** An expression is returned.
**See also:** convert(*D*), diff, D
*LI = 35*

### convert(ineq, *equality*)

Converts inequation ineq into the corresponding equation.
**Output:** An equation is returned.
**Additional information:** convert(*equality*) replaces the < or <= operators with =.
**See also:** convert(*lessthan*), convert(*lessequal*)
*LI = 37*

### convert(eqn, *lessequal*)

Converts equation (or strict inequality) eqn into the corresponding nonstrict inequality.
**Output:** A nonstrict inequality is returned.

**Additional information:** convert(*lessequal*) replaces the = or < operators with <=.
**See also:** convert(*equality*), convert(*lessthan*)
*LI = 37*

### convert(eqn, *lessthan*)

Converts equation (or nonstrict inequality) eqn into the corresponding strict inequality.
**Output:** A strict inequality is returned.
**Additional information:** convert(*lessthan*) replaces the = or <= operators with <.
**See also:** convert(*equality*), convert(*lessequal*)
*LI = 37*

### D(fnc)(num) = expr

Defines an initial value of expr for the derivative of function fnc at numeric value num.
**Output:** The initial value equation is returned.
**Argument options:** (D@@2)(fnc)(num) = expr to define an initial value for the second derivative. ♣ (D@@n)(fnc)(num) = expr to define an initial value for the $n^{th}$ derivative.
**Additional information:** The D command is also used as the differential operator for expressions and functional expressions. In its above form, it is used with the dsolve command to solve differential equations. ♣ D can also be used to represent unevaluated derivatives; but it is recommended that you use diff instead.
**See also:** dsolve, diff, convert(diff), *D*

### DESol(deqn, fnc(var))

Inert form of dsolve used to represent the solution of differential equation deqn with respect to the function fnc(var).
**Output:** An unevaluated DESol command is returned.
**Argument options:** (deqn, fnc(var), {$cond_1$, ..., $cond_n$}) to represent the solution of deqn under initial conditions $cond_1$ through $cond_n$. ♣ Sets of differential equations and functions can also be supplied.
**Additional information:** Unevaluated DESol commands can be manipulated with the commands diff, int, series, and evalf.
**See also:** dsolve, diff, *int*, *series*, *evalf*

## DEtools package

Provides specialized tools for working with differential equations and their solutions.

**Additional information:** To access the command fnc from the DEtools package, use the long form of the name DEtools[fnc], or with(DEtools, fnc) to load in a pointer for fnc only, or with(DEtools) to load in pointers to all the commands. ♣ Most of the commands deal with different aspects of plotting the results of the dsolve command. ♣ Be forewarned that the syntax for these commands can be very confusing, and the on-line help pages lengthy and obtuse.

### DEtools[Dchangevar]({eqn$_1$, ..., eqn$_n$}, deqn)

Performs a change of variables on the differential equation deqn. The new variables are defined in terms of the old ones in the eqn$_i$, usually in the form expr$_1$ := expr$_2$, where expr$_1$ is in terms of a variable in deqn and expr$_2$ is in terms of a new variable.

**Output:** A differential equation is returned.

**Argument options:** ({eqn$_1$, ..., eqn$_n$}, deqn, {var$_1$, ..., var$_m$}) to specify that variables var$_1$ through var$_m$, which occur in eqn$_1$ through eqn$_n$, are to be treated as constants by DEtools[Dchangevar].

**See also:** dsolve, *Student[changevar]*

### DEtools[DEplot](deqn, [var$_1$, var$_2$], a..b, {[expr$_1$, ..., expr$_n$]})

Displays a two-dimensional plot of the result of solving differential equation deqn for variables var$_1$ and var$_2$ and with initial conditions expressed by expr$_1$ through expr$_n$, as the independent variable ranges from a to b. The values of var$_1$ are translated to the $x$-axis and the values of var$_2$ are translated to the $y$-axis.

**Output:** A two-dimensional plot is displayed.

**Argument options:** (deqn, [var$_1$, var$_2$, var$_3$], a..b, {[expr$_1$, ..., expr$_n$]}) to display a three-dimensional plot of the result of solving the differential equation deqn. ♣ (deqn, [var$_1$, var$_2$], a..b, {[expr$_1$, ..., expr$_n$]}, option) to control standard options to the plot. For more information on available options, see the plot and plot3d commands, respectively.

**Additional information:** Many other parameter sequences provide results in DEtools[DEplot]. Their complicated descriptions are found in the on-line help page. ♣ plots[odeplot] is much more intuitive and easy to use, and while it may not provide as broad a range of services, it is recommended that you use it as much as possible.

**See also:** dsolve(*numeric*), *plots[odeplot]*, DEtools[DEplot1], DEtools[DEplot2], DEtools[dfieldplot], DEtools[phaseportrait]

### DEtools[DEplot1](deqn, [var$_1$, var$_2$], a..b, {[expr$_1$, expr$_2$]})

Displays a two-dimensional plot of the result of solving first order differential equation deqn for variables var$_1$ and var$_2$ and with initial conditions expressed by expr$_1$ and expr$_2$, as the independent variable ranges from a to b. The values of var$_1$ are translated to the $x$-axis and the values of var$_2$ are translated to the $y$-axis.

**Output:** A two-dimensional plot is displayed.

**Argument options:** (deqn, [var$_1$, var$_2$], a..b, {[expr$_1$, ..., expr$_n$]}, option) to control standard options to the plot. For more information on available options, see the plot command.

**Additional information:** Many other parameter sequences provide results in DEtools[DEplot1]. Their complicated descriptions are found in the on-line help page. ♣ plots[odeplot] is much more intuitive and easy to use, and while it may not provide as broad a range of services, it is recommended that you use it as much as possible.

**See also:** dsolve(*numeric*), *plots[odeplot]*, DEtools[DEplot], DEtools[DEplot2], DEtools[dfieldplot]

### DEtools[DEplot2]([deqn$_1$, deqn$_2$], [var$_1$, var$_2$], a..b, {[expr$_1$, expr$_2$, expr$_3$]})

Displays a two-dimensional plot of the result of solving the system of two first order differential equations, deqn$_1$ and deqn$_2$ for variables var$_1$ and var$_2$ and with initial conditions expressed by expr$_1$, expr$_2$, and expr$_3$, as the independent variable ranges from a to b. The values of var$_1$ are translated to the $x$-axis and the values of var$_2$ are translated to the $y$-axis.

**Output:** A two-dimensional plot is displayed.

**Argument options:** ([deqn$_1$, deqn$_2$], [var$_1$, var$_2$, var$_3$], a..b, {[expr$_1$, expr$_2$, expr$_3$]}) to display a three-dimensional plot of the result of solving the system of two first order differential equations deqn$_1$ and deqn$_2$. ♣ ([deqn$_1$, deqn$_2$], [var$_1$, var$_2$], a..b, {[expr$_1$, expr$_2$, expr$_3$], option}) to control standard options to the plot. For more information on available options, see the plot and plot3d commands, respectively.

**Additional information:** Many other parameter sequences provide results in DEtools[DEplot2]. Their complicated descriptions are found in the on-line help page. ♣ plots[odeplot] is much more intuitive and easy to use, and while it may not provide as broad a range of services, it is recommended that you use it as much as possible.

**See also:** dsolve(*numeric*), *plots[odeplot]*, DEtools[DEplot], DEtools[DEplot1], DEtools[dfieldplot], DEtools[phaseportrait]

## DEtools[dfieldplot](deqn, [var$_1$, var$_2$], a..b)

Displays a two-dimensional field plot of the result of solving the first order differential equation deqn for variables var$_1$ and var$_2$, as the independent variable ranges from a to b.

**Output:** A two-dimensional field plot is displayed.

**Argument options:** ([deqn$_1$, deqn$_2$], [var$_1$, var$_2$], a..b) to plot the field plot from the solution of the system of two differential equations, deqn$_1$ and deqn$_2$. The independent variable must be able to be eliminated from the system. ♣ (deqn, [var$_1$, var$_2$], a..b, {[expr$_1$, ..., expr$_n$]}, option) to control standard options to the plot. For more information on available options, see the plot command.

**Additional information:** There are other parameter sequences that provide results in DEtools[dfieldplot] and other restrictions that apply. Their descriptions are found in the on-line help page.

**See also:** dsolve(*numeric*), *plots[fieldplot]*, DEtools[DEplot], DEtools[phaseportrait]


## DEtools[PDEplot](list$_1$, list$_2$, list$_3$, var=a..b)

Displays a three-dimensional plot of the result of solving the first-order quasi-linear partial differential equation with terms represented by the elements of list$_1$, variables represented by the elements of list$_2$, and initial data represented by the elements of list$_3$, as variable var ranges from a to b.

**Output:** A three-dimensional plot is displayed.

**Argument options:** (list$_1$, list$_2$, list$_3$, var=a..b, option) to control standard options to the plot. For more information on available options, see the entry for plot3d.

**Additional information:** For more information about parameters and examples of their usage, see the lengthy on-line help page.

**See also:** *plots[odeplot]*, DEtools[DEplot]


## DEtools[phaseportrait]([deqn$_1$, deqn$_2$], [var$_1$, var$_2$], a..b, {[expr$_1$, expr$_2$, expr$_3$]})

Displays a two-dimensional phase portrait of the result of solving the system of two first-order differential equations deqn$_1$ and deqn$_2$ for variables var$_1$ and var$_2$ and with initial conditions expressed by expr$_1$, expr$_2$, and expr$_3$, as the independent variable ranges from a to b. The values of var$_1$ are translated to the $x$-axis and the values of var$_2$ are translated to the $y$-axis.

**Output:** A two-dimensional plot is displayed.

**Argument options:** ([deqn$_1$, deqn$_2$], [var$_1$, var$_2$, var$_3$], a..b, {[expr$_1$, expr$_2$, expr$_3$]}) to display a three-dimensional phase portrait of the result of solving the system of two first order differential equations deqn$_1$ and deqn$_2$. ✢ ([deqn$_1$, deqn$_2$], [var$_1$, var$_2$], a..b, {[expr$_1$, expr$_2$, expr$_3$]}, option) to control standard options to the plot. For more information on available options, see the plot and plot3d commands, respectively.

**Additional information:** Many other parameter sequences provide results in DEtools[phaseportrait]. Their complicated descriptions are found in the on-line help page.

**See also:** dsolve(*numeric*), DEtools[DEplot], DEtools[dfieldplot]

### diff(fnc(var), var)

Represents the derivative of function fnc with respect to variable var. These representations are used to build differential equations.

**Output:** A formatted version of the derivative is returned.

**Argument options:** diff(fnc(var), var$2) to represent the second derivative. ✢ diff(fnc(var), var$n) to represent the $n^{th}$ derivative. ✢ diff(fnc, var$_1$, var$_2$, ..., var$_n$) to represent the derivative of fnc with respect to var$_1$ through var$_n$.

**Additional information:** diff is also used to perform differentiations on expressions. ✢ In its above form, diff is used with dsolve to solve differential equations. ✢ D can also be used to represent differentiations; but it is recommended that you use diff instead.

**See also:** dsolve, D, convert(D), *diff*

**FL = 44** *LA = 7–8, 23–24*

### Diff(fnc, var)

Inert form of diff used to represent partial derivative of function fnc with respect to the variable var.

**Output:** An unevaluated Diff command is returned.

**Argument options:** (fnc, var$n) for $n^{th}$-order partial derivative. ✢ (fnc, var$_1$, var$_2$, ..., var$_n$) for partial differentiation of fnc with respect to var$_1$ through var$_n$, in that order.

**See also:** diff, DESol, liesymm[determine], liesymm[makeforms], liesymm[mixpar], liesymm[TD]

### dsolve(deqn, fnc(var))

Symbolically solves ordinary differential equation deqn for the function fnc(var).

**Output:** If a complete solution in closed form can be found, an equation in var and fnc(var) is returned. Otherwise, a partial result which uses calls to DESol is returned.

**Argument options:** ($expr_d$, var) to solve a differential equation of the form $expr_d = 0$. ♣ ({deqn, cond$_1$, ..., cond$_n$}, fnc(var)) to solve equation deqn with initial conditions cond$_1$ through cond$_n$. ♣ ({deqn$_1$, ..., deqn$_n$}, {fnc$_1$(var), ..., fnc$_n$(var)}) to solve the system of differential equations for the given functions of var. ♣ (deqn, fnc(var), *explicit*) to have the solution stated explicitly in terms of fnc(var), if possible. ♣ (deqn, fnc(var), *laplace*) to force the use of Laplace transforms in solving the equation. ♣ (deqn, fnc(var), *series*) to force the use of a series method in solving the equation. Set the order of the solution with Order. ♣ (deqn, fnc(var), *numeric*) to force the use of numerical methods in solving the equation. See the entry for dsolve(*numeric*) for more information. ♣ (deqn, fnc(var), *output=basis*) to specify that the solution be returned in basis form.

**Additional information:** Arbitrary constants are represented by global variables _C1 through _Cn. ♣ To read more about defining differential equations and initial conditions, see the entries for diff and D, respectively. ♣ If a NULL expression sequence is returned, this can mean that either there are no solutions or Maple was unable to find any. ♣ If your equations contain any numeric values, you must use the *numeric* option or an error message is generated.

**See also:** dsolve('*numeric*'), powseries[powsolve], diff, D, DESol, solve, fsolve, isolve, msolve, rsolve, *laplace*, *series*, subs

**FL = 43–45, 201** *LA = 23–24* **LI = 67**

### dsolve(deqn, fnc(var), *numeric*)

Numerically solves ordinary differential equation deqn for the function fnc(var).

**Output:** A Maple procedure is returned.

**Argument options:** (deqn, fnc(var), *numeric, method=rkf45*) to to force the use of a fourth-fifth order Runge-Kutta method. ♣ (deqn, fnc(var), *numeric, method=dverk78*) to to force the use of a seventh-eighth order continuous Runge-Kutta method. ♣ (deqn, fnc(var), *numeric, output=listprocedures*) to return a list of single-value output procedures instead of one multiple-output procedure. ♣ There are several other available options for tweaking the operation of dsolve(, *numeric*). See the on-line help page for more details. ♣ See the entry for dsolve for more information on alternate parameter sequences.

**Additional information:** Typically, the result of dsolve(*numeric*) is assigned to a name, say F, and then F is evaluated at specific numeric values representing values of the independent variable in the system. ♣ To plot the result, use plots[odeplot] or the commands

in the DEtools package. ♣ To read more about defining differential equations and initial conditions, see the entries for diff and D, respectively. ♣ For more information on the algorithms used and how to set tolerances, see the on-line help page for dsolve[numeric].
**See also:** dsolve, *plots[odeplot]*, DEtools[DEplot], DEtools[DEplot1], DEtools[DEplot2], diff, D
*LI = 69*

### fsolve(eqn, var)

Numerically solves equation eqn for variable var.
**Output:** An expression sequence of real solutions is returned.
**Argument options:** (expr, var) to solve the equation of the form expr = 0. ♣ ({$eqn_1$, ..., $eqn_n$}, {$var_1$, ..., $var_n$}) to solve the system of equations $eqn_1$ through $eqn_n$ for the unknown variables $var_1$ through $var_n$. The result is an expression sequence of sets, representing possible real solutions to the system. ♣ (expr) or ({$eqn_1$, ..., $eqn_n$}) to solve an equation or set of equations for all of the unknowns contained in it. Note: the n equations must contain exactly n unknowns. ♣ (eqn, var, a..b) or ({$eqn_1$, ..., $eqn_n$}, {$var_1$, ..., $var_n$}, {$var_1 = a_1..b_1$, ..., $var_n = a_n..b_n$}) to find real solutions that fall within the given ranges. ♣ (eqn, var, *'complex'*) to find all complex solutions of eqn. ♣ (eqn, var, *'maxsols'*=m) to find at most the m least roots. ♣ (eqn, var, *'fulldigits'*) to ensure that fsolve does not lower the number of digits used within its computations. The computation takes longer, but is less likely to miss roots.
**Additional information:** If a NULL expression sequence is returned, this can mean either that there are no solutions in the appropriate field or Maple was unable to find any. ♣ Sometimes, because fsolve chose poorly for a beginning interval, a solution or solutions might be missed. In this case, supplying an initial range can greatly affect the outcome. ♣ If any there are any symbolic variables that are not supplied to fsolve as variables to be solved for, then an error message is returned.
**See also:** solve, isolve, msolve, rsolve, dsolve, subs
**FL = 23−26, 204** *LA = 25 LI = 97*

### isolve(eqn)

Finds the integer solutions of equation eqn.
**Output:** An expression sequence of sets representing possible solutions is returned. Each set contains an equation for each unknown in eqn.

**Argument options:** (expr) to solve for the equation of the form expr = 0. ♣ ({eqn$_1$, ..., eqn$_n$}) to solve a set of equations eqn$_1$ through eqn$_n$. The result is an expression sequence of sets, each set representing a possible integer solution. ♣ (eqn, name) or (eqn, {name$_1$, ..., name$_n$}) to use name or name$_1$ through name$_n$ as variables in the solution, if necessary. Otherwise, global names _N1, _N2, ... are used.

**Additional information:** If a NULL expression sequence is returned, this can mean that either there are no solutions or Maple was unable to find any.

**See also:** solve, fsolve, msolve, rsolve, dsolve, subs

*FL = 198* ***LI = 120***

### liesymm package

Provides commands for manipulation of Lie symmetries.

**Additional information:** To access the command fnc from the liesymm package, use the long form of the name liesymm[fnc], or with(liesymm, fnc) to load in a pointer for fnc only, or with(liesymm) to load in pointers to all the commands. ♣ This package represents an implementation of the *Harrison-Estabrook procedure*. ♣ In most liesymm commands, a call to liesymm[setup] defines the coordinate system for expressions used in terms of all or some of the variables in the expressions. ♣ For a thorough overview of commands in this package, see the on-line help page for liesymm.

### liesymm[&^](expr$_1$, ..., expr$_n$)

Represents the wedge product of expr$_1$ through expr$_n$, expressions in differential.

**Output:** An expression in wedge products is returned.

**Argument options:** expr$_1$ &^ expr$_2$ &^ ... &^ expr$_n$ to produce the same results.

**Additional information:** Elementary simplifications are always done before a result is returned. ♣ More information about the ordering of products used in the automatic simplification can be produced with liesymm[wedgeset](*1*).

**See also:** liesymm[setup], liesymm[&mod], liesymm[wedgeset], liesymm[hook]

### expr liesymm[&mod] [expr$_1$, ..., expr$_n$]

Reduces the expression of differential form expr modulo the exterior ideal generated by expr$_1$ through expr$_n$.

**Output:** An expression in differential forms is returned.

**Argument options:** expr liesymm[&mod] {expr$_1$, ..., expr$_n$} to produce the same results.

**Additional information:** In real use of the &mod operator, the liesymm package is first loaded by with(liesymm). Subsequently, liesymm is not needed when using the operator. ♣ The form list $expr_1$ through $expr_n$ should be *closed*.
**See also:** liesymm[setup], liesymm[&^], liesymm[close], liesymm[d], liesymm[Lie]

### liesymm[annul]([$expr_1$, ..., $expr_m$], [$var_1$, ..., $var_n$])

Calculates the partial differential equations that annul $expr_1$ through $expr_m$, expressions in differential forms. The coordinates $var_1$ through $var_n$ are treated as independent variables on the solution manifold.
**Output:** A list of differential equations is returned.
**Argument options:** (\{$expr_1$, ..., $expr_m$\}, [$var_1$, ..., $var_n$]) to produce a set of differential equations.
**Additional information:** $expr_1$ through $expr_m$ are set equal to 0 and the appropriate differential equations are found.
**See also:** liesymm[setup], liesymm[makeforms], liesymm[close], liesymm[hasclosure], liesymm[determine]

### liesymm[close]([$expr_1$, ..., $expr_n$])

Computes the closure of $expr_1$ through $expr_n$, expressions in differential forms.
**Output:** A list of expressions in differential forms is returned.
**Argument options:** (\{$expr_1$, ..., $expr_n$\}) to produce a set of expressions in differential forms.
**Additional information:** The closure created is with respect to the exterior derivative liesymm[d]. ♣ The expressions returned include the input expressions plus whatever other expressions are necessary for closure. ♣ liesymm[hasclosure] can be used to determine if $expr_1$ through $expr_n$ already constitute a closure.
**See also:** liesymm[setup], liesymm[makeforms], liesymm[annul], liesymm[hasclosure], liesymm[determine]

### liesymm[d](expr)

Computes the exterior derivative of expr, an expression in differential forms, with respect to the setup coordinates.
**Output:** An expression in differential forms is returned.
**Argument options:** ([$expr_1$, ..., $expr_n$]) to return the exterior derivative of $expr_1$ through $expr_n$ in a list of length n.
**Additional information:**
**See also:** liesymm[hook], liesymm[&^], D, diff

Solving Equations    129

### liesymm[depvars](var$_1$, ..., var$_n$)

Sets the current dependent variables to var$_1$ through var$_n$.
**Output:** A list containing the dependent variables is returned.
**Argument options:** () to display the current list of dependent variables.
**Additional information:** Use liesymm[indepvars] to set the independent variables.
**See also:** liesymm[setup], liesymm[indepvars], liesymm[d]

### liesymm[determine]([expr$_1$, ..., expr$_n$], name)

Computes determining equations (expressed in terms of name.1, name.2, etc.) for the isovectors that are generators of the isogroup of expr$_1$ through expr$_n$, expressions in differential forms.
**Output:** A set of partial differential equations is returned.
**Argument options:** ([pde$_1$, ..., pde$_n$], name$_1$, [fnc$_1$, ..., fnc$_m$], name$_2$) to determine the equations for the isovectors of the partial differential equations pde$_1$ through pde$_n$ directly. fnc$_1$ through fnc$_m$ are function calls that represent which variables are dependent and independent. The determining equations are written in terms of name$_1$[1], name$_1$[2], etc. and any extended variables are written in terms of name$_2$[1], name$_2$[2], etc.
**Additional information:** If a set of differential forms is given, it must be closed. Use liesymm[close] to close it. ✦ For more information on defining partial differential equations, see the Diff command.
**See also:** liesymm[close], liesymm[hasclosure], liesymm[wedgeset], liesymm[makeforms], liesymm[wsubs], Diff

### liesymm[dvalue](expr)

Evaluates expr, an expression containing unevaluated Diff commands.
**Output:** An expression in diff commands is returned.
**Additional information:** The same results can be obtained with the convert(*diff*) command.
**See also:** Diff, diff, convert(*diff*)

### liesymm[Eta](expr, var)

Computes the coefficients of the generator of the finite point translation of expression expr in independent variable var.
**Output:** An inert expression in unevaluated Diff commands is returned.

**Argument options:** [1](expr, var), [2](expr, var$_1$, var$_2$), or [3] (expr, var$_1$, var$_2$, var$_3$) to compute the Eta value at different levels.
**Additional information:** For more information, see the on-line help page.
**See also:** liesymm[TD], liesymm[indepvars], Diff

### liesymm[extvars]()

Displays the currently assigned variable names for dynamically created partial derivatives.
**Output:** A table is returned.
**Additional information:** For more information on this command, see the on-line help page.
**See also:** liesymm[depvars], liesymm[indepvars], liesymm[makeforms]

### liesymm[getcoeff](expr)

Extracts the coefficient from expr, an expression containing one wedge product.
**Output:** An expression in differential forms is returned.
**Additional information:** Use liesymm[getform] to extract the basis element from such an expression.
**See also:** liesymm[wedgeset], liesymm[&^], liesymm[getform], liesymm[d]

### liesymm[getform](expr)

Extracts the basis element of an expression containing one wedge product.
**Output:** An expression is returned.
**Additional information:** Use liesymm[getcoeff] to extract the coefficient from such an expression.
**See also:** liesymm[wedgeset], liesymm[&^], liesymm[getcoeff], liesymm[d]

### liesymm[hasclosure]([expr$_1$, ..., expr$_n$])

Determines whether expr$_1$ through expr$_n$, expressions in differential forms, have closure.
**Output:** A boolean value of true or false is returned.
**Argument options:** ({expr$_1$, ..., expr$_n$}) to produce the same result.
**Additional information:** The closure tested for is with respect to the exterior derivative liesymm[d]. ♣ liesymm[close] can be

Solving Equations 131

used to find a closure which includes $expr_1$ through $expr_n$.
**See also:** liesymm[makeforms], liesymm[close]

### liesymm[hook](expr, name)

Computes the inner product of expr, an expression in differential forms, with respect to the unassigned vector name.
**Output:** An expression in differential forms is returned. It contains functions in the setup variables for each element of name, that is, name[1] through name[n].
**Argument options:** (expr, [$name_1$, ..., $name_n$]) to assign the names $name_1$ through $name_n$ to the functions in the setup variables present in the result.
**See also:** liesymm[setup], liesymm[&^], liesymm[d]

### liesymm[indepvars]($var_1$, ..., $var_n$)

Sets the current independent variables to $var_1$ through $var_n$.
**Output:** A list containing the independent variables is returned.
**Argument options:** () to display the current list of independent variables.
**Additional information:** Use liesymm[depvars] to set the dependent variables.
**See also:** liesymm[setup], liesymm[depvars], liesymm[d]

### liesymm[Lie](expr, name)

Computes the Lie derivative of expr, an expression in differential forms, with respect to the setup coordinates. The derivative is written in name[1] through name[n], where n is the number of setup coordinates.
**Output:** An expression in differential forms and partial derivatives is returned.
**Argument options:** (expr, [$name_1$, ..., $name_n$]) to write the derivative in terms of the n names $name_1$ through $name_n$.
**See also:** liesymm[setup], liesymm[&^], liesymm[d], liesymm[hook]

### liesymm[Lrank]([$expr_1$, ..., $expr_n$])

Determines the Lie rank of $expr_1$ through $expr_n$, expressions in differential forms.
**Output:** A list of expression in differential forms is returned.
**Argument options:** ({$expr_1$, ..., $expr_n$}) to produce a set of expressions in differential forms.

**Additional information:** The list of differential forms that is returned is equivalent to the list entered, except that any forms redundant to the generation of the determining equations are removed.
**See also:** liesymm[setup], liesymm[makeforms], liesymm[close], liesymm[determine], liesymm[d]

### liesymm[makeforms]([pde$_1$, ..., pde$_m$], [fnc$_1$, ..., fnc$_n$], name)

Constructs a list of expressions in differential forms equivalent to the partial differential equations pde$_1$ through pde$_m$. fnc$_1$ through fnc$_m$ are function calls that represent which variables are dependent and independent. Any extended variables are written in terms of name.1, name.2, etc.
**Output:** A list of expression in differential forms is returned.
**Argument options:** ({pde$_1$, ..., pde$_m$}, [fnc$_1$, ..., fnc$_n$], name) to produce the same results. ♣ ([pde$_1$, ..., pde$_m$], [fnc$_1$, ..., fnc$_n$], [name$_1$, ..., name$_p$]) to write any extended variables in terms of name$_1$ through name$_p$.
**Additional information:** The differential forms and the partial differential equations are equivalent in the *Cartan* sense. ♣ Any of the parameters of this command can be single expressions (i.e., not a list or set). ♣ For more information, see the on-line help page.
**See also:** liesymm[setup], liesymm[close], liesymm[annul], Diff

### liesymm[mixpar](pde)

Sorts the partial differentials in pde created by nested calls to the Diff or diff commands.
**Output:** An expression in unevaluated Diff commands is returned.
**Additional information:** The sorting is done in *address order* on the variables of differentiation. ♣ This command's purpose is to ensure that mixed partial derivatives are recognized as equivalent.
**See also:** liesymm[setup], Diff, diff

### liesymm[prolong](name)

Constructs equations that are used to rewrite components of isovectors. The equations are in variables name.1, name.2, etc.
**Output:** A set of equations in unevaluated Diff commands is returned.
**Additional information:** The results of liesymm[prolong] can easily be used in a subs command. ♣ For more information, see the on-line help page.
**See also:** liesymm[setup], liesymm[TD], liesymm[Eta], Diff, *subs*

### liesymm[reduce](expr)

Reduces expr, a differential form.
**Output:** A reduced differential form is returned.
**Argument options:** ($\{expr_1, ..., expr_n\}$) to reduce a set of differential forms.
**Additional information:** This command is typically called within liesymm[determine]. ♣ For more information, see the on-line help page.
**See also:** liesymm[determine]

### liesymm[setup]($var_1$, ..., $var_n$)

Defines variables $var_1$ through $var_n$ used as coordinates in Lie symmetry calculations.
**Output:** A list of n variable names is returned.
**Argument options:** ([$var_1$, ..., $var_n$]) to produce the same results.
**Additional information:** liesymm[setup] usually must be called before any of the other commands in the liesymm package are used. ♣ Each subsequent call to liesymm[setup] removes all previously defined coordinates and replaces them with the new set. Therefore, calling liesymm[setup] with no parameters erases *all* defined coordinates. ♣ Each call to liesymm[setup] erases information in several remember tables used in the liesymm package. ♣ It is recommended that liesymm[setup] be called between problems to keep overhead to a minimum.
**See also:** liesymm[d], liesymm[makeforms], liesymm[hook]

### liesymm[TD](expr, var)

Computes the extended derivative of expr, an expression in independent variable var whose other variables are explicitly either dependent or independent.
**Output:** An inert expression in unevaluated Diff commands is returned.
**Argument options:** [n](expr, var) to compute the extended derivative at level n. 1 is the default.
**Additional information:** Dependent variables are specified with the liesymm[depvars] command. ♣ Independent variables are specified with the liesymm[indepvars] command. ♣ Partial derivatives in the result are represented by the name w.1, w.2, etc. ♣ Use liesymm[dvalue] to evaluate the result.
**See also:** liesymm[setup], liesymm[depvars], liesymm[indepvars], liesymm[dvalues], liesymm[translate], liesymm[vfix], liesymm[&^], Diff

### liesymm[translate]()

Displays the currently assigned variable names for dynamically created partial derivatives.
**Output:** A translation table is returned.
**Argument options:** (pde), where pde is a currently named partial derivative, to retrieve the table entry for that partial derivative.
**Additional information:** For more information on this command, see the on-line help page.
**See also:** liesymm[depvars], liesymm[indepvars], liesymm[makeforms]

### liesymm[wdegree](expr)

Calculates the wedge degree of expr, an expression in differential forms.
**Output:** An integer value is returned.
**Additional information:** If none of the variables in expr have been included in the coordinate system, $-1$ is the result. Otherwise, the result is a non-negative integer representing the number of terms in the wedge product.
**See also:** liesymm[setup], liesymm[d], liesymm[&^]

### liesymm[wedgeset](n)

Calculates the differential forms of degree n of the current coordinate system.
**Output:** An expression sequence of differential forms is returned.
**Argument options:** (0) to return a list of the unknowns in the current coordinate system. ♣ (1) to return a list of the unknowns in the current coordinate system operated on by liesymm[d]. ♣ (2) to return a list of possible wedge products of the unknowns in the current coordinate system operated on by liesymm[d].
**See also:** liesymm[d], liesymm[&^], liesymm[Lie]

### liesymm[wsubs](eqn, expr)

Substitutes for all occurrences of the left-hand side of equation eqn with the right-hand side of equation eqn, in the expression in differential forms expr.
**Output:** An expression in differential forms is returned.
**Argument options:** ({eqn$_1$, ..., eqn$_n$}, expr) or ([eqn$_1$, ..., eqn$_n$], expr) to perform the n substitutions eqn$_1$ through eqn$_n$, in the order provided.
**Additional information:** liesymm[wsubs] differs from subs in that instead of examining the underlying structure of expr directly,

Solving Equations    135

algebraic factors are considered. Because of this, liesymm[wsubs] cannot find expressions that occur as part of a sum.
**See also:** liesymm[setup], liesymm[&^], liesymm[&mod], *subs*

### msolve(eqn, m)

Finds the integer solutions modulo m of equation eqn.
**Output:** An expression sequence of sets is returned. Each set contains an equation for each unknown in eqn.
**Argument options:** (expr, m) to solve for an equation of the form expr = 0. ♣ ({eqn$_1$, ..., eqn$_n$}, n) to solve a system of equations. The result is an expression sequence of sets, each set representing a possible integer solution modulo m. ♣ (eqn, name, m) or (eqn, {name$_1$, ..., name$_n$}, m) to use name or name$_1$ through name$_n$ as variables in the solution of eqn, if necessary. Otherwise, global names _NN1, _NN2, ... are used.
**Additional information:** If a NULL expression sequence is found, this can mean that either there are no solutions or Maple was unable to find any.
**See also:** solve, fsolve, isolve, rsolve, dsolve, *mod*, Roots, subs
*LI = 142*

### powseries[powsolve](deqn, cond$_1$, ..., cond$_n$)

Solves the linear differential equation deqn, with initial conditions cond$_1$ through cond$_n$, in terms of a power series.
**Output:** A power series structure is returned.
**Additional information:** To read more about the syntax for differential equations and initial conditions, see the entries for diff and D, respectively. ♣ All initial conditions must be defined at 0. ♣ To render a power series as a standard series, use powseries[tpsform].
**See also:** dsolve, diff, D, *powseries[powcreate]*, *powseries[tpsform]*

### rsolve(eqn, fnc(var))

Solves recurrence relation eqn for function fnc(var).
**Output:** An expression in var, which typically contains functional value holders for any or all of fnc(0), fnc(1), etc., is returned.
**Argument options:** (expr, fnc(var)) to solve for an equation of the form expr = 0. ♣ ({eqn$_1$, ..., eqn$_n$}, {fnc$_1$(var$_1$), ..., fnc$_m$(var$_m$)}) to solve a set of recurrence relations for functions fnc$_1$(var$_1$) through fnc$_m$(var$_m$). The result is a set containing equations for each function. ♣ ({eqn$_1$, ..., eqn$_n$}, {fnc$_1$, ..., fnc$_m$}, 'genfunc'(var)) to return the generating functions, expressed in variable var, of the functions fnc$_1$ through fnc$_m$. ♣ ({eqn$_1$, ..., eqn$_n$}, {fnc$_1$, ...,

$fnc_m$}, 'makeproc') to return a proc structure that returns the values of the recurrence relation for various values.
**Additional information:** If rsolve cannot find a solution, the unevaluated recurrence relation is returned. At that point, try to find an asymptotic solution with asmypt. ♣ For more information on rsolve, see the on-line help page.
**See also:** solve, fsolve, isolve, msolve, dsolve, asympt, *genfunc*
**FL = 50−51** *LA = 24−25* ***LI = 183***

### simplex package

Provides commands for performing linear optimization on systems of equalities and inequalities using the *simplex algorithm* in whole or part.
**Additional information:** To access the command fnc from the simplex package, use the long form of the name simplex[fnc], or with(simplex, fnc) to load in a pointer for fnc only, or with(simplex) to load in pointers to all the commands. ♣ The entire simplex algorithm can be performed with simplex[maximize] or simplex [minimize], or the process can be taken step-by-step with commands simplex[pivoteqn], simplex[pivotvar], and simplex[pivot].

### simplex[basis]({$eqn_1$, ..., $eqn_n$})

Computes a variable basis as used by the simplex algorithm for the equations $eqn_1$ through $eqn_n$.
**Output:** A set containing n variables is returned.
**Argument options:** ([$eqn_1$, ..., $eqn_n$]) to return a list of variables forming the basis.
**Additional information:** $eqn_1$ through $eqn_n$ should be in the form dictated by passing them through the simplex[setup] command. Otherwise, an error message is returned.
**See also:** simplex[setup], simplex[pivotvar]

### simplex[convert]({$ineq_1$, ..., $ineq_n$}, *std*)

Converts linear constraints $ineq_1$ through $ineq_n$ into an equivalent set of linear constraints in standard form. The constants (and only the constants) are gathered on the right-hand side of each constraint.
**Output:** A set of linear constraints is returned.
**Argument options:** ([$eqn_1$, ..., $eqn_n$]) to return a list of constraints in standard form.
**Additional information:** If constraints $ineq_1$ through $ineq_n$ contain strict inequalities, an error message is returned. ♣ The difference between this command and simplex[convert](*stdle*) is that

the latter transforms strict equalities into two nonstrict inequalities.
**See also:** simplex[dual], simplex[feasible], simplex[convert](*stdle*), simplex[standardize]


### simplex[convert]({ineq$_1$, ..., ineq$_n$}, *stdle*)

Converts linear constraints ineq$_1$ through ineq$_n$ into an equivalent set of nonstrict inequalities in standard form. The constants (and only the constants) are gathered on the right-hand side of each constraint.
**Output:** A set of nonstrict linear inequalities is returned.
**Argument options:** ([eqn$_1$, ..., eqn$_n$]) to return a list of nonstrict inequalities in standard form.
**Additional information:** Each strict equality encountered is converted into two nonstrict inequalities. ♣ If the constraints contain any strict inequalities, an error message is returned.
**See also:** simplex[dual], simplex[feasible], simplex[convert](*std*), simplex[standardize]


### simplex[convexhull]({pt$_1$, ..., pt$_n$})

Computes the convex hull of the set of $x, y$-points pt$_1$ through pt$_n$.
**Output:** A list of points arranged in clockwise order is returned.
**Argument options:** ([pt$_1$, ..., pt$_n$]) to return the same result.
**Additional information:** pt$_1$ through pt$_n$ must each be in the form of a list of two numeric values. ♣ An $n \log n$ algorithm is used. ♣ Each of the points in the result is from the original set.


### simplex[cterm]([ineq$_1$, ..., ineq$_n$])

Computes the constant terms of linear constraints ineq$_1$ through ineq$_n$.
**Output:** A list of constants is returned.
**Argument options:** ({ineq$_1$, ..., ineq$_n$}) to compute a set of constants.
**Additional information:** Before the constant terms are determined, ineq$_1$ through ineq$_n$ are passed through simplex[convert](*std*).
**See also:** simplex[convert](*std*)


### simplex[dual](expr, {ineq$_1$, ..., ineq$_n$}, name)

Computes the dual of the linear program defined by the objective expression expr and the nonstrict linear constraint inequalities ineq$_1$ through ineq$_n$. name is used as a base for the creation of variable names in the dual.
**Output:** An expression sequence, containing the dual's objective

function and a set of nonstrict constraint inequalities defining the dual, is returned.

**Argument options:** (expr, [eqn$_1$, ..., eqn$_n$], name) to return the constraint inequalities in a list instead of a set.

**Additional information:** ineq$_1$ through ineq$_n$ should be in the form dictated by passing them through simplex[convert](*stdle*). Otherwise, an error message is returned.

**See also:** simplex[convert](*stdle*), simplex[convert](*std*), simplex[standardize]

### simplex[feasible]({ineq$_1$, ..., ineq$_n$})

Determines whether the linear system represented by nonstrict linear constraints ineq$_1$ through ineq$_n$ has a feasible solution.

**Output:** A boolean value of true or false is returned.

**Argument options:** ([ineq$_1$, ..., ineq$_n$]) to produce the same result. ♣ ({ineq$_1$, ..., ineq$_n$}, *NONNEGATIVE*) to indicate that all variables in the system are non-negative. ♣ ({ineq$_1$, ..., ineq$_n$}, *UNRESTRICTED*) to indicate that all variables in the system are unrestricted. This is the default value. ♣ ({ineq$_1$, ..., ineq$_n$}, name$_1$, name$_2$) to assign the final feasible system found, if any, and the necessary variable transformations to name$_1$ and name$_2$, respectively.

**Additional information:** ineq$_1$ through ineq$_n$ should be in the form dictated by passing them through simplex[convert](*std*). Otherwise, an error message is returned.

**See also:** simplex[convert](*std*), simplex[maximize], simplex[minimize]

### simplex[maximize](expr, {ineq$_1$, ..., ineq$_n$})

Maximizes the linear program represented by objective expression expr and nonstrict linear constraint system ineq$_1$ through ineq$_n$.

**Output:** If a feasible solution exists, a set of equalities representing the maximizing values for the variables in the objective function is returned. If no feasible solution exists, the empty set is returned. If the solution is unbounded, NULL is returned.

**Argument options:** ([ineq$_1$, ..., ineq$_n$]) to produce the same result. ♣ (expr, {ineq$_1$, ..., ineq$_n$}, *NONNEGATIVE*) to indicate that all variables in the linear program are non-negative. ♣ (expr, {ineq$_1$, ..., ineq$_n$}, *UNRESTRICTED*) to indicate that all variables in the linear program are unrestricted. This is the default value. ♣ (expr, {ineq$_1$, ..., ineq$_n$}, name$_1$, name$_2$) to assign the optimal system found, if any, and the necessary variable transformations to name$_1$ and name$_2$, respectively.

## Solving Equations   139

**Additional information:** ineq$_1$ through ineq$_n$ should be in the form dictated by passing them through simplex[convert](*std*). Otherwise, an error message is returned.

**See also:** simplex[convert](*std*), simplex[minimize], simplex[feasible]

### simplex[minimize](expr, {ineq$_1$, ..., ineq$_n$})

Minimizes the linear program represented by objective expression expr and nonstrict linear constraint system ineq$_1$ through ineq$_n$.

**Output:** If a feasible solution exists, a set of equalities representing the minimizing values for the variables in the objective function is returned. If no feasible solution exists, the empty set is returned. If the solution is unbounded, NULL is returned.

**Argument options:** ([ineq$_1$, ..., ineq$_n$]) to produce the same result. ♣ (expr, {ineq$_1$, ..., ineq$_n$}, *NONNEGATIVE*) to indicate that all variables in the linear program are non-negative. ♣ (expr, {ineq$_1$, ..., ineq$_n$}, *UNRESTRICTED*) to indicate that all variables in the linear program are unrestricted. This is the default value. ♣ (expr, {ineq$_1$, ..., ineq$_n$}, name$_1$, name$_2$) to assign the optimal system found (if any) and the necessary variable transformations to name$_1$ and name$_2$, respectively.

**Additional information:** ineq$_1$ through ineq$_n$ should be in the form dictated by passing them through simplex[convert](*std*). Otherwise, an error message is returned.

**See also:** simplex[convert](*std*), simplex[maximize], simplex[feasible]

### simplex[pivot]({eqn$_1$, ..., eqn$_n$}, var, eqn$_p$)

Performs a simplex pivot on linear equations eqn$_1$ through eqn$_n$ with respect to pivot variable var and pivot equation eqn$_p$.

**Output:** A set of n equations is returned.

**Argument options:** ([eqn$_1$, ..., eqn$_n$], var, eqn$_p$) to return a list of equations. ♣ ({eqn$_1$, ..., eqn$_n$}, var, [eqn$_{p1}$, ..., eqn$_{pm}$]) to use one of eqn$_{p1}$ through eqn$_{pn}$ as the pivot equation. Note that eqn$_{p1}$ is always chosen.

**Additional information:** In order to reflect the pivot on the objective function, simply substitute the results of simplex[pivot] into that function. ♣ To determine a pivot variable and pivot equation, use simplex[pivotvar] and simplex[pivoteqn], respectively.

**See also:** simplex[pivotvar], simplex[pivoteqn], *subs*

### simplex[pivoteqn]({eqn₁, ..., eqn_n}, var)

Determines the subset of linear equations $eqn_1$ through $eqn_n$ that represents the pivot equation(s) with respect to variable var.
**Output:** If a pivot equation can be found, a list of linear equations is returned. Otherwise, FAIL is returned.
**Argument options:** ([eqn₁, ..., eqn_n], var) to produce the same result.
**Additional information:** $eqn_1$ through $eqn_n$ should be in the form dictated by passing them through simplex[setup]. Otherwise, an error message is returned. ♣ All equations returned achieve a minimum non-negative ratio. For more information on how that ratio is calculated, see the on-line help page. ♣ Once an equation is determined, a pivot can be performed with simplex[pivot].
**See also:** simplex[pivotvar], simplex[pivot]

### simplex[pivotvar](expr)

Determines a variable with positive coefficient from the objective expression expr.
**Output:** If a pivot variable exists, its variable name is returned. Otherwise, FAIL is returned.
**Argument options:** (expr, [var₁, ..., var_n]) to return the first variable with positive coefficient, checking in order from $var_1$ through to $var_n$.
**Additional information:** A list of variables in a linear system can be determined with simplex[basis]. ♣ Once a pivot variable is determined, simplex[pivoteqn] can be used to determine the pivot equation(s).
**See also:** simplex[basis], simplex[pivoteqn], simplex[pivot]

### simplex[ratio]({eqn₁, ..., eqn_n}, var)

Determine which of the equations $eqn_1$ through $eqn_n$ is most restrictive with respect to the simplex pivot for variable var.
**Output:** A set containing n rational values is returned.
**Argument options:** ([eqn₁, ..., eqn_n], var) to return a list of ratios.
**Additional information:** $eqn_1$ through $eqn_n$ should be in the form dictated by passing them through simplex[setup]. Otherwise, an error message is returned. ♣ The ratios computed equal, if possible, the negative of the constant term of each equation divided by the coefficient of var to the first power term. ♣ If a ratio is negative or has a zero denominator, it is reported as infinity.
**See also:** simplex[setup]

## simplex[setup]({eqn$_1$, ..., eqn$_n$})

Converts linear equations eqn$_1$ through eqn$_n$ into equivalent equations that form the basis for the corresponding linear system.
**Output:** A set of linear equations is returned. The single variables that constitute the left-hand sides of these equations do not appear in the right-hand sides of any of the same equations.
**Argument options:** ([eqn$_1$, ..., eqn$_n$]) to return a list of equations forming the basis. ♣ ({ineq$_1$, ..., ineq$_n$}) to transform any nonstrict inequalities in ineq$_1$ through ineq$_n$ into equalities. Slack variables named _SL1, _SL2, etc. are used. ♣ ({eqn$_1$, ..., eqn$_n$}, *NONNEGATIVE*) to specify that all variables are assumed nonnegative. ({eqn$_1$, ..., eqn$_n$}, name) to assign the set of transformations used for unrestricted variables to name.
**Additional information:** It is not necessary that the results of simplex[setup] correspond to a *feasible* system of equations.
**See also:** simplex[basis], simplex[feasible]

## simplex[standardize]({ineq$_1$, ..., ineq$_n$})

This command is identical to simplex[convert](*stdle*).
**See also:** simplex[convert](*stdle*), simplex[convert](*std*)

## solve(eqn, var)

Symbolically solves equation eqn for variable var.
**Output:** An expression sequence of possible solutions is returned.
**Argument options:** (expr, var) to solve the equation of the form expr = 0. ♣ ({eqn$_1$, ..., eqn$_n$}, {var$_1$, ..., var$_m$}) to solve a set of equations for a set of unknown variables. The result is an expression sequence of sets, representing possible solutions for the system. ♣ (eqn) or ({eqn$_1$, ..., eqn$_n$}) to solve an equation or set of equations for all of the unknowns contained in them. ♣ (ineq, var) to solve inequality ineq for variable var. An expression sequence of sets of inequalities is returned. ♣ (s, var) to solve series s for variable var. A series, in one of the remaining variables, is returned.
**Additional information:** If a NULL expression sequence is found, this can mean that either there are no solutions or Maple was unable to find any. ♣ If the equation(s) entered are of degree greater than four, then answers are provided in terms of RootOfs. ♣ The number of equations and variables need not be equal. That is, a system can be underdetermined or overdetermined. ♣ Both equations and inequations may be entered in a single call to solve. ♣ If floating-point values are used in solve, they are converted to their rational

counterparts, the symbolic solution is found, and all numeric values in the result are converted back to floating-point. ♣ If var is only used as a function in eqn, then a procedure representing the solution is returned.
**See also:** fsolve, isolve, msolve, rsolve, dsolve, subs
**FL = 15−19** *LA = 20, 22−24* ***LI = 196−203***

### testeq(expr$_1$ = expr$_2$)

Tests, using a random polynomial-time equivalence tester, if the expressions expr$_1$ and expr$_2$ are equivalent.
**Output:** If equivalence can be proved or disproved, a boolean value of true or false is returned. Otherwise, FAIL is returned.
**Argument options:** (expr$_1$, expr$_2$) to achieve the same result.
**Additional information:** For more information about how expressions are tested, see the on-line help page.
**See also:** *evalb*
***LI = 214***

### type(expr, '=')

Tests whether expr is an equation.
**Output:** A boolean value of either true or false is returned.
**See also:** type('<'), type('<='), type('<>')

### type(expr, '<')

Tests whether expr is a strict inequality.
**Output:** A boolean value of either true or false is returned.
**Additional information:** There is no equivalent command for >, because Maple automatically converts all greater thans to less thans.
**See also:** type('='), type('<='), type('<>')

### type(expr, '<=')

Tests whether expr is a nonstrict inequality.
**Output:** A boolean value of either true or false is returned.
**Additional information:** There is no equivalent command for >=, because Maple automatically converts all greater than or equals to less than or equals.
**See also:** type('='), type('<'), type('<>')

### type(expr, '<>')

Tests whether expr is an inequality of the <> type.
**Output:** A boolean value of either true or false is returned.
**See also:** type('='), type('<'), type('<=')

**type(expr, *equation*)**

This command is identical to type('=').
**See also:** type('=')

# Polynomials and Common Transforms

## Introduction

Polynomials are an integral part of any symbolic computation package, and Maple is no exception. Maple contains commands for their creation, manipulation, and transformation. Some of these commands are scattered throughout other chapters of *The Maple Handbook*, but the majority are listed in this chapter.

## What is a Polynomial?

At its most basic level, a polynomial is simply a sum of terms with positive integer exponents. These terms can contain almost any type of variable and coefficient. A single constant term is regarded as a polynomial even though it fails as type('+').

> x*fred - 2/Pi;

$$x\,fred - 2\,\frac{1}{\pi}$$

> type(", polynom);

$$true$$

> exp(1);

$$e$$

> type(", polynom);

$$true$$

Of course, the most frequently used polynomials are more standard. Maple uses both univariate and multivariate polynomials. The type(*polynom*) command can be used to check for polynomials with certain coefficient types and over specific variables.

> p := x^3 + 2*x^2 - 5*x +6;

$$p := x^3 + 2x^2 - 5x + 6$$

> type(p, polynom(integer));

$$true$$

> type(p, polynom(integer, y));

$$false$$

> type(p, polynom(integer, x));

$$true$$

> q := x^2*y - y^2*x + 5/3*x*y - 2/3*x + y - 1;

$$q := x^2 y - y^2 x + \frac{5}{3} x y - \frac{2}{3} x + y - 1$$

> type(q, polynom(integer));

$$false$$

> type(q, polynom(rational));

$$true$$

> type(p, polynom(anything, y));

$$true$$

> type(p, polynom(anything, [x, y]));

$$true$$

As well, there are several commands for determining other information about polynomials.

```
> lcoeff(p);
```

$$1$$

```
> degree(p);
```

$$3$$

The internal representation of a polynomial is straightforward; the terms of a polynomial directly correspond to its operands.

```
> op(p);
```

$$x^3, 2x^2, -5x, 6$$

```
> op(q);
```

$$x^2 y, -y^2 x, \frac{5}{3} x y, -\frac{2}{3} x, y, -1$$

## Standard Manipulations

There are many standard commands available for polynomial manipulation. Typically, they take as input one or more polynomials and perhaps some variable names. Instead of listing all the available commands, let's just run through examples of a representative subset of them.

```
> p1 := 3*x^3-3*x^2-15*x-9:
> p2 := x^3-7*x-6:
> p3 := x^2+3*x+2:
> expand(p1*p2);
```

$$3x^6 - 36x^4 - 6x^3 - 3x^5 + 123x^2 + 153x + 54$$

```
> divide(p2, p3, quotient);
```

$$true$$

```
> quotient;
```

$$x - 3$$

```
> gcd(p1, p2);
```

$$x^2 - 2x - 3$$

## Common Transforms

Maple contains many standard symbolic transforms, including Fourier transforms, Laplace transforms, and $Z$ transforms. Typically, the inverse of these transforms is also avaliable.

```
> readlib(fourier):
> fourier(t/(1+t^2), t, w);
```

$$I\pi\left(e^w\,Heaviside(-w) - e^{(-w)}\,Heaviside(w)\right)$$

```
> ztrans(2^n/n!, n, w);
```

$$e^{\left(2\frac{1}{w}\right)}$$

```
> invztrans(", w, n);
```

$$\frac{2^n}{n!}$$

## Parameter Types Specific to This Chapter

| | |
|---|---|
| polyp1 | a polynomial in modp1 format |
| ratpoly | a rational polynomial |

## Command Listing

### allvalues(expr)

Evaluates all possible values of expr, an expression typically containing RootOfs.
**Output:** An expression sequence is returned.
**Argument options:** (expr, *d*) to indicate that identical RootOfs in expr represent the same value and are not to be evaluated separately.

**Additional information:** allvalues uses solve to calculate the roots exactly, if possible. If not, fsolve is used. ♣ The number of values returned equals the product of the numbers of solutions to each independent RootOf expression.
**See also:** RootOf, *solve*, *fsolve*
*LI = 7*

### bernstein(n, < expr$_{var}$ | var >, name)

Calculates the n$^{th}$ degree Bernstein polynomial in name that approximates the functional operator defined by < expr$_{var}$ | var > on the [0, 1] interval.
**Output:** An polynomial in name is returned.
**Argument options:** (n, proc, name) to calculate the polynomial approximating the Maple procedure proc on the [0, 1] interval. proc typically consists of an if/then/else/fi programming structure.
**Additional information:** This command needs to be defined by readlib(bernstein) before being invoked. ♣ bernstein only works with functions of *one* variable.
**See also:** *proc*, *if/then/else/fi*, op(polynom)
*LI = 264*

### coeff(poly, var, int)

Determines the coefficient in poly of variable var raised to the power int.
**Output:** An expression is returned.
**Argument options:** (poly, expr), where expr is an expression in var, to determine the coefficient attached to expr in poly.
**Additional information:** Before calling coeff, collect should be applied to poly to collect like terms in var.
**See also:** lcoeff, tcoeff, coeffs, *collect*
**FL = 77** *LI = 19*

### coeffs(poly, var)

Determines all the coefficients in poly with respect to variable var.
**Output:** An expression sequence is returned.
**Argument options:** (poly) to determine the coefficients with respect to *all* the variables in poly. ♣ (poly, [var$_1$, ..., var$_n$]) or (poly, {var$_1$, ..., var$_n$}) to determine the coefficients with respect to variables var$_1$ through var$_n$. ♣ (poly, var, name) to assign the *terms* of poly with respect to var to name.

**Additional information:** Before calling coeffs, collect should be applied to poly to collect like terms in var.
**See also:** coeff, lcoeff, tcoeff, *indets*, *collect*
*LI = 20*


### compoly(poly, var)

Computes a polynomial composition in var that could have created the input polynomial poly.
**Output:** If a composition can be found, a two-element expression sequence is returned. The first element is the base polynomial $poly_b$, and the second element is the composition equation $eqn_c$ such that subs($eqn_c$, $poly_b$) = poly. Otherwise, FAIL is returned.
**Additional information:** Keep in mind that the polynomial composition found by compoly may not be unique.
**See also:** *subs*, *factor*
*LI = 27*


### content(poly, var)

Determines the content of poly with respect to variable var.
**Output:** An expression is returned.
**Argument options:** (poly) to determine the content with respect to *all* the variables in poly. ♣ (poly, [$var_1$, ..., $var_n$]) or (poly, {$var_1$, ..., $var_n$}) to determine the content with respect to variables $var_1$ through $var_n$. ♣ (poly, var, name) to assign the primitive part of poly with respect to var to name.
**Additional information:** The content is the greatest common divisor of the coefficients of poly with respect to var. ♣ The primitive part is simply poly divided by the content.
**See also:** primpart, icontent, coeffs, *gcd*
*LI = 28*


### Content(poly, var)

Represents the inert content command for poly with respect to variable var.
**Output:** An unevaluated Content call is returned.
**Argument options:** Content(poly, var) *mod* posint to evaluate the inert command call modulo posint.
**Additional information:** More information on other parameter sequences for Content is found in the listing for content.
**See also:** Primpart, content, evala
*LI = 27*

## convert(poly, *horner*, var)

Converts poly, a polynomial in var, into Horner form.
**Output:** A polynomial is returned.
**Argument options:** (poly, *horner*) to convert to Horner form with respect to *all* the variables in poly. ♣ (poly, *horner*, {$var_1$, ..., $var_n$}) to convert to Horner form with respect to n variables in uncontrolled order. ♣ (poly, *horner*, [$var_1$, ..., $var_n$]) to convert to Horner form with respect to n variables in specific order.
**See also:** convert(*mathorner*)
*LI = 40*

## convert(poly, *mathorner*, var)

Converts poly, a polynomial in var, into matrix Horner form.
**Output:** A polynomial is returned.
**Argument options:** (poly, *horner*, M) to convert a matrix polynomial in M into Horner form.
**Additional information:** This command is designed to perform efficient calculation with matrix polynomials ♣ For practical examples of its use, see the on-line help page.
**See also:** convert(*horner*)

## convert(poly, *sqrfree*, var)

Converts poly, a polynomial in var, into square-free form through square-free factorization.
**Output:** A factored polynomial is returned.
**Argument options:** (poly, *sqrfree*) to convert to square-free form with respect to *all* the variables in poly.
**Additional information:** Square-free form is reached by applying content(poly, var), removing the content, and then factoring the primitive part that remains. ♣ If var is not specified, then before factorization occurs content is applied recursively to poly for each variable in poly. ♣ The factors in the result of convert(*sqrfree*) are in random order.
**See also:** content, primpart
*LI = 54*

## degree(poly, var)

Determines the highest degree of poly with respect to variable var.
**Output:** An integer is returned.
**Argument options:** (poly) to determine the highest total degree with respect to *all* the variables in poly. ♣ (poly, {$var_1$, ..., $var_n$}) to determine the highest *total* degree of poly with respect to n variables. ♣ (poly, [$var_1$, ..., $var_n$]) to determine the highest *vector*

degree of poly with respect to n variables.
**Additional information:** Before calling degree, collect should be applied to poly to collect like terms in var. ♣ For details on the difference between total degree and vector degree, see the on-line help page for degree.
**See also:** ldegree, lcoeff, tcoeff, dinterp, *collect*
*LI = 61*


### dinterp(proc, $int_1$, $int_2$, $int_3$, $int_p$)

Determines probabilistically the degree of the variable in the $int_2$ position of the polynomial defined by procedure proc. proc defines a polynomial modulo $int_p$ (a prime) in $int_1$ variables with a degree bound on the desired variable of $int_3$.
**Output:** An integer is returned.
**Additional information:** This command needs to be defined by readlib(dinterp) before being invoked. ♣ To decrease the probability that dinterp returns a degree that is too low, increase the size of the modulus $int_p$. ♣ dinterp is short for *d*egree *interp*olation.
**See also:** degree
*LI = 278*


### discrim(poly, var)

Computes the discriminant of poly with respect to variable var.
**Output:** An expression is returned.
**Additional information:** The discriminant is defined in terms of the resultant, degree, and leading coefficient of poly. ♣ See the on-line help page for the exact definition.
**See also:** resultant, degree, lcoeff
*LI = 65*


### Discrim(poly, var)

Represents the inert discriminant command for poly with respect to variable var.
**Output:** An unevaluated Discrim call is returned.
**Argument options:** Discrim(poly, var) *mod* posint to evaluate the inert command call modulo posint.
**See also:** Resultant, discrim

### DistDeg(poly, var)

Represents the inert distinct degree factorization for poly with respect to variable var.

**Output:** An unevaluated DistDeg call is returned.
**Argument options:** DistDeg(poly, var) *mod* posint to evaluate the inert command call modulo posint. posint must be a prime. A list of lists, where the internal lists contain factors and their multiplicities, is returned. ♣ For more information on distinct degree factorization, see the on-line help page.
**See also:** Factors, Sqrfree
*LI = 65*

### divide(poly$_1$, poly$_2$, name)

Determines if polynomial poly$_1$ is divisible by polynomial poly$_2$. If they are divisible, then the quotient is assigned to name.
**Output:** A boolean value of true or false is returned.
**See also:** gcd, quo, rem, Divide
*LI = 67*

### Divide(poly$_1$, poly$_2$, name)

Represents the inert divisibility command for polynomials poly$_1$ and poly$_2$.
**Output:** An unevaluated Divide call is returned.
**Argument options:** Divide(poly$_1$, poly$_2$, name) *mod* posint to evaluate the inert command call modulo posint. If they are divisible, name is assigned to the quotient.
**See also:** divide, Quo, Rem, Gcd, evala, modp1
*LA = 53 LI = 66*

### Eval(poly, var=expr)

Represents the inert evaluation command for univariate polynomial poly at var=expr.
**Output:** An unevaluated Eval call is returned.
**Argument options:** (poly, {var$_1$=expr$_1$, ..., var$_n$=expr$_n$}) to represent the inert evaluation for multivariate poly at var$_1$=expr$_1$ through var$_n$=expr$_n$. ♣ Eval(poly, var) *mod* posint to evaluate the inert command call modulo posint.
**See also:** *eval, subs,* Powmod
*LI = 75*

### evala(Fnc(expr$_1$, ..., expr$_n$))

Evaluates inert command Fnc, with parameters expr$_1$ through expr$_n$, over the *algebraic closure* of its coefficient field.
**Output:** A data structure, that may contain RootOfs, is returned.
**Argument options:** (Fnc(expr$_1$, ..., expr$_n$), *independent*) to specify that no independence checking is done on the RootOf statements in the result. ♣ (*AFactor*(poly)) to compute the absolute

factorization of poly. ♣ (*AFactors*(poly)) to compute the absolute factors of poly. ♣ (*Expand*(expr)) to expand expr into a polynomial. ♣ (*Factor*(poly, expr), option), where option is one of *lenstra*, *trager*, or *linear*, to factor poly using a special method. Default is *trager*. ♣ (*Factors*(poly, expr), option), where option is one of *lenstra*, *trager*, or *linear*, to list the factors of poly derived through a special method. Default is *trager*. ♣ (*Indep*(set, name)) to determine whether the RootOf statements in set are independent. If relations are found, they are assigned to name. ♣ (*Norm*(expr, set$_1$, set$_2$)) to compute the norm of algebraic expression expr over an algebraic number field determined by set$_1$ and set$_2$. See the on-line help page for Norm to learn more about restrictions on the sets. ♣ (*Normal*(expr)) to compute the normal form of algebraic expression expr. ♣ (*Trace*(expr, set$_1$, set$_2$)) to compute the trace of algebraic expression expr over an algebraic number field determined by set$_1$ and set$_2$. See the on-line help page for Trace to learn more about restrictions on the sets. ♣ Other inert command calls that work with evala include *Content*, *Divide*, *Gcd*, *Gcdex*, *Prem*, *Primfield*, *Primpart*, *Quo*, *Rem*, *Resultant*, *Sprem*, and *Sqrfree*. For more details, see the respective command entries for these inert commands.

**Additional information:** As a general rule, the coefficients of the polynomials passed as parameters to Fnc must be algebraic numbers (i.e., they may contain RootOf commands). ♣ The inert command is performed in the smallest possible algebraic number field. ♣ If Fnc is unknown to evala, it is returned unevaluated. ♣ For more information on the possible parameter sequences and result types for specific unevaluated functions, see the on-line help for that function or its *active* counterpart (e.g., Prem and prem).

**See also:** evalgf, modp1, mod, RootOf

*LI = 76*

### evalgf(Fnc(expr$_1$, ..., expr$_n$), posint)

Evaluates the inert command Fnc, with parameters expr$_1$ through expr$_n$, in an algebraic extension of finite field $Z$ modulo posint.

**Output:** A data structure, which may contain RootOf commands, is returned.

**Argument options:** (*Expand*(expr), posint) to expand expr into a polynomial. ♣ (*Normal*(expr), posint) to compute the normal form of the algebraic expression expr. ♣ Other inert command calls that work with evala include *Content*, *Divide*, *Gcd*, *Gcdex*, *Prem*, *Primpart*, *Quo*, *Rem*, *Resultant*, and *Sprem*. For more details, see the respective command entries for these inert commands.

**Additional information:** This command needs to be defined by

readlib(evalgf) before being invoked. ♣ As a general rule, the coefficients of the polynomials passed as parameters to Fnc must be algebraic numbers (i.e., they may contain RootOf commands). ♣ If a dependency between RootOfs is encountered during computation, an error message is returned. ♣ The inert command is performed in the smallest possible algebraic extension of $Z$ mod posint. ♣ If Fnc is unknown to evalgf, it is returned unevaluated.
**See also:** evala, *GF*, modp1, mod, RootOf
*LI = 281*

### factor(poly)

Factors the multivariate polynomial poly.
**Output:** An expression is returned.
**Argument options:** (poly, expr) to factor poly over the number field defined by expr, a radical or RootOf expression, for example, sqrt(2) or RootOf(x^2-3). ♣ (poly, [expr$_1$, ..., expr$_n$]) or (poly, {expr$_1$, ..., expr$_n$}) to factor expr over the number field defined by n radicals or RootOf expressions. ♣ (rat) to factor rat, a rational polynomial or expression. rat is normalized and the numerator and denominator are factored separately. ♣ (eqn) to factor both sides of an equation. ♣ (expr) to factor an expression that is not necessarily a polynomial. ♣ ([poly$_1$, ..., poly$_n$]) or ({poly$_1$, ..., poly$_n$}) to factor poly$_1$ through poly$_n$ and return the results in a list or set, respectively.
**Additional information:** factor also works on each component of a range, series, or function. ♣ If only one parameter is given, poly is factored over the field implied by its coefficients.
**See also:** *expand*, *collect*, *combine*, *simplify*, *normal*, factors, irreduc, Factor, Factors, RootOf
**FL = 65−67** *LA = 4* *LI = 92*

### Factor(poly)

Represents the inert factoring command for multivariate polynomial poly.
**Output:** An unevaluated Factor call is returned.
**Argument options:** Factor(poly) *mod* posint to evaluate the inert command call modulo posint.
**Additional information:** For more information on valid parameter sequences, see the command listing for factor. ♣ Three special methods can be called upon when Factor is combined with evala. See the listing for evala for more information.
**See also:** factor, factors, Factors, evala, modp1
*LA = 4* *LI = 91*

### factors(poly)

Computes the factors of multivariate polynomial poly.

**Output:** A two-element list is returned. The first element is the constant factor and the second element is a list of two-element lists, where the first elements are the factors and the second elements are the corresponding multiplicities.

**Argument options:** (poly, expr) to factor poly over the number field defined by expr, a radical or RootOf expression, for example, sqrt(2) or RootOf(x^2-3)). ♣ (poly, [expr$_1$, ..., expr$_n$]) or (poly, {expr$_1$, ..., expr$_n$}) to factor poly over the number field defined by n radicals or RootOf expressions.

**Additional information:** If only one parameter is given, poly is factored over the field implied by its coefficients.

**See also:** factor, irreduc, Factor, Factors

*LI = 284*


### Factors(poly)

Represents the inert form of the factors command for multivariate polynomial poly.

**Output:** An unevaluated Factors call is returned.

**Argument options:** Factors(poly) *mod* posint to evaluate the inert command call modulo posint.

**Additional information:** For more information on valid parameter sequences, see the command listing for factors. ♣ Three special methods can be called upon when Factors is combined with evala. See the listing for evala for more information.

**See also:** factor, factors, Factor, evala, modp1

*LI = 93*


### FFT(n, A$_{re}$, A$_{im}$)

Computes the fast Fourier transform of $2^n$ complex values. The $2^n$-element arrays A$_{re}$ and A$_{im}$ represent the real and imaginary components of the complex values.

**Output:** The value $2^n$ is returned. As well, the elements in arrays A$_{re}$ and A$_{im}$ are given new values representing the transformed complex values.

**Additional information:** This command needs to be defined by readlib(FFT) before being invoked. ♣ Standard Fourier transforms can be computed with fourier. ♣ To use more efficient hardware floating-point calculations, call FFT inside evalhf. See the listing for evalhf for more information.

**See also:** iFFT, fourier, laplace, ztrans, mellin, *evalhf*

*LI = 285*

### fixdiv(poly, var)

Computes the fixed divisor of poly, a polynomial in var over the integers.
**Output:** A positive integer is returned.
**Additional information:** This command needs to be defined by readlib(fixdiv) before being invoked. ♣ The fixed divisor is the largest integer that divides poly evaluated at all the integers.
**See also:** *subs*
*LI = 288*

### fourier(expr, var, name)

Computes the Fourier transform of expr with respect to var. The resulting expression is in terms of variable name.
**Output:** An expression in name is returned.
**Additional information:** This command needs to be defined by readlib(fourier) before being invoked. ♣ The input expression may contain unevaluated calls to commands such as int, BesselJ, or exp. ♣ The resulting expression may contain unevaluated calls to commands such as fourier, Heaviside, or Dirac. ♣ Fast Fourier transforms can be computed with the FFT command. ♣ For more information, see the on-line help page.
**See also:** invfourier, FFT, laplace, ztrans, mellin

### galois(poly)

Computes the Galois group of poly, a irreducible univariate rational polynomial of degree less than eight.
**Output:** An sequence of three expressions is returned. The first expression represents the name of the Galois group and includes a + as the first character if the group is even. The second expression is an integer representing the order of the group. The third expression is a set of strings representing the generators of the group.
**Additional information:** If poly is not of the correct form, an error message is returned.
**See also:** *GF, group[permgroup]*
*LI = 98*

### gcd(poly$_1$, poly$_2$)

Computes the greatest common divisor of poly$_1$ and poly$_2$, polynomials with rational coefficients.
**Output:** A polynomial is returned.
**Argument options:** (poly$_1$, poly$_2$, name$_1$, name$_2$) to assign

poly$_1$/gcd(poly$_1$, poly$_2$) to name$_1$ and poly$_2$/gcd(poly$_1$, poly$_2$) to name$_2$.
**See also:** lcm, *igcd*, *GaussInt[GIgcd]*, gcdex, Gcd
*LA = 3−4 **LI** = 100*

### Gcd(poly$_1$, poly$_2$)

Represents the inert form of the gcd command for multivariate polynomials poly$_1$ and poly$_2$.
**Output:** An unevaluated Gcd call is returned.
**Argument options:** Gcd(poly$_1$, poly$_2$) *mod* posint to evaluate the inert command call modulo posint.
**Additional information:** For more information on other valid parameter sequences for Gcd, see the command listing for gcd.
**See also:** gcd, Gcdex, gcdex, evala, modp1
*LA = 53 **LI** = 100*

### gcdex(poly$_1$, poly$_2$, var, name$_1$, name$_2$)

Computes the greatest common divisor of poly$_1$ and poly$_2$, two univariate polynomials in var with rational coefficients, using the extended Euclidian algorithm.
**Output:** A polynomial is returned. As well, name$_1$ and name$_2$ are assigned expressions such that name$_1$ ∗ poly$_1$ + name$_2$ ∗ poly$_2$ = gcd(poly$_1$, poly$_2$).
**Argument options:** (poly$_1$, poly$_2$, poly$_3$, var, name$_1$, name$_2$) to assign to name$_1$ and name$_2$ expressions such that name$_1$ ∗ poly$_1$ + name$_2$ ∗ poly$_2$ = poly$_3$. In this invocation, no gcd of poly$_1$ and poly$_2$ is returned.
**See also:** gcd, *igcdex*, *GaussInt[GIgcdex]*, Gcdex
***LI** = 102*

### Gcdex(poly$_1$, poly$_2$, var, name$_1$, name$_2$)

Represents the inert form of the gcdex command for poly$_1$ and poly$_2$, multivariate polynomials in var.
**Output:** An unevaluated Gcdex call is returned.
**Argument options:** Gcdex(poly$_1$, poly$_2$, var, name$_1$, name$_2$) *mod* posint to evaluate the inert command call modulo posint.
**Additional information:** For more information on other valid parameter sequences for Gcdex, see the command listing for gcdex.
**See also:** gcdex, Gcd, gcd, evala, modp1
***LI** = 101*

### genpoly(n, n$_m$, var)

Generates a polynomial in var from integer n using Z-adic expansion. The value of n$_m$ controls the magnitude of the coefficients.

**Output:** A polynomial is returned.
**Argument options:** (poly, $n_m$, var) to generate the polynomial from poly, a polynomial in fully expanded form.
**Additional information:** The coefficients of the result have magnitude less than $n_m/2$. ♣ For more information on how the polynomial is generated, see the on-line help page.
*LI = 103*

### grobner package

Provides commands for performing *Gröbner basis* calculations.
**Additional information:** The two orderings available are *plex* (pure lexicographical—alphabetical by the names) and *tdeg* (total degree of the terms). ♣ To access the command fnc from the grobner package, use the long form of the name grobner[fnc], or with(grobner, fnc) to load in a pointer for fnc only, or with(grobner) to load in pointers to all the commands.

### grobner[finduni](var, [poly$_1$, ..., poly$_m$], [var$_1$, ..., var$_n$])

Constructs the univariate polynomial in var of smallest degree in the ideal generated by polynomials poly$_1$ through poly$_m$ with respect to variables var$_1$ through var$_n$.
**Output:** A polynomial in var is returned.
**Argument options:** (var, [poly$_1$, ..., poly$_m$]) to compute the Gröbner basis of poly$_1$ through poly$_m$ with respect to *all* the variables in those polynomials.
**Additional information:** Only if the polynomial system has finitely many solutions, is a polynomial solution guaranteed. Use grobner[finite] to check whether there are infinitely many solutions. ♣ If grobner[finite] returns false, don't attempt grobner[finduni]!
**See also:** grobner[finite], grobner[solvable], grobner[gsolve]

### grobner[finite]([poly$_1$, ..., poly$_m$], [var$_1$, ..., var$_n$])

Determines if the system of polynomials poly$_1$ through poly$_m$ has a finite number of solutions with respect to variables var$_1$ through var$_n$.
**Output:** A boolean value of true or false is returned.
**Argument options:** ([poly$_1$, ..., poly$_m$]) to check whether there are a finite number of solutions of poly$_1$ through poly$_m$ with respect to *all* the variables in those polynomials.
**Additional information:** A total degree Gröbner basis is used to determine the number of solutions. ♣ This command should be

called before calling grobner[finduni]. You should proceed only if the result is true.
**See also:** grobner[finduni], grobner[solvable], grobner[gsolve]

### grobner[gbasis]([poly$_1$, ..., poly$_m$], [var$_1$, ..., var$_n$])

Computes a reduced, minimal Gröbner basis for polynomials poly$_1$ through poly$_m$ with respect to variables var$_1$ through var$_n$.
**Output:** A list of polynomials in the Gröbner basis is returned.
**Argument options:** ([poly$_1$, ..., poly$_m$], [var$_1$, ..., var$_n$], *plex*) to specify that the terms in the resulting Gröbner basis polynomials be sorted in *pure lexicographical* order. ♣ ([poly$_1$, ..., poly$_m$], [var$_1$, ..., var$_n$], *tdeg*) to specify that the terms in the resulting basis polynomials be sorted in *total degree* order. This is the default. ♣ ({poly$_1$, ..., poly$_m$}, {var$_1$, ..., var$_n$}, option, name) to reorder the input variables var$_1$ through var$_n$ in an optimal manner. This reordering gets assigned to name. option is either *plex* or *tdeg*.
**Additional information:** If the variables are entered in a list, they are not reordered by grobner[gbasis]. The ordering stands as var$_1$ > var$_2$ > ... > var$_n$.
**See also:** grobner[normalf]

### grobner[gsolve]([poly$_1$, ..., poly$_m$])

Computes reduced Gröbner bases for polynomials poly$_1$ through poly$_m$ with respect to *all* the variables in these polynomials.
**Output:** A list of lists is returned. Each internal list is a reduced subsystem with roots identical to the original system.
**Argument options:** ({poly$_1$, ..., poly$_m$}) to produce the same results. ♣ ([poly$_1$, ..., poly$_m$], {poly$_{r1}$, ..., poly$_{rn}$}) to prevent the roots of poly$_{r1}$ through poly$_{rn}$ being considered in computing the Gröbner bases. Warning: this option is not always effective. ♣ ([poly$_1$, ..., poly$_m$], [var$_1$, ..., var$_n$], {poly$_{r1}$, ..., poly$_{rn}$}) to specify the permutation of variables to be used in the Gröbner bases computation.
**Additional information:** The solve command can be applied to each sublist if desired. ♣ For more information on how the Gröbner bases are computed, see the on-line help page.
**See also:** *solve*, grobner[finite], grobner[solvable]

### grobner[leadmon](poly, [var$_1$, ..., var$_n$])

Computes the leading monomial of polynomial poly with respect to variables var$_1$ through var$_n$.
**Output:** A list with two elements is returned. The first element is

the coefficient of the leading monomial. The second element is the product of the variables and their respective exponents.
**Argument options:** (poly, [$var_1$, ..., $var_n$], *plex*) to specify the leading monomial be found with respect to a *pure lexicographical* ordering of the terms of poly. ♣ (poly, [$var_1$, ..., $var_n$], *tdeg*) to specify the leading monomial be found with respect to a *total degree* ordering of the terms of poly. This is the default.
**Additional information:** If poly contains any variables not in $var_1$ through $var_n$, they are treated as constants and included in the coefficients.
**See also:** grobner[spoly]

### grobner[normalf](poly$_r$, [poly$_1$, ..., poly$_m$], [var$_1$, ..., var$_n$])

Computes the fully reduced form of polynomial poly$_r$ with respect to the Gröbner basis represented by poly$_1$ through poly$_m$ and the variables var$_1$ through var$_n$.
**Output:** A polynomial is returned.
**Argument options:** (poly$_r$, [poly$_1$, ..., poly$_m$], [var$_1$, ..., var$_n$], *plex*) to specify that the terms in the resulting polynomial be sorted in *pure lexicographical* order. ♣ (poly$_r$, [poly$_1$, ..., poly$_m$], [var$_1$, ..., var$_n$], *tdeg*) to specify that the terms in the resulting polynomial be sorted in *total degree* order. This is the default.
**Additional information:** Typically, the grobner[gbasis] command is used to compute the Gröbner basis [poly$_1$, ..., poly$_m$]. This Gröbner basis must be with respect to the given ordering of the variables var$_1$ through var$_n$.
**See also:** grobner[gbasis]

### grobner[solvable]([poly$_1$, ..., poly$_m$], [var$_1$, ..., var$_n$])

Determines if the system of polynomials poly$_1$ through poly$_m$ is solvable (i.e., algebraically consistent) with respect to variables var$_1$ through var$_n$.
**Output:** A boolean value of true or false is returned.
**Argument options:** ([poly$_1$, ..., poly$_m$]) to check the solvability of poly$_1$ through poly$_m$ with respect to *all* the variables in those polynomials. ♣ ([poly$_1$, ..., poly$_m$], [var$_1$, ..., var$_n$], *plex*) to specify a *pure lexicographical* sorted Gröbner basis be used in the computations. ♣ ([poly$_1$, ..., poly$_m$], [var$_1$, ..., var$_n$], *tdeg*) to specify a *total degree* sorted Gröbner basis be used in the computations. This is the default.
**See also:** grobner[finite], grobner[gsolve]

### grobner[spoly](poly$_1$, poly$_2$, [var$_1$, ..., var$_n$])

Computes the S-polynomial of poly$_1$ and poly$_2$ with respect to variables var$_1$ through var$_n$.
**Output:** A polynomial is returned.
**Argument options:** (poly$_1$, poly$_2$, [var$_1$, ..., var$_n$], *plex*) to specify that computations be performed with respect to a *pure lexicographical* ordering of the terms. ♣ (poly$_1$, poly$_2$, [var$_1$, ..., var$_n$], *tdeg*) to specify that computations be performed with respect to a *total degree* ordering of the terms. This is the default.
**Additional information:** The grobner[leadmon] command is used in computing the S-polynomial. ♣ For more information about S-polynomials, see the on-line help page.
**See also:** grobner[leadmon]

### Hermite(M, var)

Represents the inert Hermite normal form command, where M is a matrix of univariate polynomials in var.
**Output:** An unevaluated Hermite call is returned.
**Argument options:** Hermite(M, var) *mod* posint to evaluate the inert command call modulo posint.
**Additional information:** More information on other parameter sequences for Hermite is found in the listing for linalg[hermite].
**See also:** *linalg[hermite]*, Smith, modp1
*LI = 105*

### icontent(poly)

Determines the integer content of poly.
**Output:** A rational value is returned.
**Additional information:** The integer content is a rational value such that dividing poly by this value makes the polynomial primitive over the integers. ♣ If poly has strictly integer coefficients, icontent returns the greatest common divisor of the coefficients.
**See also:** content, *igcd*
*LI = 106*

### iFFT(n, A$_{re}$, A$_{im}$)

Computes the inverse fast Fourier transform of $2^n$ complex values. The $2^n$-element arrays A$_{re}$ and A$_{im}$ represent the real and imaginary components of the complex values.
**Output:** The value $2^n$ is returned. As well, the elements in arrays A$_{re}$ and A$_{im}$ are given new values representing the transformed complex values.

**Additional information:** This command needs to be defined by readlib(FFT) before being invoked. ♣ Standard inverse Fourier transforms can be computed with invfourier. ♣ To use more efficient hardware floating-point calculations, call iFFT inside evalhf. See the listing for evalhf for more information.
**See also:** FFT, invfourier, invlaplace, invztrans, mellin, *evalhf*
*LI = 285*

### interp([exprx$_1$, ..., exprx$_{n+1}$], [expry$_1$, ..., expry$_{n+1}$], var)

Computes a polynomial in var of degree at most n, which represents the polynomial interpolation on independent values exprx$_1$ through exprx$_{n+1}$ and dependent values expry$_1$ through expry$_{n+1}$.
**Output:** A polynomial is returned.
**Additional information:** Polynomial interpolation is performed on the points [exprx$_1$, expry$_1$] through [exprx$_{n+1}$, expry$_{n+1}$].
♣ If two or more of the exprx$_i$ values are equal, an error message is returned.
**See also:** sinterp, Interp
*LI = 117*

### Interp([exprx$_1$, ..., exprx$_{n+1}$], [expry$_1$, ..., expry$_{n+1}$], var)

Represents the inert interp command for computing a multivariate polynomial.
**Output:** An unevaluated Interp call is returned.
**Argument options:** Interp([exprx$_1$, ..., exprx$_{n+1}$], [expry$_1$, ..., expry$_{n+1}$], var) *mod* posint to evaluate the inert command call modulo posint.
**Additional information:** For more information on polynomial interpolation, see the command listing for interp.
**See also:** interp, sinterp, modp1
*LI = 116*

### invfourier(expr, var, name)

Computes the inverse Fourier transform of expr with respect to var. The resulting expression is in terms of variable name.
**Output:** An expression is returned.
**Additional information:** This command needs to be defined by readlib(fourier) before being invoked. ♣ The input expression may contain unevaluated calls to commands such as Heaviside, Dirac, or exp. ♣ The resulting expression may contain unevaluated calls to commands such as fourier, Heaviside, or Dirac. ♣ For more information, see the on-line help page.
**See also:** fourier, iFFT, invlaplace, invztrans, mellin

### invlaplace(expr, var, name)

Computes the inverse Laplace transform of expr with respect to var. The resulting expression is in terms of variable name.
**Output:** An expression is returned.
**Additional information:** The input expression may contain unevaluated calls to commands such as laplace or sin. ♣ The resulting expression may contain unevaluated calls to commands such as int, Dirac, or Heaviside. ♣ For more information on invlaplace and examples of its usage, see the on-line help page.
**See also:** laplace, invfourier, invztrans, mellin, *dsolve*
*LI = 294*


### invztrans(expr, var, name)

Computes the inverse $Z$ transform of expr with respect to var. The resulting expression is in terms of variable name.
**Output:** An expression is returned.
**Additional information:** This command needs to be defined by readlib(ztrans) or by using the ztrans command before being invoked. ♣ The input expression may contain unevaluated calls to commands such as sin or ztrans. ♣ For more information on invztrans and examples of its usage, see the on-line help page.
**See also:** ztran, invfourier, invlaplace, mellin
*LI = 295*


### irreduc(poly)

Determines whether multivariate polynomial poly is irreducible over the field implied by its coefficients.
**Output:** A boolean value of true or false is returned.
**Argument options:** (poly, expr) determine whether poly is irreducible over the number field defined by expr, a radical or RootOf expression, for example, sqrt(2) or RootOf(x^2-3). ♣ (poly, [expr$_1$, ..., expr$_n$]) or (poly, {expr$_1$, ..., expr$_n$}) to determine if expr is irreducible over the number field defined by radicals or RootOf expressions expr$_1$ through expr$_n$.
**Additional information:** A constant polynomial is irreducible by definition.
**See also:** factor, roots, Irreduc, RootOf
*LI = 119*


### Irreduc(poly)

Represents the inert irreduc command for multivariate polynomial poly.

**Output:** An unevaluated Irreduc call is returned.
**Argument options:** Irreduc(poly) *mod* posint to evaluate the inert command call modulo posint.
**Additional information:** For more information on other valid parameter sequences for Irreduc, see the command listing for irreduc.
**See also:** irreduc, factor, Primitive, evala, modp1
FL = 210−211  *LI = 118*

### laplace(expr, var, name)

Computes the Laplace transform of expr with respect to var. The resulting expression is in terms of variable name.
**Output:** An expression is returned.
**Additional information:** The input expression may contain unevaluated calls to commands such as exp, sin, or diff. ♣ The resulting expression may contain unevaluated calls to commands such as laplace or D. ♣ For more information on laplace and examples of its usage, see the on-line help page.
**See also:** invlaplace, fourier, ztrans, mellin, *dsolve*
*LI = 122*

### lcm(poly$_1$, ..., poly$_n$)

Computes the lowest common multiple of poly$_1$ through poly$_n$, polynomials with rational coefficients.
**Output:** A polynomial is returned.
**See also:** gcd, *ilcm*, *GaussInt[GIlcm]*
*LA = 4 LI = 100*

### lcoeff(poly, var)

Determines the leading coefficient in poly with respect to variable var.
**Output:** An expression is returned.
**Argument options:** (poly) to determine the leading coefficient with respect to *all* the variables in poly. ♣ (poly, [var$_1$, ..., var$_n$]) or (poly, {var$_1$, ..., var$_n$}) to determine the leading coefficient with respect to variables var$_1$ through var$_n$. ♣ (poly, var, name) to assign the leading *term* of poly with respect to var to name.
**Additional information:** Before calling lcoeff, collect should be applied to poly to collect like terms in var.
**See also:** tcoeff, coeff, coeffs, sign, degree, ldegree, *collect*
FL = 78  *LI = 125*

### ldegree(poly, var)

Determines the lowest degree of poly with respect to variable var.
**Output:** An integer is returned.

**Argument options:** (poly) to determine the lowest total degree with respect to *all* the variables in poly. ♣ (poly, {$var_1$, ..., $var_n$}) to determine the lowest *total* degree of poly with respect to variables $var_1$ through $var_n$. ♣ (poly, [$var_1$, ..., $var_n$]) to determine the lowest *vector* degree of poly with respect to variables $var_1$ through $var_n$.

**Additional information:** Before calling ldegree, collect should be applied to poly to collect like terms in var. ♣ For more details on the difference between total and vector degree, see the on-line help page for degree.

**See also:** degree, lcoeff, tcoeff, *collect*

*LI = 61*


### maxnorm(poly)

Calculates the infinity norm of poly, a fully expanded polynomial.

**Output:** A numerical value is returned.

**Additional information:** The infinity norm is equivalent to the coefficient in poly with the largest absolute value.

**See also:** norm

*LI = 135*


### maxorder(expr)

Computes a basis for maximal order of the field represented by expr, which consists of a RootOf command.

**Output:** A list containing the basis elements is returned.

**Argument options:** ({$expr_1$, ..., $expr_n$}, var) to determine a basis for maximal order of the field represented by the n independent RootOf commands $expr_1$ through $expr_n$. ♣ (expr, var) to determine a basis for maximal order of the field represented by expr, a RootOf command in the single variable var.

**Additional information:** This command needs to be defined by readlib(maxorder) before being invoked.

**See also:** Indep, evala, RootOf


### mellin(expr, var, name)

Computes the Mellin transform of expr with respect to var. The resulting expression is in terms of variable name.

**Output:** An expression is returned.

**Additional information:** The input expression may contain unevaluated calls to commands such as exp, sin, or cosh. ♣ The resulting expression may contain unevaluated calls to commands such as GAMMA or Zeta. ♣ An internal set of simplification

Polynomials 167

rules can be controlled through mellintable. ♣ For more information on mellin and examples of its usage, see the on-line help page.
**See also:** mellintable, laplace, fourier, ztrans
*LI = 137*


### mellintable(expr$_{in}$, expr$_{out}$, var, name)

Adds an entry to the internal table of Mellin transforms. The Mellin transform of expression expr$_{in}$ with respect to var is set to expr$_{out}$, an expression in name.
**Output:** A NULL expression is returned.
**Argument options:** (expr$_{in}$, proc, var, name), where proc is a procedure whose parameters are in expr$_{in}$, to allow screening of input expressions. If the parameter(s) of proc are valid for transformation, proc should return an expression in name. Otherwise, proc should return a non-null ERROR message.
**Additional information:** This command needs to be defined by readlib(mellin) or by use of mellin, before being invoked. ♣ For more information on mellin and examples of its usage, see the on-line help page.
**See also:** mellin


### minpoly(num, n)

Calculates a polynomial of degree n or less with small integer coefficients and algebraic number num as one of its roots.
**Output:** A polynomial in the variable _X is returned.
**Argument options:** (num, n, expr) to specify the desired accuracy for the calculations. The default accuracy is 10^(Digits-2).
**Additional information:** This command needs to be defined by readlib(lattice) before being invoked. ♣ The algorithm for this command is also used by the lattice command.
**See also:** *lattice, linalg[minpoly]*


### expr mod posint

Allows you to evaluate expr modulo posint.
**Output:** An expression is returned.
**Argument options:** 'mod'(expr, posint) to produce the same results. mod must be in backquotes because it is a keyword.
**Additional information:** expr can take one of many forms, including a typical arithmetic expression or an inert command call like Det, Factor, Resultant, etc. ♣ mod can be assigned to either

mods, for symmetric representation, or modp for positive representation. modp is the default. ♣ For more information on the uses of mod, see the on-line help page and the various inert command entries throughout this chapter.
**See also:** *msolve*, modp1, Det, Factor, Resultant
*LA = 52−53* **LI = 138**


## modp1(Fnc(expr$_1$, ..., expr$_n$), posint)

Uses efficient arithmetic methods to calculate the inert command Fnc modulo posint. expr$_1$ through expr$_n$ are univariate polynomials expressed in special modp1 format.

**Output:** Typically, an integer (which represents the polynomial solution in modp1 format) is returned.

**Argument options:** (*Add*(polyp1$_1$, ..., polyp1$_n$), posint) to sum the polynomials polyp1$_1$ through polyp1$_n$. ♣ (*Coeff*(polyp1, n), posint) to compute the coefficient of the $n^{th}$ power term of polyp1. ♣ (*Degree*(polyp1), posint) to compute the degree of polynomial polyp1. ♣ (*Det*(M), posint) to compute the determinant of matrix M, whose elements are polynomials in proper modp1 format. ♣ (*Diff*(polyp1), posint) to compute the derivative of polynomial polyp1. ♣ (*Gausselim*(M), posint) to perform Gaussian elimination on matrix M, whose elements are polynomials in proper modp1 format. ♣ (*Gaussjord*(M), posint) to transform to reduced row echelon form matrix M, whose elements are polynomials in proper modp1 format. ♣ (*Lcoeff*(polyp1), posint) to compute the leading coefficient of polynomial polyp1. ♣ (*Lcm*(polyp1$_1$, polyp1$_2$), posint) to compute the least common multiple of polyp1$_1$ and polyp1$_2$. ♣ (*Ldegree*(polyp1), posint) to compute the lowest degree of the polynomial polyp1. ♣ (*Multiply*(polyp1$_1$, polyp1$_2$), posint) to compute the product of polyp1$_1$ and polyp1$_2$. ♣ (*Power*(polyp1, n), posint) to compute polynomial polyp1 raised to the $n^{th}$ power. ♣ (*Subtract*(polyp1$_1$, polyp1$_2$), posint) to subtract polyp1$_2$ from polyp1$_1$. ♣ (*Tcoeff*(polyp1), posint)) to compute the trailing coefficient of polynomial polyp1. ♣ Other inert command calls that work with evala include *Chrem*, *Divide*, *Eval*, *Factors*, *Gcd*, *Gcdex*, *Interp*, *Irreduc*, *Powmod*, *Prem*, *Quo*, *Rem*, *Resultant*, *Roots*, *Smith*, and *Sqrfree*. For more details, see the respective command entries for these inert commands.

**Additional information:** Before being used by modp1(Fnc), the polynomials passed as parameters must have first been converted to proper form with modp1(*ConvertIn*). ♣ To convert from this special form back to a polynomial, use modp1(*ConvertOut*). ♣

For more information on possible parameter sequences and result types for specific unevaluated functions, see the on-line help for modp1. Information may also be available in on-line help for the inert function (e.g., Degree) or its *active* counterpart (e.g., degree).
**See also:** modp1(*ConvertIn*), modp1(*ConvertOut*), modp1(*One*), modp1(*Zero*), modp1(*Constant*), modp1(*Randpoly*), evala, evalgf, mod, RootOf
*LI = 140*

### modp1(*ConvertIn*(poly, var), posint)

Converts poly, a univariate polynomial in var with integer coefficients, into a corresponding modp1 format polynomial modulo posint.
**Output:** A special form of integer is returned.
**Additional information:** The special integer that is created allows efficient calculations to be performed on poly. ✤ To convert from this special form back to a standard polynomial, use the modp1(*ConvertOut*) command.
**See also:** modp1, modp1(*ConvertOut*)

### modp1(*ConvertOut*(polyp1, var), posint)

Converts polyp1, a modp1 format polynomial modulo posint, into a corresponding standard univariate polynomial in var with integer coefficients.
**Output:** A polynomial is returned.
**Argument options:** modp1(*ConvertOut*(polyp1), posint) to return a list of integers representing the coefficients of the univariate polynomial.
**Additional information:** To convert the polynomial back to modp1 format, use the modp1(*ConvertIn*) command.
**See also:** modp1, modp1(*ConvertIn*)

### modp1(*Constant*(expr), posint)

Represents the constant expr as a polynomial modulo posint in proper modp1 format.
**Output:** A special form of integer is returned.
**Additional information:** To convert from this special form back to a constant, use modp1(*ConvertOut*). ✤ modp1(*Constant*) is used along with modp1(*Multiply*) to perform scalar multiplication. ✤ For more information on these types of commands, see the on-line help page for modp1.
**See also:** modp1, modp1(*ConvertOut*), modp1(*One*), modp1(*Zero*), modp1(*Randpoly*)

### modp1(*One*(), posint)

Represents the 1 polynomial modulo posint in proper modp1 format.
**Output:** A special form of integer is returned.
**Additional information:** To convert from this special form back to a standard polynomial, use modp1(*ConvertOut*). ♣ For more information, see the on-line help page for modp1.
**See also:** modp1, modp1(*ConvertOut*), modp1(*Zero*), modp1(*Constant*), modp1(*Randpoly*)


### modp1(*Randpoly*(n), posint)

Creates a random polynomial modulo posint of degree n in proper modp1 format.
**Output:** A special form of integer is returned.
**Additional information:** To convert from this special form back to a standard polynomial, use modp1(*ConvertOut*). ♣ For more information, see the on-line help page for modp1.
**See also:** modp1, modp1(*ConvertOut*), modp1(*One*), modp1(*Zero*), modp1(*Constant*)


### modp1(*Zero*(), posint)

Represents the 0 polynomial modulo posint in proper modp1 format.
**Output:** A special form of integer is returned.
**Additional information:** To convert from this special form back to a standard polynomial, use modp1(*ConvertOut*). ♣ For more information, see the on-line help page for modp1.
**See also:** modp1, modp1(*ConvertOut*), modp1(*One*), modp1(*Constant*), modp1(*Randpoly*)


### modpol(expr, poly, var, int$_p$)

Evaluates expr, a rational expression in var over $Q$, with respect to the quotient field represented by $Z\text{int}_p[\text{var}]/\text{poly}(\text{var})$. int$_p$ is a prime integer and poly is a polynomial in var over $Q$.
**Output:** An expression in var is returned.
**Additional information:** This command needs to be defined by readlib(modpol) before being invoked.
**See also:** evala, mod, Powmod
*LI = 302*

### Nextprime(poly, var)

Represents the inert command for determining the irreducible polynomial in var that is next highest to poly.
**Output:** An unevaluated Nextprime call is returned.
**Argument options:** Nextprime(poly, var, $expr_a$) to represent next highest irreducible polynomial over the algebraic extension $expr_a$.
♣ Nextprime(poly, var) *mod* $int_p$ to evaluate the inert command call modulo $int_p$. $int_p$ must be a prime integer.
**See also:** Prevprime, Randprime

### norm(poly, n, var)

Calculates the $n^{th}$ norm of poly, a polynomial in var.
**Output:** A numerical value is returned.
**Argument options:** (poly, n) to calculate the norm of poly with respect to the result of indet(poly). ♣ (poly, *infinity*, var) to calculate the *infinity norm* of poly.
**Additional information:** The norm of a polynomial is defined as sum(abs(c)^n for c in [coeffs(poly,var)])^(1/n). This previous ''command'' is not in valid Maple syntax, so you cannot enter it as such.
**See also:** maxnorm, *linalg[norm]*, *indets*, evala
*LI = 143*

### op(polynom)

Displays the internal structure of polynomial data structure polynom and allows you access to its operands.
**Output:** An expression sequence containing $n$ operands, where n is the number of terms in the polynomial, is returned. The $i^{th}$ operand is the $i^{th}$ term of the polynomial.
**Additional information:** The terms with zero coefficients are eliminated and not represented in the data structure. ♣ More information about the features of a polynomial can be determined with the type(*polynom*) command. ♣ To substitute for an operand in a polynom structure, use subsop.
**See also:** type(*polynom*), *subsop*

### op(ratpoly)

Displays the internal structure of rational polynomial data structure ratpoly and allows you access to its operands.
**Output:** An expression sequence is returned.
**Argument options:** (*1*, ratpoly) to extract the numerator polyno-

mial. ♣ (2, ratpoly) to extract the *reciprocal* of the denominator polynomial.

**Additional information:** The terms with zero coefficients are eliminated and not represented in the data structure. ♣ More information about the features of a rational polynomial can be determined with the type(ratpoly) command. ♣ To substitute for an operand in a ratpoly structure, use subsop.

**See also:** type(*ratpoly*), *subsop*

### orthopoly package

Provides commands for calculating orthogonal polynomials.

**Additional information:** To access the command fnc from the orthopoly package, use the long form of the name orthopoly[fnc], or with(orthopoly, fnc) to load in a pointer for fnc only, or with(orthopoly) to load in pointers to all the commands. ♣ For more information on the recurrence relations that define specific orthogonal polynomials in this package and the conditions for their othogonality, see the individual on-line help pages.

### orthopoly[G](n, expr$_1$, expr$_2$)

Calculates the $n^{th}$ Gegenbauer (ultraspherical) polynomial with parameter expr$_1$ and evaluated at expr$_2$.

**Output:** A polynomial in expr$_1$ and expr$_2$ is returned.

**Additional information:** If expr$_1$ is a rational number, it must have a value greater than $-1/2$.

**See also:** orthopoly[H], orthopoly[L], orthopoly[P], orthopoly[T], orthopoly[U]

### orthopoly[H](n, expr)

Calculates the $n^{th}$ Hermite polynomial evaluated at expr.

**Output:** A polynomial in expr is returned.

**See also:** orthopoly[G], orthopoly[L], orthopoly[P], orthopoly[T], orthopoly[U], *linalg[hermite]*, *Hermite*

### orthopoly[L](n, expr$_1$, expr$_2$)

Calculates the $n^{th}$ generalized Laguerre polynomial with parameter expr$_1$ and evaluated at expr$_2$.

**Output:** A polynomial in expr$_1$ and expr$_2$ is returned.

**Argument options:** (n, expr) to set the parameter of the Laguerre polynomial equal to 0. That is, a call equivalent to orthopoly[L](n, 0, expr) is calculated.

**Additional information:** If expr$_1$ is a rational number, it must

have a value greater than $-1$.
**See also:** orthopoly[G], orthopoly[H], orthopoly[P], orthopoly[T], orthopoly[U]

### orthopoly[P](n, expr$_1$, expr$_2$, expr$_3$)

Calculates the $n^{th}$ Jacobi polynomial with parameters expr$_1$ and expr$_2$, evaluated at expr$_3$.
**Output:** A polynomial in expr$_1$, expr$_2$, and expr$_3$ is returned.
**Argument options:** (n, expr) to calculate the $n^{th}$ Legendre spherical polynomial. This is to the Jacobi polynomial defined by orthopoly[P](n, 0, 0, expr).
**Additional information:** If expr$_1$ or expr$_2$ are rational numbers, they each must have a value greater than $-1$.
**See also:** orthopoly[G], orthopoly[H], orthopoly[L], orthopoly[T], orthopoly[U], *linalg[jacobian]*

### orthopoly[T](n, expr)

Calculates the $n^{th}$ Chebyshev polynomial of the first kind evaluated at expr.
**Output:** A polynomial in expr is returned.
**See also:** orthopoly[G], orthopoly[H], orthopoly[L], orthopoly[P], orthopoly[U], *chebyshev*

### orthopoly[U](n, expr)

Calculates the $n^{th}$ Chebyshev polynomial of the second kind evaluated at expr.
**Output:** A polynomial in expr is returned.
**See also:** orthopoly[G], orthopoly[H], orthopoly[L], orthopoly[P], orthopoly[T], *chebyshev*

### Power(poly, n)

Represents the inert power operator for polynomial poly to the $n^{th}$ power.
**Output:** An unevaluated Power call is returned.
**Argument options:** Power(poly, n) *mod* posint to evaluate the inert command call modulo posint.
**See also:** Powmod, modp1
*LI = 164*

## Powmod(poly$_1$, n, poly$_2$, var)

Represents the inert remainder command Rem(poly$_1$^n, poly$_2$) for poly$_1$ and poly$_2$, polynomials in var.
**Output:** An unevaluated Powmod call is returned.
**Argument options:** (poly$_1$, n, poly$_2$, var) *mod* posint to evaluate the inert command call modulo posint.
**See also:** Power, Rem, Gcdex, modp1
**FL = 210−211** *LI = 165*

## prem(poly$_1$, poly$_2$, var)

Computes the pseudo-remainder when polynomial poly$_1$ is divided by polynomial poly$_2$. Both poly$_1$ and poly$_2$ are polynomials in variable var.
**Output:** An expression is returned.
**Argument options:** (poly$_1$, poly$_2$, var, name$_m$, name$_q$) to assign the multiplier and quotient to name$_m$ and name$_q$, respectively.
**Additional information:** The remainder rem is an expression such that name$_m$ * poly$_1$ = poly$_2$ * name$_q$ + rem.
**See also:** sprem, rem, Prem, *irem, GaussInt[GIrem]*
*LI = 167*

## Prem(poly$_1$, poly$_2$, var)

Represents the inert pseudo-remainder command for polynomials poly$_1$ and poly$_2$ with respect to variable var.
**Output:** An unevaluated Prem call is returned.
**Argument options:** Prem(poly$_1$, poly$_2$, var) *mod* posint to evaluate the inert command call modulo posint.
**Additional information:** More information on other parameter sequences for Prem is found in the listing for prem.
**See also:** prem, Sprem, evala, evalgf

## Prevprime(poly, var)

Represents the inert command for determining the irreducible polynomial in var that is next lowest to poly.
**Output:** An unevaluated Prevprime call is returned.
**Argument options:** Prevprime(poly, var, expr$_a$) to represent next lowest irreducible polynomial over the algebraic extension expr$_a$.
♣ Prevprime(poly, var) *mod* int$_p$ to evaluate the inert command call modulo int$_p$. int$_p$ must be a prime integer.
**See also:** Nextprime, Randprime

## Prime(poly)

This command is identical to Irreduc.
**See also:** Irreduc

## Primfield({expr$_1$, ..., expr$_n$})

Represents the inert primitive description of the algebraic extension represented by expr$_1$ through expr$_n$, expressions consisting of RootOf commands.
**Output:** An unevaluated Primfield call is returned.
**Argument options:** Primfield({expr$_1$, ..., expr$_n$}) *mod* posint to evaluate the inert command call modulo posint. ♣ (set$_1$, set$_2$), where set$_1$ and set$_2$ are both sets of RootOf expressions, to represent a primitive element of the field defined by set$_1$ over the field defined by set$_2$.
**Additional information:** By default, the primitive description is found over the smallest possible transcendental extension of the rational numbers.
**See also:** RootOf, evala

## Primitive(poly)

Represents the inert command to determine whether univariate polynomial poly is primitive.
**Output:** An unevaluated Primitive call is returned.
**Argument options:** Primitive(poly) *mod* posint to evaluate the inert command call modulo posint.
**Additional information:** For more information on what makes a polynomial primitive, see the on-line help page.
**See also:** Irreduc, Powmod
*LI = 167*

## primpart(poly, var)

Determines the primitive part of poly with respect to variable var.
**Output:** An polynomial is returned.
**Argument options:** (poly) to determine the primitive part with respect to *all* the variables in poly. ♣ (poly, [var$_1$, ..., var$_n$]) or (poly, {var$_1$, ..., var$_n$}) to determine the primitive part with respect to variables var$_1$ through var$_n$. ♣ (poly, var, name) to assign the content of poly with respect to var to name.
**Additional information:** The content is the greatest common divisor of the coefficients of poly with respect to var. ♣ The primitive part is simply poly divided by the content.
**See also:** content, icontent
*LI = 28*

### Primpart(poly, var)

Represents the inert primitive part command for poly with respect to variable var.
**Output:** An unevaluated Primpart call is returned.
**Argument options:** Primpart(poly, var) *mod* posint to evaluate the inert command call modulo posint.
**Additional information:** More information on other parameter sequences for Primpart is found in the listing for primpart.
**See also:** Content primpart, evala
*LI = 27*


### proot(poly, n)

Computes the $n^{th}$ root of polynomial poly.
**Output:** If the $n^{th}$ root of poly exists, it is returned. Otherwise, the value _NOROOT is returned.
**Additional information:** This command needs to be defined by readlib(proot) before being invoked. ♣ poly must have rational coefficients.
**See also:** psqrt, *iroot*
*LI = 314*


### psqrt(poly)

Computes the square root of polynomial poly.
**Output:** If the square root of poly exists, it is returned. Otherwise, the value _NOSQRT is returned.
**Additional information:** This command needs to be defined by readlib(psqrt) before being invoked. ♣ poly must have rational coefficients.
**See also:** proot, *isqrt*
*LI = 314*


### quo(poly$_1$, poly$_2$, var)

Computes the quotient when polynomial poly$_1$ is divided by polynomial poly$_2$. Both poly$_1$ and poly$_2$ are polynomials in variable var.
**Output:** A polynomial is returned.
**Argument options:** (poly$_1$, poly$_2$, var, name) to assign the remainder of the same division to name.
**See also:** rem, Quo, divide, *iquo, GaussInt[GIquo]*
*LI = 177*

## Polynomials 177

### Quo(poly$_1$, poly$_2$, var)

Represents the inert quotient command for poly$_1$ and poly$_2$ with respect to variable var.
**Output:** An unevaluated Quo call is returned.
**Argument options:** Quo(poly$_1$, poly$_2$, var) *mod* posint to evaluate the inert command call modulo posint.
**Additional information:** More information on other parameter sequences for Quo is found in the listing for quo.
See also: Rem, quo, evala, modp1
*LI = 176*

### randpoly(var)

Creates a random univariate polynomial in var with six terms.
**Output:** A polynomial is returned.
**Argument options:** ({var$_1$, ..., var$_n$}) or ([var$_1$, ..., var$_n$]) to return a random multivariate polynomial in variables var$_1$ through var$_n$. ✦ (var, *coeffs*= *rand*(a..b)) to compute coefficients using the random generator created by rand(a..b). ✦ (var, *expons*= *rand*(n)) to compute exponents using the random generator created by rand(n). ✦ (var, *terms*=int) to specify that the random polynomial has int terms. ✦ (var, *dense*) to specify that a *dense* polynomial is to be created. ✦ (var, *degree*=int) to specify the degree of the dense polynomial.
**Additional information:** If the *terms* and *degree* options are in conflict, the *degree* option takes precedence.
See also: *rand*
*LI = 173*

### Randpoly(n, var)

Represents the inert command for creating a random polynomial of degree n in var.
**Output:** An unevaluated Randpoly call is returned.
**Argument options:** Randpoly(n, var) *mod* posint to evaluate the inert command call modulo posint.
**Additional information:** More information on the parameters for Randpoly is found in the on-line help page.
See also: randpoly, Randprime, modp1
*LI = 172*

### Randprime(n, var)

Represents the inert command for creating a random monic irreducible polynomial of degree n in var.
**Output:** An unevaluated Randprime call is returned.
**Argument options:** (n, var, expr$_a$) to represent a random monic

irreducible polynomial over algebraic extension $expr_a$. ♣ Randprime(n, var) *mod* $int_p$ to evaluate the inert command call modulo $int_p$. $int_p$ must be a prime integer.
**See also:** Randpoly, Nextprime, Prevprime
*LI = 172*

### ratrecon($poly_1$, $poly_2$, var, $int_1$, $int_2$, $name_1$, $name_2$)

Reconstructs a signed rational function from $poly_1$, its image modulo $poly_2$. (Both polynomials are in var.) $name_1$ and $name_2$ are assigned integers, such that $name_1/name_2$ = $poly_1$ mod $poly_2$.
**Output:** If a reconstruction can be found, a boolean value of true is returned. Otherwise, a boolean value of false is returned.
**Additional information:** This command needs to be defined by readlib(ratrecon) before being invoked. ♣ The bounds on the values assigned to $name_1$ and $name_2$ are such that degree($name_1$) <= $int_1$ and degree($name_2$) <= $int_2$.
**See also:** degree, mod, *iratrecon*

### Ratrecon($poly_1$, $poly_2$, var, $int_1$, $int_2$, $name_1$, $name_2$)

Represents the inert rational reconstruction command for $poly_1$ and $poly_2$ with respect to variable var.
**Output:** An unevaluated Ratrecon call is returned.
**Argument options:** Ratrecon($poly_1$, $poly_2$, var, $int_1$, $int_2$, $name_1$, $name_2$) *mod* posint to evaluate the inert command call modulo posint.
**Additional information:** More information on the parameters for Ratrecon is found in the listing for ratrecon.
**See also:** ratrecon, iratrecon, evala, modp1

### realroot(poly)

Computes intervals in which lie the real roots of poly, a univariate polynomial with integer coefficients.
**Output:** A list of lists is returned, where each internal list contains two values representing a range.
**Argument options:** (poly, num) to specify that the ranges returned should be no wider than positive value num.
**Additional information:** This command needs to be defined by readlib(realroot) before being invoked. ♣ The intervals are real and open. ♣ No multiplicity information is returned.
**See also:** roots, *solve*, *fsolve*
*LI = 314*

### recipoly(poly, var)

Determine whether poly, a polynomial in var, is self-reciprocal.
**Output:** A boolean value of true or false is returned.
**Argument options:** (poly, var, name) to assign the polynomial p of degree degree(poly,var)/2, such that var^(degree(poly,var)/2) * p(var+1/var) = poly, to name.
**Additional information:** This command needs to be defined by readlib(recipoly) before being invoked.
**See also:** degree
*LI = 315*

### rem(poly$_1$, poly$_2$, var)

Computes the remainder when polynomial poly$_1$ is divided by polynomial poly$_2$. Both poly$_1$ and poly$_2$ are polynomials in var.
**Output:** A polynomial is returned.
**Argument options:** (poly$_1$, poly$_2$, var, name) to assign the quotient of the same division to name.
**See also:** quo, Rem, divide, *irem*, *GaussInt[GIrem]*
*LI = 177*

### Rem(poly$_1$, poly$_2$, var)

Represents the inert remainder command for poly$_1$ and poly$_2$ with respect to variable var.
**Output:** An unevaluated Rem call is returned.
**Argument options:** Rem(poly$_1$, poly$_2$, var) *mod* posint to evaluate the inert command call modulo posint.
**Additional information:** More information on other parameter sequences for Rem is found in the listing for rem.
**See also:** Quo, rem, evala, modp1
*LI = 176*

### resultant(poly$_1$, poly$_2$, var)

Computes the resultant of poly$_1$ and poly$_2$ with respect to var.
**Output:** An expression is returned.
**Additional information:** The resultant is defined in terms of products and is calculated either with the Euclidean algorithm or as the determinant of Sylvester or Bezout matrices. ♣ See the on-line help page for the exact definition.
**See also:** Resultant, discrim, gcd, *linalg[sylvester]*, *linalg[bezout]*
*LI = 178*

### Resultant(poly$_1$, poly$_2$, var)

Represents the inert resultant command for poly$_1$ and poly$_2$ with respect to variable var.
**Output:** An unevaluated Resultant call is returned.
**Argument options:** Resultant(poly$_1$, poly$_2$, var) *mod* posint to evaluate the inert command call modulo posint.
**See also:** Discrim, resultant, evala, modp1
*LI = 178*

### RootOf(eqn, var)

Represents the roots of equation eqn with respect to var.
**Output:** An unevaluated RootOf expression in _Z is returned.
**Argument options:** (expr, var) to represent the roots of the equation expr = 0 with respect to var. ♣ (expr, var, num) to represent the individual root of expr closest to the value num. ♣ (eqn) to represent the roots of the univariate equation eqn.
**Additional information:** RootOf expressions can be found in the results of many commands, such as solve, linalg[eigenvalues], int, etc. ♣ RootOf expressions can also represent solutions of transcendental equations. ♣ RootOf expressions are used as input to commands like Roots, Factor, Resultant, etc., where they represent algebraic extensions over which computations are made. It is often convenient to alias these algebraic extensions to shorter, simpler names. ♣ To evaluate a RootOf expression, try the allvalues command.
**See also:** allvalues, evala, solve, Roots, mod, *alias*, *convert(RootOf)*
*LI = 180*

### roots(poly)

Computes the roots of univariate polynomial poly.
**Output:** A list of lists is returned, where the first elements are the roots and the second elements are the corresponding multiplicities.
**Argument options:** (poly, expr) to find the roots of poly over the number field defined by expr, a radical or RootOf expression, for example, sqrt(2) or RootOf(x^2-3). ♣ (poly, [expr$_1$, ..., expr$_n$]) or (poly, {expr$_1$, ..., expr$_n$}) to find the roots over the number field defined by radicals or RootOf expressions expr$_1$ through expr$_n$.
**Additional information:** roots finds roots over the field implied by the coefficients of poly. ♣ If no roots are found, then an empty list is returned.
**See also:** realroot, solve, factors, Roots, RootOf
*LI = 182*

### Roots(poly)

Represents the inert form of the roots command for univariate polynomial poly.
**Output:** An unevaluated Roots call is returned.
**Argument options:** Roots(poly) *mod* posint to evaluate the inert command call modulo posint.
**Additional information:** For more information on other valid parameter sequences for Roots, see the command listing for roots.
**See also:** roots, msolve, Factors, modp1
*LI = 180*


### sign(poly)

Computes the sign of the leading coefficient of poly, a multivariate polynomial.
**Output:** If poly is positive, 1 is returned. If poly is negative, $-1$ is returned. If the sign of poly is indeterminate, an error message is returned.
**Argument options:** (poly, var) to treat poly as a univariate polynomial in var. ♣ (poly, [$var_1$, ..., $var_n$]) to treat poly as a multivariate polynomial in $var_1$ through $var_n$. ♣ (poly, var, name) to assign the leading term of poly to name.
**Additional information:** This command works equally well for numeric values. ♣ The leading coefficient may depend on the ordering of the terms in poly.
**See also:** lcoeff, indets, *signum*


### sinterp(proc, [$var_1$, ..., $var_n$], posint, $int_p$)

Computes a polynomial, in $var_1$ through $var_n$ with degree bound posint, whose values modulo $int_p$ equal the values of procedure proc when given corresponding values.
**Output:** A polynomial is returned.
**Additional information:** This command needs to be defined by readlib(sinterp) before being invoked. ♣ sinterp uses *sparse interp*olation. ♣ If $int_p$ is not sufficiently large, FAIL is returned. ♣ For an example of sinterp, see the on-line help page.
**See also:** interp
*LI = 319*


### Smith(M, var)

Represents the inert Smith normal form command, where M is a matrix of univariate polynomials in var.
**Output:** An unevaluated Smith call is returned.

**Argument options:** Smith(M, var) *mod* posint to evaluate the inert command call modulo posint.
**Additional information:** More information on other parameter sequences for Smith is found in the listing for linalg[smith].
**See also:** *linalg[smith]*, Hermite, modp1

### sort(poly)

Sort the terms of poly in total degree order over all the indeterminates of poly.
**Output:** A polynomial is returned.
**Argument options:** (poly, *plex*) to sort poly in pure lexicographic ordering. ♣ (poly, *tdeg*) to sort poly in total degree order. This is the default. ♣ (poly, option, [$var_1$, ..., $var_n$]) or (poly, option, {$var_1$, ..., $var_n$}) to sort a polynomial in order option over the variables $var_1$ through $var_n$. option is either *plex* or *tdeg*.
**Additional information:** sort actually changes the input poly to the new sorted version. This behavior is different than the vast majority of Maple commands. ♣ For information on how sort works on lists, see the entry for sort(list).
**See also:** *sort(list)*
**FL = 60−62  *LI = 204***

### split(poly, var)

Computes the complete factorization of poly, a univariate polynomial in var.
**Output:** An expression containing linear factors is returned.
**Argument options:** (poly, var, name) to assign the extensions used in the complete representation to name.
**Additional information:** This command needs to be defined by readlib(split) before being invoked. ♣ If the factorization contains RootOf commands, they are included appropriately in linear factors. ♣ See the on-line help page for examples of split.
**See also:** factor, Factor, Sqrfree, RootOf

### sprem($poly_1$, $poly_2$, var)

Computes the sparse pseudo-remainder when polynomial $poly_1$ is divided by polynomial $poly_2$. Both $poly_1$ and $poly_2$ are polynomials in variable var.
**Output:** An expression is returned.
**Argument options:** ($poly_1$, $poly_2$, var, $name_m$, $name_q$) to assign the multiplier and quotient to $name_m$ and $name_q$, respectively.
**Additional information:** The workings of sprem are identical to those of prem, except that sprem uses as a multiplier the smallest

possible power of lcoeff(poly$_2$) such that the division does not introduce fractions into the quotient or the remainder. ♣ If it can be used, sprem is more efficient than prem.
**See also:** prem, rem, Sprem, *irem*, *GaussInt[GIrem]*
*LI = 167*

### Sprem(poly$_1$, poly$_2$, var)

Represents the inert sparse pseudo-remainder command for polynomials poly$_1$ and poly$_2$ with respect to variable var.
**Output:** An unevaluated Sprem call is returned.
**Argument options:** Sprem(poly$_1$, poly$_2$, var) *mod* posint to evaluate the inert command call modulo posint.
**Additional information:** More information on other parameter sequences for Sprem is found in the listing for sprem.
**See also:** sprem, Prem, evala, evalgf

### sqrfree(poly)

Computes the square free factors of multivariate polynomial poly.
**Output:** A two-element list is returned. The first element is the constant factor and the second element is a list of two-element lists, where the first elements are the factors and the second elements are their corresponding multiplicities.
**Argument options:** (poly, var) to factor poly, a polynomial in var.
**Additional information:** poly must have rational coefficients.
**See also:** factor, Sqrfree, *isqrfree*

### Sqrfree(poly)

Represents the inert form of the sqrfree command for multivariate polynomial poly.
**Output:** An unevaluated Sqrfree call is returned.
**Argument options:** Sqrfree(poly) *mod* posint to evaluate the inert command call modulo posint.
**Additional information:** For more information on other valid parameter sequences for Sqrfree, see the command listing for sqrfree.
**See also:** sqrfree, Factors, evala, modp1
*LI = 205*

### sturm(expr$_s$, var, a, b)

Computes the number of real roots of the polynomial represented by the Sturm sequence expr$_s$ in the interval (a, b].
**Output:** An integer is returned.
**Additional information:** This command needs to be defined by readlib(sturm) before being invoked. ♣ expr$_s$ can be created with

the sturmseq command. ♣ If b is less than a, then the result is the corresponding negative integer.
**See also:** sturmseq, roots, solve
*LI = 320*

### sturmseq(poly, var)

Computes a Sturm sequence from poly, a univariate polynomial in var.
**Output:** A list of polynomials representing the Sturm sequence is returned.
**Additional information:** This command needs to be defined by readlib(sturm) before being invoked.
**See also:** sturm, roots, solve
*LI = 320*

### tcoeff(poly, var)

Determines the trailing coefficient in poly with respect to var.
**Output:** An expression is returned.
**Argument options:** (poly) to determine the trailing coefficient with respect to *all* variables in poly. ♣ (poly, [$var_1$, ..., $var_n$]) or (poly, {$var_1$, ..., $var_n$}) to determine the trailing coefficient with respect to variables $var_1$ through $var_n$. ♣ (poly, var, name) to assign the trailing *term* of poly with respect to var to name.
**Additional information:** Before calling tcoeff, collect should be applied to poly to collect like terms in var.
**See also:** lcoeff, coeff, coeffs, degree, ldegree, *collect*
*LI = 125*

### translate(poly, var, num)

Translates the polynomial poly by replacing every occurrence of var by var + num.
**Output:** An expression is returned.
**Additional information:** This command needs to be defined by readlib(translate) before being invoked. ♣ poly does *not* need to be a univariate polynomial. ♣ The result is automatically expanded.
**See also:** subs, *expand*
*LI = 325*

### type(expr, *cubic*)

Tests whether expr is a cubic polynomial.
**Output:** A boolean value of either true or false is returned.
**Argument options:** (expr, *cubic*({$var_1$, ..., $var_n$})) to determine if expr is a cubic polynomial in variables $var_1$ through $var_n$. By

default, the result of indet(expr) is used for the variables.
**Additional information:** A cubic polynomial is of degree 3.
**See also:** type(*polynom*), type(*linear*), type(*quadratic*), type(*quartic*), indets, degree
*LI = 231*


**type(expr,** *expanded***)**

Tests whether expr is an expanded polynomial.
**Output:** A boolean value of either true or false is returned.
**See also:** type(*polynom*), *expand*
*LI = 228*


**type(expr,** *linear***)**

Tests whether expr is a linear polynomial.
**Output:** A boolean value of either true or false is returned.
**Argument options:** (expr, *linear*({var$_1$, ..., var$_n$})) to determine if expr is a linear polynomial in variables var$_1$ through var$_n$. By default, the result of indet(expr) is used for the variables.
**Additional information:** A linear polynomial is of degree 1.
**See also:** type(*polynom*), type(*quadratic*), type(*cubic*), type(*quartic*), indets, degree
*LI = 231*


**type(expr,** *monomial***)**

Tests whether expr is a monomial.
**Output:** A boolean value of either true or false is returned.
**Additional information:** A monomial is a polynomial that contains no sums (i.e., a one-term polynomial). ♣ In all other ways, this command is identical to the type(*polynom*) command.
**See also:** type(*polynom*), op(polynom)
*LI = 234*


**type(expr,** *polynom***)**

Tests whether expr is a polynomial.
**Output:** A boolean value of either true or false is returned.
**Argument options:** (expr, *polynom*(typename)) to determine if expr is a polynomial with all coefficients in domain typename. ♣ (expr, *polynom*(anything , var)) to determine if expr is a polynomial in var. ♣ (expr, *polynom*(typename, var)) to determine if expr is a polynomial in var with all coefficients in domain typename. ♣ (expr, *polynom*(typename, [var$_1$, var$_2$, ..., var$_n$])) to determine if expr is a multivariate polynomial in var$_1$ through var$_n$

with all coefficients in domain typename.
**Additional information:** If typename is not specified, then the default is type constant.
**See also:** type(*ratpoly*), type(*linear*), type(*quadratic*), type(*cubic*), type(*quartic*), type(*expanded*), op(polynom)
**FL = 132** *LA = 86* **LI = 237**


### type(expr, *quadratic*)

Tests whether expr is a quadratic polynomial.
**Output:** A boolean value of either true or false is returned.
**Argument options:** (expr, *quadratic*({var$_1$, ..., var$_n$})) to determine if expr is a quadratic polynomial in the variables var$_1$ through var$_n$. By default, the result of indet(expr) is used for the variables.
**Additional information:** A quadratic polynomial is of degree 2.
**See also:** type(*polynom*), type(*linear*), type(*cubic*), type(*quartic*), indets, degree
*LI = 231*


### type(expr, *quartic*)

Tests whether expr is a quartic polynomial.
**Output:** A boolean value of either true or false is returned.
**Argument options:** (expr, *quartic*({var$_1$, ..., var$_n$})) to determine if expr is a quartic polynomial in the variables var$_1$ through var$_n$. By default, the result of indet(expr) is used for the variables.
**Additional information:** A quartic polynomial is of degree 4.
**See also:** type(*polynom*), type(*linear*), type(*quadratic*), type(*cubic*), indets, degree
*LI = 231*


### type(expr, *ratpoly*)

Tests whether expr is a rational polynomial. In all other ways, this command is identical to the type(*polynom*) command.
**Output:** A boolean value of either true or false is returned.
**See also:** type(*polynom*), op(ratpoly)
*LI = 245*


### ztrans(expr, var, name)

Computes the $Z$ transform of expr with respect to var. The resulting expression is in terms of variable name.
**Output:** An expression is returned.
**Additional information:** The input expression may contain unevaluated calls to commands such as sin or invztrans. ♣ For more

information on the ztrans command and examples of its usage, see the on-line help page.
**See also:** invztrans, fourier, laplace, mellin, *rsolve*
*LI = 263*

# Geometry

## Introduction

The commands dealing with special types of geometries are unique in that they are almost completely restricted to functions in packages, and have very little to do with the standard and miscellaneous libraries.

## Three Packages—Three Geometries

The geometric knowledge contained in Maple can be found in the three packages: geometry, geom3d, and projgeom. Each of these packages consists of commands that create geometric objects (e.g., lines, circles, spheres, etc.) and commands that operate upon these objects to compute standard values (e.g., midpoint, area, volume, etc.) and not-so-standard values (e.g., Euler lines, Gergonne points, etc.). All of the basic geometric structures are useful only within their respective packages; you cannot use a point created in the geometry package in a command from geom3d nor plot a sphere in three-dimensions by passing the result of geom3d[sphere] to plot3d. (Note: you can plot a sphere with plots[sphereplot].)

Throughout the three packages a similar calling structure is used in commands that define geometric objects. The first parameter to these commands is the name to which you wish to assign the completed object. After that, defining information is provided. More complex data types are constructed from their simpler cousins. This design differs fundamentally from most Maple commands, where you must use the assignment operator, :=, to assign results to a name.

All objects in these packages are stored internally as tables of indexed information. Because they follow the same evaluation rules as other Maple tables (i.e., *first name evaluation*), in order to examine individual elements of an object you must use either the op command or table indexing. The following example should illustrate this.

```
> with(geometry):
> point(pt, [2,3]):
> pt;
```

$$pt$$

```
> op(pt);
```

$$table([\\ form = point\\ x = 2\\ y = 3\\ ])$$

```
> pt[x], pt[y];
```

$$2, 3$$

The *form* entry is an element common to all geometric objects. Other types of entries vary widely depending on how an object is defined. Often, certain important values that are not provided by you are automatically calculated when an object is defined. For example, when you define a circle by providing the center point and the radius, the equation of the circle is automatically computed.

```
> point(pt, [4,5]):
> circle(circ, [pt, 6]):
> op(");
```

$$table([\\ form = circle\\ given = [pt, 6]\\ center = pt\\ radius = 6\\ equation = (x^2 + y^2 + 5 - 8x - 10y = 0)\\ ])$$

In most cases, geometric objects can be defined with either numeric or symbolic values. Because of the fact that many of the elements of data type tables refer to $x$, $y$, and $z$ coordinates, it is a good idea not to use the variables x, y, or z in the actual expressions that define geometric objects.

```
> with(geom3d):
> point3d(pt3d, [x,y,z]);
Error, (in point3d) wrong number or type of arguments
> point3d(pt3d, [a,b,c]);
```

$$pt3d$$

```
> radius := t*sqrt(2):
> sphere(s, [pt3d,radius]):
> volume(s);
```

$$\frac{8}{3}\pi t^3 \sqrt{2}$$

## Two-Dimensional Geometry

The geometry package contains many well- and lesser-known commands for manipulating geometrical objects in two-dimensional Euclidean space (i.e., the $x, y$-plane). Points are the most basic type of structures in this geometry. Other objects include lines, triangles, squares, circles, ellipses, and conics.

The following example rotates a triangle, with vertices at $(0,0)$, $(2,0)$, and $(1,3)$, $\pi/4$ radians (90 degrees) clockwise about its own centroid.

```
> with(geometry):
> point(A, [0,0]), point(B, [2,0]), point(C, [1,3]):
> triangle(ABC, [A,B,C]):
> centroid(ABC, cent_ABC):
> rotate(ABC[1], Pi/4, clockwise, v_rot_1, cent_ABC):
> rotate(ABC[2], Pi/4, clockwise, v_rot_2, cent_ABC):
> rotate(ABC[3], Pi/4, clockwise, v_rot_3, cent_ABC):
> coordinates(v_rot_1), coordinates(v_rot_2),
> coordinates(v_rot_3);
```

$$\left[-\sqrt{2}+1, 1\right], \left[1, -\sqrt{2}+1\right], \left[\sqrt{2}+1, \sqrt{2}+1\right]$$

```
> triangle(ABC_rot, [v_rot_1, v_rot_2, v_rot_3]):
> op(ABC_rot);
```

$$table([$$
$$form = triangle$$
$$given = [v\_rot\_1, v\_rot\_2, v\_rot\_3]$$
$$1 = v\_rot\_1$$
$$2 = v\_rot\_2$$
$$3 = v\_rot\_3$$
$$])$$

## Three-Dimensional Geometry

The geom3d package contains many well-known and lesser-known commands for manipulating geometrical objects in three-dimensional Euclidean space (i.e., $x, y, z$-space). Points are the most basic type of structures in this geometry. Other objects include lines, triangles, planes, spheres, and tetrahedra.

The following example calculates a plane that is parallel with the $x, y$-plane and goes through the point of intersection of two given lines.

```
> with(geom3d):
> point3d(AA, [0,0,0]), point3d(BB, [2,2,2]):
> point3d(CC, [3/5,1/3,7/8]), point3d(DD, [9/5,7/3,5/4]):
> line3d(AB, [AA,BB]), line3d(CD, [CC,DD]):
> inter(AB, CD, AB_CD):
> op(AB_CD);
```

$$table([$$
$$form = point3d$$
$$x = 1$$
$$y = 1$$
$$z = 1$$
$$])$$

```
> plane(P, [z=0]):
> parallel(AB_CD, P, para_plane):
> op(para_plane);
```

$$table([$$
$$normal\_vector = [0, 0, 1]$$
$$form = plane$$
$$given = [AB\_CD, [0, 0, 1]]$$
$$equation = (z - 1 = 0)$$
$$])$$

## Projective Geometry

The projgeom package contains many commands for manipulating geometrical objects in three-dimensional projective geometry. Points are the most basic type of structures in this geometry. Other objects include lines and conics.

## Parameter Types Specific to This Chapter

| | |
|---|---|
| point, point2d | a two-dimensional point |
| line, line2d | a two-dimensional line |
| tri, triangle2d | a two-dimensional triangle |
| square, square2d | a two-dimensional square |
| circle, circle2d | a two-dimensional circle |
| ellipse, ellipse2d | a two-dimensional ellipse |
| conic, conic2d | a two-dimensional conic |
| point3d | a three-dimensional point |
| line3d | a three-dimensional line |
| triangle3d | a three-dimensional triangle |
| plane | a three-dimensional plane |
| sphere | a three-dimensional sphere |
| tetra, tetrahedron | a three-dimensional tetrahedron |
| ptpj, pointpj | a three-dimensional projective point |
| linepj | a three-dimensional projective line |
| conicpj | a three-dimensional projective conic |

## Command Listing

### ellipsoid(expr$_1$, expr$_2$, expr$_3$)

Calculates the surface area of the ellipsoid with semi-axes defined by expr$_1$, expr$_2$, and expr$_3$.

**Output:** If it is possible to determine the exact surface area, an exact expression is returned. Otherwise, an unevaluated integral is returned.

**Additional information:** This command needs to be defined by readlib(ellipsoid) before being invoked. ♣ The evalf command can sometimes be used to find a numerical approximation to an unevaluated integral.

**See also:** *evalf(Int)*, geometry(ellipse)

*LI = 280*

### geom3d package

Provides commands for creating objects from three-dimensional Euclidean geometry and calculating important values concerning them. The basic objects available are point3d, line3d, triangle3d, plane, sphere, and tetrahedron.

**Additional information:** To access the command fnc from the geom3d package, use with(geom3d, fnc) to load in a pointer for fnc only or with(geom3d) to load in pointers to all the commands. Unfortunately, the long form geom3d[fnc] does not work. ♣ The design of several commands in the geom3d package differs from most of the Maple library of commands in that the result is assigned to a name supplied as one of the parameters of the command itself.

### geom3d[angle](line3d$_1$, line3d$_2$)

Computes the smallest angle between line3d$_1$ and line3d$_2$.

**Output:** An expression representing the angle (in radians) is returned.

**Argument options:** (plane$_1$, plane$_2$) to compute the angle between two planes.

**See also:** geom3d[line3d], geom3d[plane], geom3d[distance], geometry[find_angle]

### geom3d[area](tri3d)

Computes the area of triangle tri3d.

**Output:** An expression representing the area is returned.

**Argument options:** (sphere) to compute the surface area of sphere.
**See also:** geom3d[triangle3d], geom3d[sphere], geom3d[volume], geometry[area]

### geom3d[are_collinear](pt3d$_1$, pt3d$_2$, pt3d$_3$)

Determines whether points pt3d$_1$, pt3d$_2$, and pt3d$_3$ lie on the same line.
**Output:** If it can be determined completely, a boolean value of true or false is returned. Otherwise, a condition that must be met for a result of true is returned.
**Additional information:** This function cannot be called with any more or less than three points.
**See also:** geom3d[point3d], geom3d[inter], geom3d[are_concurrent], geom3d[coplanar], geometry[are_collinear]

### geom3d[are_concurrent](line3d$_1$, line3d$_2$, line3d$_3$)

Determines whether line3d$_1$, line3d$_2$, and line3d$_3$ pass through a common point.
**Output:** If it can be determined completely, a boolean value of true or false is returned. Otherwise, a condition that must be met for a result of true is returned.
**Additional information:** This function cannot be called with any more or less than three lines.
**See also:** geom3d[line3d], geom3d[inter], geom3d[are_collinear], geometry[are_concurrent]

### geom3d[are_parallel](line3d$_1$, line3d$_2$)

Determines whether line3d$_1$ and line3d$_2$ are parallel.
**Output:** If it can be determined completely, a boolean value of true or false is returned. Otherwise, a condition that must be met for a result of true is returned.
**Argument options:** (plane$_1$, plane$_2$) to see if two planes are parallel. ♣ (plane, line) to see if a plane and a line are parallel.
**Additional information:** This function cannot be called with any more or less than two lines or planes.
**See also:** geom3d[line3d], geom3d[inter], geom3d[find_angle], geom3d[are_perpendicular], geometry[are_parallel]

### geom3d[are_perpendicular](line3d$_1$, line3d$_2$)

Determines whether line3d$_1$ and line3d$_2$ are perpendicular.
**Output:** If it can be determined completely, a boolean value of true

or false is returned. Otherwise, a condition that must be met for a result of true is returned.
**Argument options:** (plane$_1$, plane$_2$) to see if two planes are perpendicular. ♣ (plane, line) to see if a plane and a line are perpendicular.
**Additional information:** This function cannot be called with any more or less than two lines or planes.
**See also:** geom3d[line], geom3d[inter], geom3d[find_angle], geom3d[are_parallel], geometry[are_perpendicular]

### geom3d[are_tangent](line3d, sphere)

Determines whether line3d is tangent to sphere.
**Output:** If it can be determined completely, a boolean value of true or false is returned. Otherwise, a condition that must be met for a result of true is returned.
**Argument options:** (plane, sphere) to determine if a plane is tangent to a sphere. ♣ (sphere$_1$, sphere$_2$) to determine if two spheres are tangent to one another.
**Additional information:** A line/plane and a sphere (or two spheres) are tangent if they intersect at one and only one point.
**See also:** geom3d[line3d], geom3d[sphere], geom3d[inter], geometry[are_tangent]

### geom3d[center](sphere)

Computes the center of sphere and assigns it to the point *center_of_sphere*.
**Output:** The center point's name, *center_of_sphere*, is returned.
**See also:** geom3d[sphere], geom3d[point3d], geom3d[radius], geometry[center]

### geom3d[centroid](tri3d, name)

Computes the centroid of the triangle tri3d and assigns it to name.
**Output:** The centroid point's name is returned.
**Argument options:** (tetra, name) to compute the centroid of a tetrahedron. ♣ ([pt$_1$, ..., pt$_n$]) to compute the centroid of the points pt$_1$ through pt$_n$.
**Additional information:** There must be no symbolic variables in the coordinates of tri3d's points (vertices).
**See also:** geom3d[point3d], geom3d[triangle3d], geom3d[tetrahedron], geometry[centroid]

### geom3d[coordinates](pt3d)

Displays the coordinates of pt3d, a point in three-dimensional space.
**Output:** A list containing the $x$, $y$, and $z$ coordinates of the point (in that order) is returned.
**See also:** geom3d[point3d], geometry[coordinates]

### geom3d[coplanar](pt3d$_1$, pt3d$_2$, pt3d$_3$, pt3d$_4$)

Determines whether points pt3d$_1$ through pt3d$_4$ lie on the same plane.
**Output:** If it can be determined completely, a boolean value of true or false is returned. Otherwise, a condition that must be met for a result of true is returned.
**Argument options:** (line3d$_1$, line3d$_2$) to determine if two lines are coplanar.
**Additional information:** This function cannot be called with any more or less than four points or two lines.
**See also:** geom3d[point3d], geom3d[line3d], geom3d[inter], geom3d[are_collinear]

### geom3d[distance](pt3d$_1$, pt3d$_2$)

Computes the distance between points pt3d$_1$ and pt3d$_2$.
**Output:** An expression is returned.
**Argument options:** (pt3d, line3d) to compute the distance between a point and a line. ♣ (pt3d, plane) to compute the distance between a point and a plane. ♣ (line3d$_1$, line3d$_2$) to compute the distance between two lines.
**See also:** geom3d[point3d], geom3d[line3d], geom3d[plane], geometry[distance]

### geom3d[inter](line3d$_1$, line3d$_2$, name)

Calculates the point of intersection of line3d$_1$ and line3d$_2$ and assigns it to name.
**Output:** The intersection point's name is returned.
**Argument options:** (plane$_1$, plane$_2$, name) to assign to name the point (or plane) of intersection of two planes. ♣ (line3d, plane, name) to assign to name the point of intersection of a line and a plane.
**Additional information:** If the two objects have no intersection, no point is returned. ♣ If the two objects are the same, the first object is returned.
**See also:** geom3d[point3d], geom3d[line3d], geometry[inter]

### geom3d[line3d](name, [pt3d$_1$, pt3d$_2$])

Defines the line in three-dimensional Euclidean space that passes through points pt3d$_1$ and pt3d$_2$, and assigns it to name.
**Output:** The line's name is returned.
**Argument options:** (name, [pt3d, [expr$_x$, expr$_y$, expr$_z$]]) to assign the line defined by point pt3d and a directional vector defined by $x$, $y$, and $z$ coordinates expr$_x$, expr$_y$, and expr$_z$ to name.
**Additional information:** Using unknowns x, y, or z for anything other than global variables representing the Euclidean dimensions $x$, $y$, and $z$ is not allowed. As well, use of _t as a variable is restricted. ♣ Like tables, spheres follow the rule of *first name evaluation*. To examine the actual equation of a line, use the command op(line3d).
**See also:** geom3d[point3d], geom3d[triangle3d], geom3d[plane], geom3d[sphere], geom3d[tetrahedron], op(line3d), geometry[line]

### geom3d[midpoint](pt3d$_1$, pt3d$_2$, name)

Calculates the midpoint of the line segment joining points pt3d$_1$ and pt3d$_2$, and assigns it to name.
**Output:** The midpoint's name is returned.
**Additional information:** If the two object are the same, the midpoint is equivalent to those points.
**See also:** geom3d[point3d], geom3d[onsegment], geometry[midpoint]

### geom3d[onsegment](pt3d$_1$, pt3d$_2$, expr, name)

Calculates the point that divides the line segment joining pt3d$_1$ to pt3d$_2$ proportionately to the ratio expr, and assigns it to name.
**Output:** The dividing point's name is returned.
**Additional information:** The ratio expr can be numeric or symbolic, floating point or rational, but must not be equal to $-1$.
**See also:** geom3d[point3d], geom3d[midpoint], geometry[onsegment]

### geom3d[on_plane](pt3d, plane)

Determines whether pt3d lies on plane.
**Output:** If it can be determined completely, a boolean value of true or false is returned. Otherwise, a condition that must be met for a result of true is returned.
**Argument options:** ({pt3d$_1$, ..., pt3d$_n$}, plane) or ([pt3d$_1$, ..., pt3d$_n$], plane) to determine whether *all* the points pt3d$_1$ through

pt3d$_n$ lie on plane.
**See also:** geom3d[point3d], geom3d[plane], geom3d[coplanar], geom3d[on_sphere], geometry[on_line]

### geom3d[on_sphere](pt3d, sphere)

Determines whether pt3d lies on sphere.
**Output:** If it can be determined completely, a boolean value of true or false is returned. Otherwise, a condition that must be met for a result of true is returned.
**Argument options:** ({pt3d$_1$, ..., pt3d$_n$}, sphere) or ([pt3d$_1$, ..., pt3d$_n$], sphere) to determine whether all the points pt3d$_1$ through pt3d$_n$ lie on sphere.
**See also:** geom3d[point3d], geom3d[sphere], geom3d[on_plane], geometry[on_plane]

### geom3d[parallel](pt3d, plane, name)

Calculates the plane that goes through point pt3d and is parallel to plane, and assigns it to name.
**Output:** The plane's name is returned.
**Argument options:** (pt3d, line3d, name) to assign the line going through pt3d and parallel to line3d to name. ♣ (line3d$_1$, line3d$_2$, name) to assign the plane going through line3d$_1$ and parallel to line3d$_2$ to name.
**See also:** geom3d[point3d], geom3d[plane], geom3d[perpendicular], geom3d[are_parallel], geometry[parallel]

### geom3d[perpendicular](pt3d, plane3d, name)

Calculates the plane that goes through point pt3d and is perpendicular to plane, and assigns it to name.
**Output:** The plane's name is returned.
**Argument options:** (pt3d, line3d, name) to assign the line going through pt3d and perpendicular to line3d to name. ♣ (line3d$_1$, line3d$_2$, name) to assign the plane going through line3d$_1$ and perpendicular to line3d$_2$ to name.
**See also:** geom3d[point3d], geom3d[plane], geom3d[parallel], geom3d[are_perpendicular], geometry[perpendicular]

### geom3d[plane](name, [pt3d$_1$, pt3d$_2$, pt3d$_3$])

Defines the plane in three-dimensional Euclidean space that passes through the points pt3d$_1$, pt3d$_2$, and pt3d$_3$, and assigns it to name.
**Output:** The plane's name is returned.

**Argument options:** (name, [eqn$_{x,y,z}$]) to assign the plane defined by eqn$_{x,y,z}$, a linear equation in x, y, and z, to name. ♣ (name, [expr$_{x,y,z}$]) to assign the plane defined by expr$_{x,y,z}$ = 0 to name. ♣ (name, [pt3d, [expr$_x$, expr$_y$, expr$_z$]]) to assign the plane that passes through pt3d and has a normal vector represented by [expr$_x$, expr$_y$, expr$_z$] to name. ♣ (name, [pt3d, [expr$_{x1}$, expr$_{y1}$, expr$_{z1}$], [expr$_{x2}$, expr$_{y2}$, expr$_{z2}$]]) to assign the plane defined by a point and two vectors to name.

**Additional information:** Like tables, planes follow the rule of *first name evaluation*. ♣ To examine the actual details of a plane, use the command op(plane).

**See also:** geom3d[point3d], geom3d[line3d], geom3d[sphere], geom3d[triangle3d], geom3d[tetrahedron], op(plane)

### geom3d[point3d](name, expr$_x$, expr$_y$, expr$_z$)

Defines a three-dimensional point, [expr$_x$, expr$_y$, expr$_z$], and assigns it to name.

**Output:** The point's name is returned.

**Argument options:** (name, [expr$_x$, expr$_y$, expr$_z$]) to achieve the same results.

**Additional information:** Using unknowns x, y, or z for anything other than the global variables that represent the Euclidean dimensions $x$, $y$, and $z$ is not allowed. Also, use of _t as a variable is restricted. ♣ Like tables, points follow the rule of *first name evaluation*. ♣ To examine the actual coordinates of a point, use geom3d[coordinates] or op(pt3d).

**See also:** geom3d[line3d], geom3d[sphere], geom3d[plane], geom3d[triangle3d], geom3d[tetrahedron], geom3d[coordinates], op(point3d), geometry[point]

### geom3d[powerps](pt3d, sphere)

Calculates the power of pt3d with respect to sphere.
**Output:** An expression is returned.
**See also:** geometry[point3d], geometry[sphere], geometry[powerpc]

### geom3d[projection](pt3d, plane, name)

Calculates the projection point of pt3d on plane, and assigns it to name.
**Output:** The projection point's name is returned.
**Argument options:** (pt3d, line3d, name) to assign the projection point of pt3d on line3d to name. ♣ (line3d, plane, name) to assign the projection line of line3d on plane to name.

**Additional information:** The projection of pt3d on plane is the point on plane that, when joined to pt3d, forms a perpendicular.
**See also:** geom3d[point3d], geom3d[line3d], geom3d[parallel], geom3d[perpendicular], geometry[projection]

### geom3d[radius](sphere)

Computes the radius of sphere.
**Output:** An expression is returned.
**See also:** geom3d[sphere], geometry[radius]

### geom3d[rad_plane](sphere$_1$, sphere$_2$, name)

Computes the radical axis plane of two spheres, sphere$_1$ and sphere$_2$, and assigns it to name.
**Output:** The radical axis plane's name is returned.
**See also:** geom3d[point3d], geom3d[plane], geom3d[sphere], geometry[rad_axis]

### geom3d[reflect](pt3d, plane, name)

Computes the point of reflection of pt3d with respect to plane, and assigns it to name.
**Output:** The point of reflection's name is returned.
**Argument options:** (line3d, plane, name) to compute the line of reflection of line3d with respect to plane.
**See also:** geom3d[point3d], geom3d[line3d], geom3d[symmetric], geometry[reflect]

### geom3d[sphere](name, [pt3d, expr])

Defines the sphere in three-dimensional Euclidean space with center at point pt3d and radius of length expr, and assigns it to name.
**Output:** The sphere's name is returned.
**Argument options:** (name, [pt3d$_1$, pt3d$_2$, pt3d$_3$, pt3d$_4$]) to assign the sphere passing through points pt3d$_1$, pt3d$_2$, pt3d$_3$, and pt3d$_4$ to name. ♣ (name, [pt3d$_1$, pt3d$_2$, *diameter*]) to assign the sphere with points pt3d$_1$ and pt3d$_2$ as two ends of its diameter to name. ♣ (name, [eqn$_{x,y,z}$]) to assign the sphere defined by eqn$_{x,y,z}$, an equation in x, y, and z, to name. ♣ (name, [expr$_{x,y,z}$]) to assign the sphere defined by expr$_{x,y,z}$ = 0 to name.
**Additional information:** Like tables, spheres follow the rule of *first name evaluation*. ♣ To examine the actual equation of a sphere, use the command op(sphere).
**See also:** geom3d[center], geom3d[radius], geom3d[area], geom3d[volume], geom3d[point3d], geom3d[line3d], geom3d[triangle3d], geom3d[plane], geom3d[tetrahedron], op(sphere), geometry[circle]

### geom3d[symmetric](pt3d$_1$, pt3d$_2$, name)

Computes the symmetric point of point pt3d$_1$ with respect to point pt3d$_2$, and assigns it to name.
**Output:** The symmetric point's name is returned.
**Additional information:** The symmetric point of pt3d$_1$ with respect to pt3d$_2$ is the point equally as far away from pt3d$_2$ but on its opposite side.
**See also:** geom3d[point3d], geom3d[reflect], geometry[symmetric]


### geom3d[tangent](pt3d, sphere, name)

Calculates the plane that goes through pt3d and is tangent to sphere, and assigns it to name.
**Output:** The tangent plane's name is returned.
**Additional information:** The point pt3d must be on the surface of sphere.
**See also:** geom3d[point3d], geom3d[sphere], geom3d[plane], geom3d[are_tangent], geometry[tangent]


### geom3d[tetrahedron](name, [pt3d$_1$, pt3d$_2$, pt3d$_3$, pt3d$_4$])

Defines the tetrahedron in three-dimensional Euclidean space with vertices at points pt3d$_1$ through pt3d$_4$, and assigns it to name.
**Output:** The tetrahedron's name is returned.
**Argument options:** (name, [plane$_1$, plane$_2$, plane$_3$, plane$_4$]) to assign the tetrahedron defined by plane$_1$ through plane$_4$ to name.
**Additional information:** Like tables, tetrahedra follow the rule of *first name evaluation*. ♣ To examine the actual equation of a tetrahedron, use the command op(tetra).
**See also:** geom3d[volume], geom3d[point3d], geom3d[line3d], geom3d[triangle3d], geom3d[plane], geom3d[sphere], op(tetra)


### geom3d[triangle3d](name, [pt3d$_1$, pt3d$_2$, pt3d$_3$])

Defines the triangle in three-dimensional Euclidean space with vertices at points pt3d$_1$, pt3d$_2$, and pt3d$_3$, and assigns it to name.
**Output:** The triangle's name is returned.
**Additional information:** Like tables, triangles follow the rule of *first name evaluation*. ♣ To examine the actual details of a triangle, use the command op(triangle3d).
**See also:** geom3d[area], geom3d[point3d], geom3d[line3d], geom3d[plane], geom3d[sphere], geom3d[tetrahedron], op(triangle3d), geometry[triangle]

## geom3d[type](expr, option)

Tests whether expr is of type option. option may be any of the following three-dimensional geometric types: *point3d*, *line3d*, *triangle3d*, *plane*, *sphere*, or *tetrahedron*.
**Additional information:** When this function is loaded, it updates the standard type command to recognize the six new types. Until then, *point3d*, *line3d*, *triangle3d*, *plane*, *sphere*, and *tetrahedron* are *not* valid Maple types.
**See also:** geometry[point3d], geometry[line3d], geometry[triangle3d], geometry[circle], *type*

## geom3d[volume](sphere)

Computes the volume of sphere.
**Output:** An expression is returned.
**Argument options:** (tetra) to compute the volume of tetrahedron tetra.
**See also:** geom3d[sphere], geom3d[tetrahedron], geom3d[area]

## geometry package

Provides commands for creating objects from two-dimensional Euclidean geometry and calculating important values concerning them. The basic objects available are point, line, triangle, square, circle, ellipse, and conic.
**Additional information:** To access the command fnc from the geometry package, use with(geometry, fnc) to load in a pointer for fnc only or with(geometry) to load in pointers to all the commands. Unfortunately, the long form geometry[fnc] does not work.
✱ The design of several commands in the geometry package differs from most of the Maple library of commands in that the result is assigned to a name supplied as one of the parameters of the command itself.

## geometry[altitude](tri, pt, name)

Computes the line from vertex pt of triangle tri that creates a line perpendicular to the opposite side, and assigns it to name.
**Output:** The altitude's name is returned.
**See also:** geometry[point], geometry[line], geometry[triangle], geometry[bisector], geometry[distance]

## geometry[Appolonius](circle$_1$, circle$_2$, circle$_3$)

Calculates the *Appolonius circles* of circle$_1$, circle$_2$, and circle$_3$.
**Output:** Typically, a list containing eight circles, named *Ap1_of_*circle$_1$_circle$_2$_circle$_3$ through *Ap8_of_*circle$_1$_circle$_2$_circle$_3$, is returned.
**Additional information:** There must be no symbolic variables in the coordinates of the centers or the radii of the three circles. ♣ If the three circles have no radical center, a warning is issued.
**See also:** geometry[circle], geometry[rad_center], geometry[rad_axis]

## geometry[area](tri)

Computes the area of triangle tri.
**Output:** An expression is returned.
**Argument options:** (circle) to compute the area of circle. ♣ (square) to compute the area of square.
**See also:** geometry[triangle], geometry[circle], geometry[square], geometry[sides], geom3d[area]

## geometry[are_collinear](pt$_1$, pt$_2$, pt$_3$)

Determines whether points pt$_1$, pt$_2$, and pt$_3$ lie on the same line.
**Output:** If it can be determined completely, a boolean value of true or false is returned. Otherwise, a condition that must be met for a result of true is returned.
**Additional information:** This function cannot be called with any more or less than three points.
**See also:** geometry[point], geometry[inter], geometry[are_concurrent], geometry[on_line], geom3d[are_collinear]

## geometry[are_concurrent](line$_1$, line$_2$, line$_3$)

Determines whether line$_1$, line$_2$, and line$_3$ pass through a common point.
**Output:** If it can be determined completely, a boolean value of true or false is returned. Otherwise, a condition that must be met for a result of true is returned.
**Additional information:** This function cannot be called with any more or less than three lines.
**See also:** geometry[line], geometry[inter], geometry[are_collinear], geom3d[are_concurrent]

Geometry    205

### geometry[are_harmonic](pt$_1$, pt$_2$, pt$_3$, pt$_4$)

Determines whether points pt$_1$ and pt$_2$ are harmonic conjugates to points pt$_3$ and pt$_4$.

**Output:** If it can be determined completely, a boolean value of true or false is returned. Otherwise, a condition that must be met for a result of true is returned.

**Additional information:** This function cannot be called with any more or less than four points.

**See also:** geometry[point], geometry[harmonic], geometry[are_collinear]


### geometry[are_orthogonal](circle$_1$, circle$_2$)

Determines whether circle$_1$ and circle$_2$ are orthogonal to one another.

**Output:** If it can be determined completely, a boolean value of true or false is returned. Otherwise, a condition that must be met for a result of true is returned.

**Additional information:** This function cannot be called with any more or less than two circles.

**See also:** geometry[circle]


### geometry[are_parallel](line$_1$, line$_2$)

Determines whether line$_1$ and line$_2$ are parallel.

**Output:** If it can be determined completely, a boolean value of true or false is returned. Otherwise, a condition that must be met for a result of true is returned.

**Additional information:** This function cannot be called with any more or less than two lines.

**See also:** geometry[line], geometry[inter], geometry[find_angle], geometry[are_perpendicular], geom3d[are_parallel]


### geometry[are_perpendicular](line$_1$, line$_2$)

Determines whether line$_1$ and line$_2$ are perpendicular.

**Output:** If it can be determined completely, a boolean value of true or false is returned. Otherwise, a condition that must be met for a result of true is returned.

**Additional information:** This function cannot be called with any more or less than two lines.

**See also:** geometry[line], geometry[inter], geometry[find_angle], geometry[are_parallel], geom3d[are_perpendicular]

## geometry[are_similar](tri$_1$, tri$_2$)

Determines whether triangles tri$_1$ and tri$_2$ are similar.

**Output:** If it can be determined completely, a boolean value of true or false is returned. Otherwise, a condition that must be met for a result of true is returned.

**Additional information:** This function cannot be called with any more or less than two triangles. ♣ Two triangles are similar if they contain the same three internal angles.

**See also:** geometry[triangle], geometry[find_angle]


## geometry[are_tangent](line, circle)

Determines whether line is tangent to circle.

**Output:** If it can be determined completely, a boolean value of true or false is returned. Otherwise, a condition that must be met for a result of true is returned.

**Argument options:** (circle$_1$, circle$_2$) to determine if two circles are tangent to one another.

**Additional information:** A line and a circle (or two circles) are tangent to one another if they intersect at *one and only one* point.

**See also:** geometry[line], geometry[circle], geometry[inter], geom3d[are_tangent]


## geometry[bisector](tri, pt, name)

Computes the line from vertex pt of triangle tri that bisects the opposite side, and assigns it to name.

**Output:** The bisector line's name is returned.

**See also:** geometry[point], geometry[line], geometry[triangle], geometry[altitude], geometry[distance]


## geometry[center](circle)

Computes the center of circle and assigns it to the point *center_*circle.

**Output:** The point's name, *center_*circle, is returned.

**See also:** geometry[circle], geometry[point], geometry[radius], geom3d[center]


## geometry[centroid](tri, name)

Computes the centroid of triangle tri and assigns it to name.

**Output:** The centroid point's name is returned.

**Argument options:** ([pt$_1$, ..., pt$_n$]) or ({pt$_1$, ..., pt$_n$}) to compute the centroid of points pt$_1$ through pt$_n$.

### geometry[circle](name, [pt, expr])

Defines the circle in two-dimensional Euclidean space that has center at pt and radius of expr, and assigns it to name.
**Output:** The circle's name is returned.
**Argument options:** (name, [$pt_1$, $pt_2$, $pt_3$]) to assign the circle passing through points $pt_1$, $pt_2$, and $pt_3$ to name. ♣ (name, [$pt_1$, $pt_2$, *diameter*]) to assign the circle with points $pt_1$ and $pt_2$ as ends of its diameter to name. ♣ (name, [$eqn_{x,y}$]) to assign the circle defined by $eqn_{x,y}$, an equation in x and y, to name. (name, [$expr_{x,y}$]) to assign the circle defined by $expr_{x,y} = 0$ to name.
**Additional information:** Using unknowns x or y for anything other than global variables representing Euclidean dimensions $x$ and $y$ is not allowed. ♣ Like tables, circles follow the rule of *first name evaluation*. ♣ To examine the actual equation of a circle, use geometry[detailf] or op(circle).
**See also:** geometry[center], geometry[radius], geometry[area], geometry[diameter], geometry[ellipse], geometry[point], geometry[line], geometry[triangle], geometry[square], geometry[detailf], op(circle), geometry[circle]

### geometry[circumcircle](tri, name)

Calculates the circumcircle of triangle tri and assigns it to name.
**Output:** The circumcircle's name is returned.
**Additional information:** The circumcircle of a triangle is the smallest circle that fits the triangle entirely within it. This circle passes through each vertex of the triangle.
**See also:** geometry[triangle], geometry[circle], geometry[incircle], geometry[excircle]

### geometry[concyclic]($pt_1$, $pt_2$, $pt_3$, $pt_4$)

Determines whether points $pt_1$, $pt_2$, $pt_3$, and $pt_4$ lie on a single circle.
**Output:** If it can be determined completely, a boolean value of true or false is returned. Otherwise, a condition that must be met for a result of true is returned.
**Additional information:** This function cannot be called with any more or less than four points.
**See also:** geometry[circle], geometry[point]

### geometry[conic](name, [pt$_1$, pt$_2$, pt$_3$, pt$_4$, pt$_5$])

Defines the conic in two-dimensional Euclidean space that passes through the five points pt$_1$ through pt$_5$, and assigns it to name.
**Output:** The conic's name is returned.
**Additional information:** Using unknowns x or y for anything other than global variables representing Euclidean dimensions $x$ and $y$ is not allowed. ♣ Like tables, conics follow the rule of *first name evaluation*. ♣ To examine the actual equation of a conic, use geometry[detailf] or op(conic).
**See also:** geometry[point], geometry[ellipse], geometry[detailf], op(conic)

### geometry[convexhull]([pt$_1$, ..., pt$_n$])

Computes the convex hull that encloses points pt$_1$ through pt$_n$.
**Output:** A list with three elements is returned. The elements are the three points (always from the input) that define the convex hull.
**Argument options:** ({pt$_1$, ..., pt$_n$}) to compute the convex hull of a set of points.
**Additional information:** The convex hull of pt$_1$ through pt$_n$ is the smallest circle that encloses all the points. ♣ There must be no symbolic variables in the coordinates of pt$_1$ through pt$_n$.
**See also:** geometry[point], geometry[circle], geometry[diameter], geometry[on_circle]

### geometry[coordinates](pt)

Displays the coordinates of point pt.
**Output:** A list containing the $x$ and $y$ coordinates (in that order) is returned.
**See also:** geometry[point], geom3d[coordinates]

### geometry[detailf](pt)

Displays the details of point pt.
**Output:** A list with two elements is returned. The elements are the $x$ and $y$ coordinates of pt, respectively.
**Argument options:** (line) to display a list containing the equation that defines line. ♣ (circle) to display a list containing the center and radius of circle in the form *center* = [x$_c$, y$_c$], *radius* = [expr$_r$].
**Additional information:** This command does not work for other types of geometric entities (e.g., triangles, squares, etc.). ♣ In all cases, the op command can be used to find out all the known information about a geometric object.
**See also:** geometry[point], geometry[line], geometry[circle], op(point), op(line), op(circle), op(triangle), op(square), op(ellipse), op(conic)

## geometry[diameter]([pt$_1$, ..., pt$_n$])

Computes the diameter of the smallest circle that encompasses all the points pt$_1$ through pt$_n$.
**Output:** A list with three elements is returned. The elements are the two endpoints of the diameter (always two points from the input) and the length of the diameter, in that order.
**Argument options:** ({pt$_1$, ..., pt$_n$}) to compute the diameter of a set of points.
**Additional information:** There must be no symbolic variables in the coordinates of pt$_1$ through pt$_n$.
**See also:**  geometry[point], geometry[circle], geometry[convexhull], geometry[on_circle]


## geometry[distance](pt, line)

Computes the distance between the point pt and line.
**Output:** An expression is returned.
**Argument options:** (pt$_1$, pt$_2$) to compute the distance between two points.
**See also:**  geometry[point], geometry[line], geometry[perpen_bisector], geom3d[distance]


## geometry[ellipse](name, [pt, expr$_1$, expr$_2$, x_axis])

Defines the ellipse in two-dimensional Euclidean space that has center at point pt, major axis of length expr$_1$ on the $x$-axis, and minor axis of length expr$_2$ on the $y$-axis, and assigns it to name.
**Output:** The ellipse's name is returned.
**Argument options:** (name, [pt, expr$_1$, expr$_2$, y_axis]) to assign the ellipse with center at pt, and major axis on the $y$-axis to name.
♣ (name, [pt$_1$, pt$_2$, pt$_3$, pt$_4$, pt$_5$]) to assign the ellipse passing through points pt$_1$, pt$_2$, pt$_3$, and pt$_4$ and with midpoint at point pt$_5$ to name. ♣ (name, [eqn$_{x,y}$]) to assign the ellipse defined by eqn$_{x,y}$, an equation in x and y, to name. ♣ (name, [expr$_{x,y}$]) to assign the ellipse defined by expr$_{x,y} = 0$ to name.
**Additional information:** Using unknowns x or y for anything other than global variables representing Euclidean dimensions $x$ and $y$ is not allowed. ♣ Like tables, ellipses follow the rule of *first name evaluation*. ♣ To examine the actual equation of an ellipse, use op(ellipse).
**See also:**  geometry[point], geometry[circle], geometry[conic], op(ellipse)

### geometry[Eulercircle](tri, name)

Computes the Euler circle of triangle tri and assigns it to name.
**Output:** The Euler circle's name is returned.
**Additional information:** There must be no symbolic variables in the coordinates of tri's vertices.
**See also:** geometry[triangle], geometry[circle], geometry[Eulerline]

### geometry[Eulerline](tri, name)

Computes the Euler line of triangle tri and assigns it to name.
**Output:** The Euler line's name is returned.
**Additional information:** There must be no symbolic variables in the coordinates of tri's vertices.
**See also:** geometry[triangle], geometry[line], geometry[Eulerline]

### geometry[excircle](tri)

Calculates the three *excircles* of triangle tri and assigns them to names *excircle_of_tri_A*, *excircle_of_tri_B*, and *excircle_of_tri_C*.
**Output:** An expression sequence containing the names of the three excircles is returned.
**Additional information:** An excircle of a triangle is a circle passing through one vertex and tangent to the side opposite that vertex.
**See also:** geometry[triangle], geometry[circle], geometry[incircle], geometry[circumcircle], geometry[tangent]

### geometry[find_angle](line$_1$, line$_2$)

Computes the smallest angle between line$_1$ and line$_2$.
**Output:** An expression representing the angle (in radians) is returned.
**Argument options:** (circle$_1$, circle$_2$) to compute the angle between two circles. The angle is always in the range $[0, 2\pi]$.
**See also:** geometry[line], geometry[circle], geom3d[angle]

### geometry[Gergonnepoint](tri, name)

Computes the Gergonne point of triangle tri and assigns it to name.
**Output:** The Gergonne point's name is returned.
**Additional information:** There must be no symbolic variables in the coordinates of tri's vertices.
**See also:** geometry[point], geometry[triangle], geometry[Nagelpoint], geometry[orthocenter]

### geometry[harmonic](pt$_1$, pt$_2$, pt$_3$, name)

Calculates the harmonic conjugate to pt$_1$ with respect to the points pt$_2$ and pt$_3$, and assigns it to name.
**Output:** The harmonic point's name is returned.
**See also:** geometry[point], geometry[are_harmonic]

### geometry[incircle](tri, name)

Calculates the *incircle* of triangle tri and assigns it to name.
**Output:** The incircle's name is returned.
**Additional information:** The incircle of a triangle is the largest circle that fits entirely within the triangle. This circle touches each side of the triangle.
**See also:** geometry[triangle], geometry[circle], geometry[excircle], geometry[circumcircle]

### geometry[inter](line$_1$, line$_2$)

Calculates the point of intersection of line$_1$ and line$_2$ and assigns it to the name line$_1$_*intersect*_line$_2$.
**Output:** The intersection point's name is returned.
**Argument options:** (circle$_1$, circle$_2$) to find the point(s) of intersection of two circles. ♣ (circle, line) to find the point(s) of intersection of a circle and a line.
**Additional information:** If two objects have no intersection, a warning message is issued and no point is returned. ♣ If two object are the same, a warning is issued and the first object is returned.
**See also:** geometry[point], geometry[line], geometry[circle], geometry[detailf], geom3d[inter]

### geometry[inversion](pt, circle, name)

Calculates the inversion point of pt with respect to circle, and assigns it to name.
**Output:** The inversion point's name is returned.
**Argument options:** (line, circle, name) to calculate the inversion of line with respect to circle. ♣ (circle$_1$, circle$_2$, name) to calculate the inversion of circle$_1$ with respect to circle$_2$.
**See also:** geometry[point], geometry[line], geometry[circle]

### geometry[is_equilateral](tri)

Determines whether tri is an equilateral triangle.
**Output:** If it can be determined completely, a boolean value of true or false is returned. Otherwise, a condition that must be met for a result of true is returned.

**Additional information:** A triangle is equilateral if all internal angles are the same or, equally, if all sides are of the same length.
**See also:** geometry[triangle], geometry[is_right], geometry[find_angle]

### geometry[is_right](tri)

Determines whether tri is a right triangle.
**Output:** If it can be determined completely, a boolean value of true or false is returned. Otherwise, a condition that must be met for a result of true is returned.
**Additional information:** A triangle is right if one of its angles is equal to $\pi/2$ radians.
**See also:** geometry[triangle], geometry[is_equilateral], geometry[find_angle]

### geometry[line](name, [pt$_1$, pt$_2$])

Defines the line in two-dimensional Euclidean space that passes through points pt$_1$ and pt$_2$, and assigns it to name.
**Output:** The line's name is returned.
**Argument options:** (name, [eqn$_{x,y}$]) to assign the line defined by eqn$_{x,y}$, an equation in x and y, to name. ♣ (name, [expr$_{x,y}$]) to assign the line defined by expr$_{x,y} = 0$ to name.
**Additional information:** Using unknowns x or y for anything other than global variables representing Euclidean dimensions $x$ and $y$ is not allowed. ♣ Like tables, lines follow the rule of *first name evaluation*. ♣ To examine the actual equation of a line, use geometry[detailf] or op(line).
**See also:** geometry[point], geometry[circle], geometry[square], geometry[triangle], geometry[detailf], op(line), geom3d[line3d]

### geometry[make_square](name, [pt$_1$, pt$_2$, option])

Defines the square in two-dimensional Euclidean space with vertices at points pt$_1$ and pt$_2$, and assigns it to name. option specifies the relationship between the given vertices and can equal *adjacent* or *opposite*.
**Output:** If *adjacent* is used, an expression sequence containing the names of two possible squares, name_1 and name_2, is returned. If *opposite* is used, the name of one possible square, name, is returned.
**Argument options:** (name, [pt$_v$, pt$_c$]) to define the square with one vertex at pt$_v$ and center at pt$_c$.
**Additional information:** Like tables, squares follow the rule of

*first name evaluation.* ♣ To examine the actual details of a square, use op(square).
**See also:** geometry[square], geometry[point], op(square)

### geometry[median](tri, pt, name)

This command is the same as geometry[bisector].
**See also:** geometry[bisector]

### geometry[midpoint](pt$_1$, pt$_2$, name)

Calculates the midpoint of the line segment joining points pt$_1$ and pt$_2$, and assigns it to name.
**Output:** The midpoint's name is returned.
**Additional information:** If the two object are the same, the midpoint is equivalent to those points.
**See also:** geometry[point], geometry[onsegment], geom3d[midpoint]

### geometry[Nagelpoint](tri, name)

Computes the Nagel point of triangle tri and assigns it to name.
**Output:** The Nagel point's name is returned.
**Additional information:** There must be no symbolic variables in the coordinates of tri's vertices.
**See also:** geometry[point], geometry[triangle], geometry[Gergonnepoint], geometry[orthocenter]

### geometry[onsegment](pt$_1$, pt$_2$, expr, name)

Calculates the point that divides the line segment joining pt$_1$ to pt$_2$ proportionate to the ratio expr, and assigns it to name.
**Output:** The dividing point's name is returned.
**Additional information:** The ratio expr can be numeric or symbolic, floating point or rational, but must not be equal to $-1$.
**See also:** geometry[point], geometry[midpoint], geom3d[onsegment]

### geometry[on_circle](pt, circle)

Determines whether pt lies on circle.
**Output:** If it can be determined completely, a boolean value of true or false is returned. Otherwise, a condition that must be met for a result of true is returned.

**Argument options:** ($\{pt_1, ..., pt_n\}$, circle) or ([$pt_1, ..., pt_n$], circle) to determine if the points $pt_1$ through $pt_n$ all lie on circle.
**See also:** geometry[point], geometry[circle], geometry[on_line], geom3d[on_sphere]

### geometry[on_line](pt, line)

Determines whether pt lies on line.
**Output:** If it can be determined completely, a boolean value of true or false is returned. Otherwise, a condition that must be met for a result of true is returned.
**Argument options:** ($\{pt_1, ..., pt_n\}$, line) or ([$pt_1, ..., pt_n$], line) to determine if the points $pt_1$ through $pt_n$ all lie on line.
**See also:** geometry[point], geometry[line], geometry[are_collinear], geometry[on_circle], geom3d[on_plane]

### geometry[orthocenter](tri, name)

Computes the orthocenter of triangle tri and assigns it to name.
**Output:** The orthocenter point's name is returned.
**Additional information:** There must be no symbolic variables in the coordinates of tri's vertices.
**See also:** geometry[point], geometry[triangle], geometry[centroid], geometry[Nagelpoint], geometry[Gergonnepoint]

### geometry[parallel](pt, line, name)

Calculates the line that goes through pt and is parallel to line, and assigns it to name.
**Output:** The line's name is returned.
**See also:** geometry[point], geometry[line], geometry[perpendicular], geometry[projection], geometry[are_parallel], geom3d[parallel]

### geometry[perpendicular](pt, line, name)

Calculates the line that goes through pt and is perpendicular to line, and assigns it to name.
**Output:** The line's name is returned.
**See also:** geometry[point], geometry[line], geometry[parallel], geometry[projection], geometry[are_perpendicular], geometry[perpen_bisector], geom3d[perpendicular]

### geometry[perpen_bisector](pt$_1$, pt$_2$, name)

Computes the line through the midpoint of the line segment from pt$_1$ to pt$_2$ and perpendicular to that segment, and assigns it to name.

**Output:** The perpendicular bisector's name is returned.

**See also:** geometry[point], geometry[distance], geometry[perpendicular]

### geometry[point](name, expr$_x$, expr$_y$)

Defines a two-dimensional point [expr$_x$, expr$_y$] and assigns it to name.

**Output:** The point's name is returned.

**Argument options:** (name, [expr$_x$, expr$_y$]), (name, x=expr$_x$, y=expr$_y$), or (name, [x=expr$_x$, y=expr$_y$]) to achieve the same results.

**Additional information:** Using unknowns x or y for anything other than global variables representing Euclidean dimensions $x$ and $y$ is not allowed. ♣ Like tables, points follow the rule of *first name evaluation*. ♣ To examine the actual coordinates of a point, use geometry[coordinates], geometry[detailf], or op(pt).

**See also:** geometry[line], geometry[circle], geometry[square], geometry[triangle], geometry[ellipse], geometry[coordinates], geometry[detailf], op(point), geom3d[point3d]

### geometry[polar_point](pt, conic, name)

Calculates the polar line of pt with respect to conic, and assigns it to name.

**Output:** The polar line's name is returned.

**Argument options:** (pt, circle, name) to assign the polar line with respect to circle to name. ♣ (pt, ellipse, name) to assign the polar line with respect to ellipse to name.

**Additional information:** If no polar line can be determined, NULL is returned.

**See also:** geometry[point], geometry[line], geometry[circle], geometry[ellipse], geometry[conic], geometry[pole_line]

### geometry[pole_line](line, conic, name)

Calculates the pole of line with respect to conic, and assigns it to name.

**Output:** The pole point's name is returned.

**Argument options:** (line, circle, name) to assign the pole with respect to circle to name. ♣ (line, ellipse, name) to assign the

pole with respect to ellipse to name.
**Additional information:** If no pole can be determined, NULL is returned.
**See also:** geometry[point], geometry[line], geometry[circle], geometry[ellipse], geometry[conic], geometry[polar_point]

### geometry[powerpc](pt, circle)

Calculates the power of pt with respect to circle.
**Output:** An expression is returned.
**See also:** geometry[point], geometry[circle], geometry[tangentpc], geom3d[powerps]

### geometry[projection](pt, line, name)

Calculates the projection point of pt on line, and assigns it to name.
**Output:** The projection point's name is returned.
**Additional information:** The projection of pt on line is the point on line that, when joined to pt, forms a perpendicular.
**See also:** geometry[point], geometry[line], geometry[parallel], geometry[perpendicular], geom3d[projection]

### geometry[radius](circle)

Computes the radius of circle.
**Output:** An expression is returned.
**See also:** geometry[circle], geom3d[radius]

### geometry[rad_axis](circle$_1$, circle$_2$, name)

Computes the radical axis line of circle$_1$ and circle$_2$, and assigns it to name.
**Output:** The radical axis line's name is returned.
**See also:** geometry[point], geometry[line], geometry[circle], geometry[rad_center], geometry[Appolonius], geom3d[rad_plane]

### geometry[rad_center](circle$_1$, circle$_2$, circle$_3$, name)

Computes the radical center of circle$_1$, circle$_2$, and circle$_3$, and assigns it to name.
**Output:** The radical center point's name is returned.
**See also:** geometry[circle], geometry[point], geometry[rad_axis], geometry[Appolonius]

## geometry[randpoint](line, name)

Calculates a random point on line and assigns it to name.
**Output:** The random point's name is returned.
**Argument options:** (line, a..b, name) to find a random point on line with $x$ value in range a..b. ♣ (circle, name) to find a random point on circle. ♣ ($a_x$..$b_x$, $a_y$..$b_y$, name) to find a random point with $x$ value in range $a_x$..$b_x$ and $y$ value in range $a_y$..$b_y$.
**See also:** geometry[point], geometry[line], geometry[circle]

## geometry[reflect](pt, line, name)

Computes the reflection point of pt with respect to line, and assigns it to name.
**Output:** The point of reflection's name is returned.
**See also:** geometry[point], geometry[line], geometry[rotate], geometry[symmetric], geom3d[reflect]

## geometry[rotate](pt, expr, option, name)

Computes the point of rotation of pt when rotated expr radians about the origin, and assigns it to name. option specifies the direction of rotation and can equal *clockwise* or *counterclockwise*.
**Output:** The point of rotation's name is returned.
**Argument options:** ($pt_1$, expr, option, name, $pt_2$) to rotate point $pt_1$ about the point $pt_2$.
**See also:** geometry[point], geometry[reflect], geometry[symmetric]

## geometry[sides](tri)

Computes the lengths of the sides of triangle tri.
**Output:** A list containing the lengths of the triangle's sides is returned.
**Argument options:** (square) to compute the length of sides of square. A single value is returned.
**See also:** geometry[triangle], geometry[square], geometry[area]

## geometry[similitude]($circle_1$, $circle_2$)

Calculates the *insimilitude* and *exsimilitude* of the $circle_1$ and $circle_2$, and names them *in_similitude_of_$circle_1$_$circle_2$* and *ex_similitude_of_$circle_1$_$circle_2$*, respectively.
**Output:** An expression sequence containing the names of the two similitude points is returned.
**Additional information:** Occasionally, geometry[similitude] only assigns an insimilitude.
**See also:** geometry[point], geometry[circle]

### geometry[Simsonline](tri, pt, name)

Calculates the Simson line of triangle tri with respect to pt, and assigns it to name. pt must lie on tri's circumcircle.

**Output:** The Simson line's name is returned.

**Additional information:** If pt is not on tri's circumcircle, a warning is displayed. ♣ In order for this command to work, the vertices of tri must contain only known values.

**See also:** geometry[point], geometry[triangle], geometry[line], geometry[circumcircle], geometry[on_circle]

### geometry[square](name, [$pt_1$, $pt_2$, $pt_3$, $pt_4$])

Defines the square in two-dimensional Euclidean space that has vertices at points $pt_1$, $pt_2$, $pt_3$, and $pt_4$, and assigns it to name.

**Output:** The square's name is returned.

**Additional information:** The vertices of the square must be provided such that drawing lines between them in order creates a square. ♣ Like tables, squares follow the rule of *first name evaluation*. ♣ To examine the actual details of a square, use op(square).

**See also:** geometry[area], geometry[point], geometry[line], geometry[triangle], geometry[circle], op(square)

### geometry[symmetric]($pt_1$, $pt_2$, name)

Computes the symmetric point of $pt_1$ with respect to $pt_2$, and assigns it to name.

**Output:** The symmetric point's name is returned.

**Additional information:** The symmetric point of $pt_1$ with respect to $pt_2$ is the point equally as far away from $pt_2$ but on the opposite side.

**See also:** geometry[point], geometry[rotate], geometry[reflect], geom3d[symmetric]

### geometry[tangent](pt, circle, $name_1$, $name_2$)

Calculates the two lines that go through pt and are tangent to circle, and assigns them to $name_1$ and $name_2$.

**Output:** An expression sequence containing the names of the two tangent lines is returned.

**Additional information:** If pt is within circle, then no lines are assigned. ♣ If pt is on circle, then the one line tangent at that point is assigned to $name_1$.

**See also:** geometry[point], geometry[circle], geometry[tangentpc], geometry[are_tangent], geom3d[tangent]

### geometry[tangentpc](pt, circle, name)

Calculates the line through pt and tangent to circle, and assigns it to name. pt must lie on circle.
**Output:** The tangent line's name is returned.
**Additional information:** If pt is not on circle, an error message is displayed.
**See also:** geometry[point], geometry[circle], geometry[line], geometry[tangent], geometry[are_tangent], geometry[on_circle], geom3d[tangentpc]

### geometry[triangle](name, [pt$_1$, pt$_2$, pt$_3$])

Defines the triangle in two-dimensional Euclidean space that has vertices at points pt$_1$, pt$_2$, and pt$_3$, and assigns it to name.
**Output:** The triangle's name is returned.
**Argument options:** (name, [expr$_1$, expr$_2$, expr$_3$]) to assign the triangle with sides of length expr$_1$, expr$_2$, and expr$_3$ to name. ♣ (name, [expr$_a$, expr$_b$, angle$_{a,b}$]) to assign the triangle with two sides of length expr$_1$ and expr$_2$ and an angle (in radians) between those sides of angle$_{a,b}$ to name. ♣ (name, [line$_1$, line$_2$, line$_3$]) to assign the triangle defined by the lines line$_1$, line$_2$, and line$_3$ to name.
**Additional information:** Using unknowns x or y for anything other than global variables representing Euclidean dimensions $x$ and $y$ is not allowed. ♣ Like tables, triangles follow the rule of *first name evaluation*. ♣ To examine the actual details of a triangle, use op(triangle).
**See also:** geometry[area], geometry[point], geometry[line], geometry[square], geometry[circle], op(triangle), geom3d[triangle3d]

### geometry[type](expr, option)

Tests whether expr is of type option. option may be any of the following two-dimensional geometric types: *point2d*, *line2d*, or *circle2d*.
**Additional information:** When this function is loaded, it updates the standard type command to recognize the three new types. Until that time, *point2d*, *line2d*, and *circle2d* are not valid Maple types.
**See also:** geometry[point], geometry[line], geometry[circle], *type*

### op(circle)

Displays the internal structure of the table that represents the two-dimensional circle, circle.

**Output:** A table data structure is displayed.
**Argument options:** The individual elements of the table can be accessed in typical table style: circle*[form]* to extract the type of geometric structure, circle*[center]* to extract the center of the circle, circle*[radius]* to extract the radius of the circle, circle*[equation]* to extract the equation of the circle, and circle*[given]* to extract the defining facts.
**Additional information:** Special functions geometry[center] and geometry[radius] also return information about a circle.
**See also:** geometry[circle], op(point), op(ellipse), *table*

### op(conic)

Displays the internal structure of the table that represents the two-dimensional conic, conic.
**Output:** A table data structure is displayed.
**Argument options:** The individual elements of the table can be accessed in typical table style: conic*[form]* to extract the type of geometric structure, and conic*[equation]* to extract the equation of the conic.
**See also:** geometry[conic], *table*

### op(conicpj)

Displays the internal structure of the table that represents the projective geometry conic, conicpj.
**Output:** A table data structure is displayed.
**Argument options:** The individual elements of the table can be accessed in typical table style: conicpj*[form]* to extract the type of geometric structure and conicpj*[coords]* to extract a list containing the (six) coordinates.
**See also:** projgeom[conic], op(pointpj), op(linepj), *table*

### op(ellipse)

Displays the internal structure of the table that represents the two-dimensional ellipse, ellipse.
**Output:** A table data structure is displayed. Depending on how the ellipse was defined, there may be elements for *eccentricity*, *major_axis*, *minor_axis*, *focus_1*, and *focus_2*.
**Argument options:** The individual elements of the table can be accessed in typical table style: ellipse*[form]* to extract the type of geometric structure, ellipse*[center]* to extract the center of the ellipse, ellipse*[equation]* to extract the defining equation, ellipse*[eccentricity]* to extract the eccentricity of the ellipse, ellipse*[given]* to extract the defining facts, ellipse*[major_axis]* or ellipse*[minor_axis]*

Geometry   221

to extract the axes, and ellipse*[focus_1]* or ellipse*[focus_2]* to extract the foci.

**See also:** geometry[ellipse], op(point), op(circle), *table*

## op(line)

Displays the internal structure of the table that represents the two-dimensional line, line.

**Output:** A table data structure is displayed.

**Argument options:** The individual elements of the table can be accessed in typical table style: line*[form]* to extract the type of geometric structure, line*[equation]* to extract the equation of the line, and line*[given]* to extract the defining facts.

**See also:** geometry[line], op(point), op(triangle), *table*

## op(line3d)

Displays the internal structure of the table that represents the three-dimensional line, line3d.

**Output:** A table data structure is displayed.

**Argument options:** The individual elements of the table can be accessed in typical table style: line3d*[form]* to extract the type of geometric structure, line3d*[equation]* to extract the equation of the line, line3d*[direction_vector]* to extract the line's directional vector (as a list of three elements), and line3d*[given]* to extract the defining facts.

**See also:** geom3d[line3d], op(point3d), op(triangle3d), *table*

## op(linepj)

Displays the internal structure of the table that represents the projective geometry line, linepj.

**Output:** A table data structure is displayed.

**Argument options:** The individual elements of the table can be accessed in typical table style: linepj*[form]* to extract the type of geometric structure and linepj*[coords]* to extract a list containing the three coordinates.

**See also:** projgeom[line], op(pointpj), op(conicpj), *table*

## op(plane)

Displays the internal structure of the table that represents the three-dimensional plane, plane.

**Output:** A table data structure is displayed.

**Argument options:** The individual elements of the table can be accessed in typical table style: plane*[form]* to extract the type of

geometric structure, plane*[normal_vector]*] to extract the normal vector to the plane, plane*[equation]* to extract the equation of the plane, and plane*[given]* to extract the defining facts.
**See also:** geom3d[plane], *table*

### op(point)

Displays the internal structure of the table that represents the two-dimensional point, point.
**Output:** A table data structure is displayed.
**Argument options:** The individual elements of the table can be accessed in typical table style: point*[form]* to extract the type of geometric structure, point*[x]* to extract the $x$ coordinate, and point*[y]* to extract the $y$ coordinate.
**Additional information:** geometry[coordinates] returns the $x$ and $y$ coordinates in a list.
**See also:** geometry[point], geometry[coordinates], op(line), op(triangle), op(circle), op(ellipse), op(square), *table*

### op(point3d)

Displays the internal structure of the table that represents the three-dimensional point, point3d.
**Output:** A table data structure is displayed.
**Argument options:** The individual elements of the table can be accessed in typical table style: point3d*[form]* to extract the type of geometric structure, point3d*[x]* to extract the $x$ coordinate, point3d*[y]* to extract the $y$ coordinate, and point3d*[z]* to extract the $z$ coordinate.
**Additional information:** geom3d[coordinates] returns the $x$, $y$, and $z$ coordinates in a list.
**See also:** geom3d[point3d], geom3d[coordinates], op(line3d), op(triangle3d), op(sphere), op(tetrahedron), *table*

### op(pointpj)

Displays the internal structure of the table that represents the projective geometry point, pointpj.
**Output:** A table data structure is displayed.
**Argument options:** The individual elements of the table can be accessed in typical table style: pointpj*[form]* to extract the type of geometric structure and pointpj*[coords]* to extract a list containing the three coordinates.
**See also:** projgeom[point], op(linepj), op(conicpj), *table*

## op(sphere)

Displays the internal structure of the table that represents the three-dimensional sphere, sphere.

**Output:** A table data structure is displayed.

**Argument options:** The individual elements of the table can be accessed in typical table style: sphere*[form]* to extract the type of geometric structure, sphere*[center]* to extract the center of the sphere, sphere*[radius]* to extract the radius, sphere*[equation]* to extract the equation of the sphere, and sphere*[given]* to extract the defining facts.

**Additional information:** geom3d[center] and geom3d[radius] also return information about a sphere.

**See also:** geom3d[sphere], op(point3d), *table*


## op(square)

Displays the internal structure of the table that represents the two-dimensional square, square.

**Output:** A table data structure is displayed.

**Argument options:** The individual elements of the table can be accessed in typical table style: square*[form]* to extract the type of geometric structure, square*[diagonal]* to extract the length of the diagonal, and square*[given]* to extract the defining facts.

**Additional information:** The *given* element contains four points, even if only two were given in the definition.

**See also:** geometry[square], geometry[make_square], op(point), *table*


## op(tetrahedron)

Displays the internal structure of the table that represents the three-dimensional tetrahedron, tetrahedron.

**Output:** A table data structure is displayed.

**Argument options:** The individual elements of the table can be accessed in typical table style: tetrahedron*[form]* to extract the type of geometric structure, tetrahedron*[given]* to extract the defining facts, and tetrahedron*[vertex_1]*, tetrahedron*[vertex_2]*, tetrahedron*[vertex_3]*, or tetrahedron*[vertex_4]* to extract the vertices.

**See also:** geom3d[tetrahedron], op(point3d), *table*


## op(triangle)

Displays the internal structure of the table that represents the two-dimensional triangle, triangle.

**Output:** A table data structure is displayed.
**Argument options:** The individual elements of the table can be accessed in typical table style: triangle*[form]* to extract the type of geometric structure, triangle*[given]* to extract the defining facts, and triangle*[1]*, triangle*[2]*, or triangle*[3]* to extract the vertices.
**See also:** geometry[triangle], op(point), op(line), *table*

## op(triangle3d)

Displays the internal structure of the table that represents the three-dimensional triangle, triangle3d.
**Output:** A table data structure, that includes entries for *form*, the facts that were *given* when the triangle was defined, and the three vertices labelled *vertex_1*, *vertex_2*, and *vertex_3*, is displayed.
**Argument options:** The individual elements of the table can be accessed in typical table style: triangle3d*[form]* to extract the type of geometric structure, triangle3d*[given]* to extract the defining facts, and triangle3d*[vertex_1]*, triangle3d*[vertex_2]*, or triangle3d*[vertex_3]* to extract the vertices.
**See also:** geom3d[triangle3d], op(point3d), *table*

## projgeom package

Provides commands for creating objects from three-dimensional projective geometry and calculating important values concerning them. The basic objects available are points (ptpj), lines (linepj), and conics (conicpj).
**Additional information:** To access the command fnc from the projgeom package, use with(projgeom, fnc) to load in a pointer for fnc only or with(projgeom) to load in pointers to all the commands. Unfortunately, the long form projgeom[fnc] does not work.
♣ The design of several commands in the projgeom package differs from most of the Maple library of commands in that the result is assigned to a name supplied as one of the parameters of the command itself.

## projgeom[collinear](ptpj$_1$, ptpj$_2$, ptpj$_3$)

Determines whether points ptpj$_1$, ptpj$_2$, and ptpj$_3$ lie on the same line.
**Output:** If it can be determined completely, a boolean value of true or false is returned. Otherwise, a condition that must be met for a result of true is returned.
**Additional information:** This function cannot be called with any more or less than three points.
**See also:** projgeom[point], projgeom[inter], projgeom[concur]

## projgeom[concur](linepj$_1$, linepj$_2$, linepj$_3$)

Determines whether linepj$_1$, linepj$_2$, and linepj$_3$ pass through a common point.

**Output:** If it can be determined completely, a boolean value of true or false is returned. Otherwise, a condition that must be met for a result of true is returned.

**Additional information:** This function cannot be called with any more or less than three lines.

**See also:** projgeom[line], projgeom[inter], projgeom[collinear], projgeom[linemeet]

## projgeom[conic](name, [expr$_a$, expr$_b$, expr$_c$, expr$_d$, expr$_e$, expr$_f$])

Defines the conic in three-dimensional Euclidean space whose equation has coefficients expr$_a$ through expr$_f$, and assigns it to name.

**Output:** The conic's name is returned.

**Additional information:** The formula for the equation of the conic is $ax^2 + by^2 + cz^2 + dyz + ezx + fxy$. ♣ Like tables, conic follow the rule of *first name evaluation*. ♣ To examine the actual equation of a conic, use op(conicpj).

**See also:** projgeom[point], projgeom[line], projgeom[fpconic], op(conicpj)

## projgeom[conjugate](ptpj$_1$, ptpj$_2$, conicpj)

Determines whether points ptpj$_1$ and ptpj$_2$ are conjugate with respect to conicpj.

**Output:** If it can be determined completely, a boolean value of true or false is returned. Otherwise, a condition that must be met for a result of true is returned.

**See also:** projgeom[point], projgeom[conic]

## projgeom[ctangent](ptpj, conicpj, name$_1$, name$_2$)

Calculates the lines that go through ptpj and are tangent to conicpj in the real complex plane, and assigns them to name$_1$ and name$_2$.

**Output:** An expression sequence containing the two tangent line names is returned.

**Additional information:** There may be only one tangent line. In that case, only name$_1$ is assigned to a line.

**See also:** projgeom[point], projgeom[conic], projgeom[line], projgeom[rtangent], projgeom[ptangent]

### projgeom[fpconic](name, ptpj$_1$, ptpj$_2$, ptpj$_3$, ptpj$_4$, ptpj$_5$)

Defines the conic in three-dimensional Euclidean space that passes through the points ptpj$_1$ through ptpj$_5$, and assigns it to name.
**Output:** The conic's name is returned.
**Additional information:** Like tables, conic follows the rule of *first name evaluation*. ♣ To examine the actual equation of a conic, use op(conicpj).
**See also:** projgeom[point], projgeom[conic], op(conicpj)

### projgeom[harmonic](ptpj$_1$, ptpj$_2$, ptpj$_3$, name)

Calculates the harmonic conjugate to ptpj$_1$ with respect to the points ptpj$_2$ and ptpj$_3$, and assigns it to name.
**Output:** The harmonic point's name is returned.
**See also:** projgeom[point], projgeom[tharmonic]

### projgeom[inter](linepj$_1$, linepj$_2$, name)

Calculates the point of intersection of linepj$_1$ and linepj$_2$, and assigns it to name.
**Output:** The intersection point's name is returned.
**Additional information:** If the two objects have no intersection, no point is returned. ♣ If the two objects are the same, the origin is returned.
**See also:** projgeom[point], projgeom[line], projgeom[join]

### projgeom[join](ptpj$_1$, ptpj$_2$, name)

Calculates the line joining the points ptpj$_1$ and ptpj$_2$, and assigns it to name.
**Output:** The joining line's name is returned.
**See also:** projgeom[point], projgeom[line]

### projgeom[lccutc](conicpj, linepj, name$_1$, name$_2$)

Calculates the points of intersection of conicpj and linepj in the complex plane, and assigns them to name$_1$ and name$_2$.
**Output:** An expression sequence containing the point names is returned.
**Additional information:** If the two objects have no intersection, no point is returned.
**See also:** projgeom[point], projgeom[line], projgeom[conic], projgeom[inter], projgeom[lccutr], projgeom[lccutr2p]

Geometry 227

### projgeom[lccutr](conicpj, linepj, name$_1$, name$_2$)

Calculates the points of intersection of conicpj and linepj in the real extended plane, and assigns them to name$_1$ and name$_2$.

**Output:** An expression sequence containing the point names is returned.

**Additional information:** If the two objects have no intersection, no point is returned.

**See also:** projgeom[point], projgeom[line], projgeom[conic], projgeom[inter], projgeom[lccutp], projgeom[lccutr2p]

### projgeom[lccutr2p](conicpj, ptpj$_1$, ptpj$_2$, name$_1$, name$_2$)

Calculates the points of intersection of conicpj and the line joining the points ptpj$_1$ and ptpj$_2$ in the real extended plane, and assigns them to name$_1$ and name$_2$.

**Output:** An expression sequence containing the point names is returned.

**Additional information:** If the two objects have no intersection, no point is returned.

**See also:** projgeom[point], projgeom[line], projgeom[conic], projgeom[inter], projgeom[lccutp], projgeom[lccutc]

### projgeom[line](name, [expr$_x$, expr$_y$, expr$_z$])

Defines the line in three-dimensional Euclidean space whose homogeneous equation has coefficients expr$_x$, expr$_y$, and expr$_z$, and assigns it to name.

**Output:** The line's name is returned.

**Additional information:** Like tables, lines follow the rule of *first name evaluation*. ♣ To examine the actual equation of a line, use op(linepj).

**See also:** projgeom[point], projgeom[conic], op(linepj)

### projgeom[linemeet](linepj$_1$, linepj$_2$, expr, name)

Calculates the line concurrent to linepj$_1$ and linepj$_2$ with respect to the ratio expr, and assigns it to name.

**Output:** The concurrent line's name is returned.

**Additional information:** the formula for the concurrent line is linepj$_1$*[coords]* + expr * linepj$_2$*[coords]* = name*[coords]*.

**See also:** projgeom[line], projgeom[concur], projgeom[onsegment]

### projgeom[midpoint](ptpj$_1$, ptpj$_2$, name)

Calculates the midpoint of the line joining points ptpj$_1$ and ptpj$_2$, and assigns it to name.

**Output:** The midpoint's name is returned.

**Additional information:** The midpoint is defined as the point that is harmonic to the intersection of the line at infinity and the line joining ptpj$_1$ and ptpj$_2$.

**See also:** projgeom[point], projgeom[onsegment], projgeom[tharmonic]


### projgeom[onsegment](ptpj$_1$, ptpj$_2$, expr, name)

Calculates the point that is collinear to the points ptpj$_1$ to ptpj$_2$ and divides them by the ratio expr, and assigns it to name.

**Output:** The dividing point's name is returned.

**Additional information:** The ratio expr can be numeric or symbolic, floating point or rational.

**See also:** projgeom[point], projgeom[midpoint], projgeom[linemeet]


### projgeom[point](name, [expr$_x$, expr$_y$, expr$_z$])

Defines a three-dimensional point with homogeneous coordinates expr$_x$, expr$_y$, and expr$_z$, and assigns it to name.

**Output:** The point's name is returned.

**Additional information:** Like tables, points follow the rule of *first name evaluation*. ♣ To examine the actual coordinates of a point, use op(pointpj).

**See also:** projgeom[line], projgeom[conic], geom3d[coordinates], op(pointpj)


### projgeom[polarp](ptpj, conicpj, name)

Calculates the polar line of ptpj with respect to conicpj, and assigns it to name.

**Output:** The polar line's name is returned.

**See also:** projgeom[point], projgeom[conic], projgeom[line], projgeom[poleline]


### projgeom[poleline](linepj, conicpj, name)

Calculates the pole of linepj with respect to conicpj, and assigns it to name.

**Output:** The polar point's name is returned.

**See also:** projgeom[line], projgeom[conic], projgeom[point], projgeom[polarp]

## projgeom[ptangent](ptpj, conicpj, name)

Calculates the tangent line that goes through ptpj and is tangent to conicpj, and assigns it to name. ptpj must lie on the conicpj.
**Output:** The tangent line's name is returned.
**See also:** projgeom[point], projgeom[conic], projgeom[line], projgeom[rtangent], projgeom[ctangent]

## projgeom[rtangent](ptpj, conicpj, name$_1$, name$_2$)

Calculates the lines that go through ptpj and are tangent to conicpj in the real extended plane, and assigns them to name$_1$ and name$_2$.
**Output:** An expression sequence containing the two tangent line names is returned.
**Additional information:** It is possible that there may be one or zero tangent lines. In the former case, only name$_1$ is assigned.
**See also:** projgeom[point], projgeom[conic], projgeom[line], projgeom[ctangent], projgeom[ptangent]

## projgeom[tangentte](linepj, conicpj)

Determines whether linepj is tangent to conicpj.
**Output:** If it can be determined completely, a boolean value of true or false is returned. Otherwise, a condition that must be met for a result of true is returned.
**See also:** projgeom[line], projgeom[conic], projgeom[tharmonic]

## projgeom[tharmonic](ptpj$_1$, ptpj$_2$, ptpj$_3$, ptpj$_4$)

Determines whether points ptpj$_1$ and ptpj$_2$ are harmonic conjugates to points ptpj$_3$ and ptpj$_4$.
**Output:** If it can be determined completely, a boolean value of true or false is returned. Otherwise, a condition that must be met for a result of true is returned.
**Additional information:** This function cannot be called with any more or less than four points.
**See also:** projgeom[point], projgeom[harmonic], projgeom[collinear], geometry[are_harmonic]

# Combinatorics and Graph Theory

## Introduction

The commands that deal with combinatorics and graph theory reside in the combinat and networks packages, respectively.

## Combinatorics

Input and output for commands in the combinat package follow the standard forms for Maple commands. There are few examples of returning values through names passed as parameters, as you find in the three geometry packages.

Many of the commands in the combinat package revolve around taking a finite set of objects and examining all possible ways to combine its elements to make unique subsets of that set. These subsets are then listed, counted, or perhaps even put into alternative representations, such as matrices or polynomials. Also, there are similar commands for dealing with permutations (different ways of ordering a finite set of objects) and partitions (different ways of choosing positive integer values that sum to a particular value), both of which are common concepts in combinatorial computation.

The most basic commands simply list the available combinations, permutations, and partitions or return the number of possible elements.

```
> with(combinat):
> choose([a, b, c]);
```

$$[[\,],[a],[b],[a,b],[c],[a,c],[b,c],[a,b,c]]$$

> numbcomb([a, b, c]);

$$8$$

> permute([a, b, c]);

$$[[a,b,c],[a,c,b],[b,a,c],[b,c,a],[c,a,b],[c,b,a]]$$

> numbperm([a, b, c]);

$$6$$

> partition(6);

$$[[1,1,1,1,1,1],[1,1,1,1,2],[1,1,2,2],[2,2,2],[1,1,1,3],$$
$$[1,2,3],[3,3],[1,1,4],[2,4],[1,5],[6]]$$

> numbpart(6);

$$11$$

Clearly, the numbers calculated by numbcomb, numbperm, and numbpart can also be found by applying the nops command on the results from choose, permute, and partition, respectively.

More complicated routines allow you to set up while loops that iterate through all possible combinations (subsets) or partitions (firstpart, nextpart, etc.) available. The following simple example calculates the probability of finding a 1 in a partition of 20.

```
> count := 0:
> temp := firstpart(20):
> while temp <> FAIL do
>    if member(1, temp) then
>       count := count + 1;
>    fi;
>    temp := nextpart(temp);
> od:
> evalf(count/numbpart(20), 5);
```

$$.78150$$

One thing of which you must be aware is that it is very simple to create computations in combinat that take *huge* amounts of computing time. The above example, when dealing with the integer 20, takes under 20 seconds on some machines. The same example with integer 40 takes over 1800 seconds. Imagine how long you would have to wait at 80! It is *always* a wise idea to have a basic grasp of how difficult a question is before you pose it to Maple.

To finish, many common values are available within combinat. They include binomial coefficients, Bell numbers, Fibonacci numbers, and Stirling numbers of the first and second kind.

> binomial(10, 4);

$$210$$

> bell(6);

$$203$$

> seq(fibonacci(i), i=0..10);

$$0, 1, 1, 2, 3, 5, 8, 13, 21, 34, 55$$

> stirling1(12, 3);

$$-150917976$$

> stirling2(12, 3);

$$86526$$

# Graphs and Networks

The major concerns in the networks package are the creation, modification, manipulation, and computation of graphs (also referred to as networks). A graph in this sense is very different from a graph in the plotting sense.

A graph in the networks package is created from a non-negative integer number of vertices connected pairwise by a non-negative integer number of edges. In the simplest graphs, edges are undirected (i.e., they have no specified head or tail), no edge has a single vertex as both of its endpoints (a loop), and no distinct pair of vertices has more than one edge connecting them. In more complex graphs, all of these possibilities may arise.

There are commands that allow you to create common structures within a graph quickly and efficiently. In the following example, a graph with five vertices (labelled one, two, three, four, and five) is created. The five vertices are connected in a cycle, and then edges are added connecting one to four and two to five. All edges are assigned default names by the system.

```
> with(networks):
> G := new():
> addvertex({one, two, three, four, five}, G);
```

$$two, three, one, five, four$$

```
> addedge(Cycle(one, two, three, four, five), G);
```

$$e1, e2, e3, e4, e5$$

```
> addedge([{one, four}, {two, five}], G);
```

$$e6, e7$$

Each graph is represented internally by a table containing detailed information. Some of this information is available via specialized commands that access the table, while all of the information can be viewed by using networks[show].

```
> vertices(G);
```

$$\{\, two, three, one, five, four \,\}$$

```
> edges(G);
```

$$\{\, e7, e6, e5, e1, e2, e3, e4 \,\}$$

```
> mindegree(G), maxdegree(G);
```

$$2, 3$$

```
> eweight(e2,G);
```

$$1$$

In addition to multiple edges, loops, and directed edges, another level of complexity can be added by assigning a *weight* to any edge or vertex. In some cases, these weights can be represented by unknown variables, but for many computations (such as shortest path) the weights must be numeric values (though they can be either rational, floating-point, or irrational). If no weight is specified for an edge or vertex, then a default value of 1 is used.

As in the combinat package, many of the commands in networks take an existing structure (i.e., a graph) and compute values that are then listed, counted, or put into alternative representations, such as matrices and polynomials. In the following example, we use the complete graph on seven vertices (automatically labelled 1 through 7) minus the two edges connecting 2 to 5 and 6 to 7, then perform three common graph theory calculations.

```
> H := complete(7):
> delete(incident(2, H) intersect incident(5, H), H):
> delete(incident(6, H) intersect incident(7, H), H):
> girth(H);
```

$$3$$

```
> adjacency(H);
```

$$\begin{bmatrix} 0 & 1 & 1 & 1 & 1 & 1 & 1 \\ 1 & 0 & 1 & 1 & 0 & 1 & 1 \\ 1 & 1 & 0 & 1 & 1 & 1 & 1 \\ 1 & 1 & 1 & 0 & 1 & 1 & 1 \\ 1 & 0 & 1 & 1 & 0 & 1 & 1 \\ 1 & 1 & 1 & 1 & 1 & 0 & 0 \\ 1 & 1 & 1 & 1 & 1 & 0 & 0 \end{bmatrix}$$

```
> mincut(H, 2, 5);
```

$$\{e7, e11, e10, e8, e1\}$$

While the input and output to most commands in networks follow the standards set by the majority of Maple commands, there are more than a few commands that pass information back through unassigned names passed as parameters.

## Parameter Types Specific to This Chapter

| | |
|---|---|
| G | a graph or network |
| v | a vertex |
| e | an edge |
| part | an integer partition |

## Command Listing

### binomial(n, r)

Computes the binomial coefficient, referred to as n *choose* r (the number of ways to choose r objects from n distinct objects).
**Output:** If n and r are both positive integers such that $r \leq n$, an integer is returned. If other types of numeric values are given, the result can be a fraction, a float, or an integer. If symbolic arguments are used, binomial typically returns unevaluated.
**Additional information:** In the common case, where $0 \leq r \leq n$, the formula used is n!/(r!(n-r)!). In the more general case, it is GAMMA(n+1)/(GAMMA(r+1)/GAMMA(n-r+1)).
**See also:** convert(*binomial*), combinat[choose], combinat[multinomial], combinat[bell]
*LA = 7 LI = 16*

### combinat package

Provides commands for combinatorial operations, including standard combining, permuting, and partitioning computations.
**Additional information:** To access the command fnc from the combinat package, use the long form of the name combinat[fnc], or with(combinat, fnc) to load in a pointer for fnc only, or with (combinat) to load in pointers to all the commands. ♣ The basic commands of these types include choose, permute, and partition (which list all possible combinations) and numbcomb, numbperm, and numbpart (which determine how many such combinations there are).

## combinat[bell](n)

Computes the $n^{th}$ Bell number.

**Output:** If n is an integer, an integer is returned. If n is not an integer, an unevaluated command call is returned.

**Additional information:** Bell numbers are defined by the generating function exp(exp(x)-1) = sum(bell(n)/n!*x^n, n=0..infinity).
* One interpretation of combinat[bell](n) is the number of distinct ways a product of n distinct primes can be factored.

**See also:** combinat[binomial]

## combinat[binomial](n, r)

This command is the same as binomial.
**See also:** binomial

## combinat[cartprod]([list$_1$, list$_2$])

Allows iteration over the cartesian product of the elements of list$_1$ and list$_2$.

**Output:** A special table structure is returned.

**Argument options:** ([list$_1$, ..., list$_n$]) to create the iteration table for list$_1$ through list$_n$. * ([list$_1$, list$_2$])*[finished]* to determine whether all iterations have been completed. * ([list$_1$, list$_2$])*[nextvalue]()* to determine the next iteration in the cartesian product and update [finished], if necessary.

**Additional information:** Typically, the results of this command are assigned to a variable (say S) and then used in combination with a while loop. For example, while not S[finished] do ... S[nextvalue]; ... od; . * This command can be very useful in applications of programming other than combinatorics.

**See also:** combinat[subsets], *while*

## combinat[character](n)

Computes the command Chi(part$_\lambda$, part$_\rho$) for all partitions part$_\lambda$ and part$_\rho$ of positive integer n.

**Output:** A two-dimensional square matrix is returned, where the $[i,j]^{th}$ entry is given by Chi(part[n-i+1], part[j]). part[k] is the $k^{th}$ partition of n, sorted in ascending order.

**Additional information:** The result represents the character of all conjugacy classes for all irreducible representations of the symmetric group containing n elements. * The dimension of the matrix is determined by combinat[numbpart](n). * For more information, see the on-line help page.

**See also:** combinat[Chi], combinat[partition], combinat[numbpart]

### combinat[Chi](part$_\lambda$, part$_\rho$)

Computes the trace of the matrices contained in the conjugacy class corresponding to partition part$_\rho$ in irreducible partition part$_\lambda$.
**Output:** A positive integer is returned. If part$_\lambda$ and part$_\rho$ are not the proper type or form of partitions, an error message is displayed.
**Additional information:** part$_\lambda$ and part$_\rho$ must be partitions of the same integer. ♣ The elements of these partitions must be in numerically ascending order (e.g., [1,3,2] is out). ♣ For more information, see the on-line help page.
**See also:** combinat[character], combinat[partition], combinat[numbpart]

### combinat[choose]([expr$_1$, ..., expr$_n$], r)

Generates all the distinct subsets of size r (r must be an integer) of elements expr$_1$ through expr$_n$.
**Output:** A list of lists is returned. Each internal list represents a distinct subset of the original n elements. An empty subset is represented by an empty list, [].
**Argument options:** ([expr$_1$, ..., expr$_n$]) to generate all distinct subsets of all sizes of expr$_1$ through expr$_n$. ♣ (int$_1$, int$_2$) to generate all distinct subsets of size int$_2$ of the set of integers 1 through int$_1$. ♣ (int) to generate a list of lists containing all distinct subsets of all sizes of the integers 1 through int.
**See also:** binomial, combinat[numbcomb], combinat[permute]

### combinat[conjpart](part)

Generates the conjugate partition of part.
**Output:** A partition is returned. If part is not a valid partition, then an error message is returned.
**Additional information:** If part is a partition of $n$, there are $m$ partitions of $n$, and part is the $i^{th}$ partition, then the conjugate partition is represented by combinat[encodepart](m-i+1).
**See also:** binomial, combinat[encodepart], combinat[decodepart], combinat[partition], combinat[numbpart]

### combinat[decodepart](n, m)

Generates the m$^{th}$ partition of the non-negative integer n.
**Output:** A partition of n is returned.
**Additional information:** The ordering of the partitions used is as listed by combinat[partition](n). ♣ To return a partition, m must

be an integer between 1 and combinat[numbpart](n). Otherwise, an error message is displayed.

**See also:** combinat[encodepart], combinat[firstpart], combinat[lastpart], combinat[nextpart], combinat[prevpart], combinat[conjpart], combinat[partition], combinat[numbpart]

### combinat[encodepart]([int$_1$, ..., int$_x$])

Calculates which partition of n ($=$ int$_1$+...+int$_x$) the given partition is.

**Output:** A positive integer between 1 and combinat[numbpart](n), uniquely representing the partition, is returned.

**Additional information:** The ordering of the partitions used is as listed by combinat[partition](n). ♣ If an invalid partition is entered, an error message is returned.

**See also:** combinat[decodepart], combinat[firstpart], combinat[lastpart], combinat[nextpart], combinat[prevpart], combinat[conjpart], combinat[partition], combinat[numbpart]

### combinat[fibonacci](n)

Computes the $n^{th}$ Fibonacci number. n must be an integer.

**Output:** An integer is returned.

**Argument options:** (n, var) to compute the $n^{th}$ Fibonacci polynomial in variable var.

**Additional information:** The combinat[fibonacci](n, x) formula is $F(n,x) = xF(n-1,x) + F(n-2,x)$, where $F(0,x) = 0$ and $F(1,x) = 1$.

**See also:** *linalg[fibonacci]*

### combinat[firstpart](n)

Generates the first partition of the integer n.

**Output:** A partition of n is returned.

**Additional information:** If n is not an integer, then it is rounded down to the first integer lower than itself. ♣ The ordering of the partitions used is as listed by combinat[partition](n). ♣ This command together with combinat[nextpart] and the while statement can be used to loop through all partitions of n.

**See also:** combinat[encodepart], combinat[decodepart], combinat[lastpart], combinat[nextpart], combinat[prevpart], combinat[partition], combinat[numbpart], *while*

### combinat[graycode](n)

Constructs the list of all $2^n$ n-bit integers sorted in *graycode* order. n must be a non-negative integer.

**Output:** A list containing the integers 0 through $2^n - 1$ is returned.

The integers are not ordered in the standard sequential way.

**Additional information:** Each successive pair of integers in a list in *graycode* order have binary representations that differ in one and only one bit. ♣ Use convert(*binary*) to convert an integer into its binary representation.

**See also:** *convert(binary)*

### combinat[inttovec](m, n)

Calculates the list of length n of non-negative integers corresponding one-to-one with the non-negative integer m.

**Output:** A list of n non-negative integers is returned.

**Additional information:** For more information on how the one-to-one correspondence works, consult the on-line help page.

**See also:** combinat[vectoint], combinat[decodepart]

### combinat[lastpart](n)

Generates the last partition of the integer n.

**Output:** A partition of n is returned.

**Additional information:** If n is not an integer, then n is returned as the only element of the list. ♣ The ordering of the partitions used is as listed by the combinat[partition](n). ♣ This command together with combinat[prevpart] and the while statement can be used to loop through all partitions of n.

**See also:** combinat[encodepart], combinat[decodepart], combinat[firstpart], combinat[nextpart], combinat[prevpart], combinat[partition], combinat[numbpart], *while*

### combinat[multinomial](n, $int_1$, $int_2$, ..., $int_m$)

Computes the multinomial coefficient of integer n and integers $int_1$ through $int_m$.

**Output:** If the parameters are valid integers, an integer is returned.

**Additional information:** The integers $int_1$ through $int_m$ must sum to n. ♣ The coefficient is equal to $n!/(int_1!*...*int_m!)$. ♣ If symbolic values or other types of numeric values are given, then combinat[multinomial] typically returns unevaluated.

**See also:** binomial

### combinat[nextpart](part)

Generates the next partition after part.

**Output:** A partition is returned. If part is the last partition, then FAIL is returned.

**Additional information:** If part is not a valid partition, then an error message is returned. ♣ The ordering of the partitions as listed by combinat[partition](n) is used. ♣ This command together with combinat[firstpart] and the while statement can be used to loop through all the partitions of n.

**See also:** combinat[encodepart], combinat[decodepart], combinat[firstpart], combinat[lastpart], combinat[prevpart], combinat[partition], combinat[numbpart], *while*

### combinat[numbcomb]([expr$_1$, ..., expr$_n$], r)

Calculates the number of distinct subsets of size r (r must be an integer) of n elements expr$_1$ through expr$_n$.

**Output:** An integer is returned.

**Argument options:** ([expr$_1$, ..., expr$_n$]) to calculate the number of distinct subsets of all sizes of expr$_1$ through expr$_n$. ♣ (int$_1$, int$_2$) to calculate the number of distinct subsets of size int$_2$ of the integers 1 through int$_1$. ♣ (int) to calculate the number of distinct subsets of all sizes of the integers 1 through int.

**Additional information:** The result of combinat[numbcomb] is always equivalent to the result of nops(combinat[choose]) given the same set of parameters.

**See also:** binomial, combinat[choose], combinat[numbperm]

### combinat[numbpart](n)

Calculates the number of integer partitions of non-negative integer n.

**Output:** An integer is returned.

**Additional information:** The result of combinat[numbpart](n) is always equivalent to the result of nops(combinat[partition](n)). ♣ Warning: large values of n can cause memory faults on some platforms.

**See also:** combinat[partition], combinat[Chi], combinat[character]

### combinat[numbperm]([expr$_1$, ..., expr$_n$], r)

Calculates the number of permutations of size r (r must be an integer) of the n elements expr$_1$ through expr$_n$.

**Output:** An integer is returned.

**Argument options:** ([expr$_1$, ..., expr$_n$]) to calculate the number of permutations of size n of expr$_1$ through expr$_n$. ♣ (int$_1$, int$_2$) to calculate the number of permutations of size int$_2$ of the integers 1 through int$_1$. ♣ (int) to calculate the number of permutations of size int of the integers 1 through int.

**Additional information:** The result of combinat[numbperm] is always equivalent to the result of nops(combinat[permute]) given

the same set of parameters.
**See also:** binomial, combinat[permute], combinat[numbcomb]

### combinat[partition](n)

Generates all the integer partitions of non-negative integer n.
**Output:** A list of lists is returned. Each internal list represents an integer partition of n.
**Additional information:** Warning: large values of n can cause memory faults on some platforms.
**See also:** combinat[numbpart], combinat[Chi], combinat[decodepart], combinat[encodepart]

### combinat[permute]([expr$_1$, ..., expr$_n$], r)

Generates all the permutations of size r (r must be an integer) of the n elements expr$_1$ through expr$_n$.
**Output:** A list of lists is returned. Each internal list represents a permutation of the original elements.
**Argument options:** ([expr$_1$, ..., expr$_n$]) to generate all permutations of size n of expr$_1$ through expr$_n$. ♣ (int$_1$, int$_2$) to generate all permutations of size int$_2$ of the integers 1 through int$_1$. ♣ (int) to generate all permutations of size int of the integers 1 through int.
**See also:** combinat[numbperm], combinat[choose]

### combinat[prevpart](part)

Generates the previous partition before part.
**Output:** A partition is returned. If part is the first partition, then FAIL is returned.
**Additional information:** The ordering of the partitions used is as listed by combinat[partition](n). ♣ If part is not a valid partition, an error message is returned. ♣ This command together with combinat[lastpart] and the while statement can be used to loop through all the partitions of n.
**See also:** combinat[encodepart], combinat[decodepart], combinat[firstpart], combinat[lastpart], combinat[nextpart], combinat[partition], combinat[numbpart], *while*

### combinat[powerset]({expr$_1$, ..., expr$_n$})

Generates all subsets of expressions expr$_1$ through expr$_n$.
**Output:** A set of sets is returned. Each internal set represents a unique subset of the given expressions. An empty subset is represented by an empty set, {}.
**Argument options:** ([expr$_1$, ..., expr$_n$]) to generate a list of lists containing all subsets of expr$_1$ through expr$_n$. ♣ (n) to generate

a set of sets containing all subsets of the integers 1 through n.
**Additional information:** The power set has $2^n$ elements.
**See also:** combinat[subsets], combinat[choose], combinat[permute]

### combinat[randcomb]([expr$_1$, ..., expr$_n$], r)

Generates a list containing a random subset of size r (r must be an integer) of elements expr$_1$ through expr$_n$.
**Output:** A list is returned.
**Argument options:** (int$_1$, int$_2$) to generate a set containing a random subsets of size int$_2$ of the integers 1 through int$_1$.
**See also:** combinat[choose], combinat[randpart], combinat[randperm]

### combinat[randpart](n)

Generates a random partition of non-negative integer n.
**Output:** A partition of n is returned.
**Additional information:** If n is not a non-negative integer, an error message is displayed. ♣ Warning: large values of n can cause memory faults on some platforms.
**See also:** combinat[partition], combinat[randperm], combinat[randcomb]

### combinat[randperm]([expr$_1$, ..., expr$_n$])

Generates a random permutation of the n elements expr$_1$ through expr$_n$.
**Output:** A list is returned.
**Argument options:** ({expr$_1$, ..., expr$_n$}) to create a set containing a random permutation of expr$_1$ through expr$_n$. ♣ (n) to generate a list containing a random permutation of the integers 1 through n.
**See also:** combinat[permute], combinat[randpart], combinat[randcomb]

### combinat[stirling1](n, m)

Calculates the Stirling number of the first kind, commonly denoted as $S(n, m)$, for integers n and m.
**Output:** An integer is returned.
**Additional information:** If n < m, then the answer is zero. ♣ If any parameters are symbolic, then combinat[stirling1] returns unevaluated. ♣ For information on the generating function for Stirling numbers of the first kind, see the on-line help page.
**See also:** combinat[stirling2]

### combinat[stirling2](n, m)

Calculates the Stirling number of the second kind, commonly denoted as $S(n, m)$, for integers n and m.
**Output:** An integer is returned.
**Additional information:** If n < m then the answer is zero. ♣ If any parameters are symbolic, then combinat[stirling2] returns unevaluated. ♣ For information on the generating function for Stirling numbers of the second kind, see the on-line help page.
**See also:** combinat[stirling1]


### combinat[subsets]({expr$_1$, ..., expr$_n$})

Allows iteration over subsets of expressions expr$_1$ through expr$_n$.
**Output:** A special table structure is returned.
**Argument options:** ([expr$_1$, ..., expr$_n$]) to iterate over subsets of a list of expressions. ♣ ({expr$_1$, ..., expr$_n$})*[finished]* to determine whether all iterations have been completed. ♣ ({expr$_1$, ..., expr$_n$})*[nextvalue]()* to determine the next iteration of the subsets.
**Additional information:** Typically, the results of a call to this command are assigned to a variable (say S), and then used in combination with a while loop. For example, while not S[finished] do ... S[nextvalue] ... od; . ♣ This command can be very useful in applications of programming other than combinatorics.
**See also:** combinat[cartprod], combinat[choose], combinat[powerset], *while*


### combinat[vectoint]([int$_1$, ..., int$_n$])

Calculates the non-negative integer value corresponding (one-to-one) with the list of non-negative integers int$_1$ through int$_n$.
**Output:** A non-negative integer is returned.
**Additional information:** For more information on how the one-to-one correspondence works, consult the on-line help page.
**See also:** combinat[inttovec], combinat[encodepart]


### convert(expr, *binomial*)

Converts all GAMMA functions and factorials in expr into terms of the binomial function.
**Output:** An expression in the binomial function is returned.
**Argument options:** ([expr$_1$, ..., expr$_n$], *binomial*) or ({expr$_1$, ..., expr$_n$}, *binomial*) to create a list or set of converted expressions, respectively.
**See also:** *convert(factorial)*, *convert(GAMMA)*, binomial, *GAMMA*

## networks package

Provides commands for creating standard graph theoretical objects and performing well-known calculations.
**Additional information:** To access the command fnc from the networks package, use the long form of the name networks[fnc], or with(networks, fnc) to load in a pointer for fnc only, or with(networks) to load in pointers to all the commands. ♣ Graphs can be defined with commands such as new, advertex, and addedge. ♣ Standard operations include complement, mincut, girth, and spantree.

### networks[acycpoly](G, $var_p$)

Computes the acyclicity polynomial in variable $var_p$ of undirected graph G.
**Output:** A polynomial in $var_p$ is returned.
**Additional information:** The variable $var_p$ represents edge probability. ♣ The resulting polynomial represents the probability that G is acyclic when each edge of G has probability $var_p$.
**See also:** networks[flowpoly], networks[tuttepoly], networks[rankpoly], networks[spanpoly], networks[chrompoly], networks[charpoly]

### networks[addedge]([$v_{tail}$, $v_{head}$], *names*=**ename, G**)

Adds an edge (of weight 1) joining vertices $v_{tail}$ to $v_{head}$ to graph G, and labels it ename.
**Output:** The edge name is returned.
**Argument options:** ({$v_1$, $v_2$}, *names*=ename, G) to add an undirected edge without tail or head. ♣ ([$v_{tail}$, $v_{head}$], *names*=ename, *weights*=expr, G) to add an edge ename with weight expr. ♣ ([[$v_{1tail}$, $v_{1head}$], ..., [$v_{ntail}$, $v_{nhead}$]], *names*=[$ename_1$, ..., $ename_n$], G) to add edges $ename_1$ through $ename_n$. ♣ (*Path*($v_1$, ..., $v_n$), *weights*=[$expr_1$, ..., $expr_{n-1}$], G) to add the edge path defined by vertices $v_1$ through $v_n$. ♣ (*Cycle*($v_1$, ..., $v_n$), *weights*=[$expr_1$, ..., $expr_n$], G) to add the edge cycle defined by vertices $v_1$ through $v_n$.
**Additional information:** All vertices used in networks[addedge] must be previously defined in G.
**See also:** networks[addvertex], networks[delete], networks[edges], networks[eweight], networks[ends], networks[head], networks[tail]

### networks[addvertex](v, G)

Adds a vertex (of 0 weight) to graph G, and labels it v.
**Output:** The vertex name is returned.
**Argument options:** (v, *weights*=expr, G) to add vertex v with weight expr. ♣ ($v_1$, $v_2$, ..., $v_n$, G), ({$v_1$, ..., $v_n$}, G), or ([$v_1$, ..., $v_n$], G) to add vertices $v_1$ through $v_n$. ♣ ([$v_1$, ..., $v_n$], *weights*=[$expr_1$, ..., $expr_n$], G) to add a vertices $v_1$ through $v_n$ with corresponding weights $expr_1$ through $expr_n$.
**Additional information:** If G has not been defined, be sure to call networks[new](G) to define it.
**See also:** networks[new], networks[addedge], networks[delete], networks[vertices], networks[vweight]

### networks[adjacency](G)

Computes the adjacency matrix of graph G.
**Output:** A square matrix of dimension equal to the number of vertices of G is returned. All elements are non-negative integers.
**Additional information:** The value of the $[i,j]^{th}$ element represents the number of edges connecting vertex $v_i$ to vertex $v_j$. ♣ If the edge connecting vertex $v_i$ and vertex $v_j$ is undirected, both the $[i,j]^{th}$ and the $[j,i]^{th}$ elements are incremented by one. ♣ If the edge is directed towards $v_j$, only the $[i,j]^{th}$ element is incremented.
**See also:** networks[incidence], networks[addedge], networks[charpoly]

### networks[allpairs](G)

Computes the *all-pairs shortest path* distances for every pair of vertices in graph G.
**Output:** A table containing the path distances is returned.
**Argument options:** (G)[$v_1$, $v_2$] to access the shortest path distance from vertex $v_1$ to vertex $v_2$. ♣ (G, name) to assign a table of ancestors to name. Each element [u, v] corresponds to the parent of vertex $v$ in the shortest path tree rooted at vertex $u$.
**Additional information:** Edge weights are assumed to represent distances. ♣ Undirected edges are assumed bidirectional.
**See also:** networks[shortpathtree], networks[spantree], networks[ancestor], networks[eweight], networks[diameter]

### networks[ancestor](v, Tree)

Displays the ancestor of vertex v in directed tree Tree.
**Output:** A set of vertices is returned.

## Combinatorics

**Argument options:** (Tree) to return the ancestor table for the entire directed tree. ♣ ([$v_1$, ..., $v_n$], Tree) or ({$v_1$, ..., $v_n$}, Tree) to return the list or set of ancestors of vertices $v_1$ through $v_n$.

**Additional information:** An ancestor of v is a vertex connected to it by an edge with v as its head. ♣ Undirected edges are not considered. ♣ In a directed tree, v can have either one or zero (if v is the root) ancestors.

**See also:** networks[daughter], networks[shortpathtree], networks[path], networks[neighbors], networks[arrivals]

### networks[arrivals](v, G)

Displays the neighbors, found along *incoming* edges, of vertex v in graph G.

**Output:** A set of vertices is returned.

**Argument options:** (G) to return the arrivals table for graph G.

**Additional information:** An arrival vertex of v is defined as any vertex that is joined to it by a undirected edge or by an edge that is directed towards it (i.e., with v as its head).

**See also:** networks[neighbors], networks[departures], networks[indegree]

### networks[bicomponents](G)

Computes the biconnected subgraphs of graph G and the bridges that span them.

**Output:** A list containing two sets is returned. The first set represents edges that are bridges in G. The second set contains sets of edges that represent biconnected subgraphs in G.

**Additional information:** A biconnected subgraph is a subgraph that is 2-connected. ♣ A bridge is an edge connecting two biconnected subgraphs.

**See also:** networks[components], networks[connect], networks[connectivity]

### networks[charpoly](G, var)

Computes the characteristic polynomial in variable var of undirected graph G.

**Output:** A polynomial in var is returned.

**Additional information:** The resulting polynomial represents the characteristic polynomial of the adjacency matrix of G.

**See also:** networks[adjacency], networks[flowpoly], networks[tuttepoly], networks[rankpoly], networks[spanpoly], networks[acycpoly], networks[chrompoly]

### networks[chrompoly](G, $\text{var}_c$)

Computes the chromatic polynomial in variable $\text{var}_c$ of undirected graph G.
**Output:** An unexpanded polynomial is returned.
**Argument options:** (G, n), where n is a positive integer, to determine the number of proper vertex colorings of G with n colors.
**Additional information:** The variable $\text{var}_c$ represents number of colors. ♣ The resulting polynomial represents the number of proper vertex colorings of G with $\text{var}_c$ colors.
**See also:** networks[flowpoly], networks[tuttepoly], networks[rankpoly], networks[spanpoly], networks[acycpoly], networks[charpoly]

### networks[complement](G)

Computes the complement of graph G with respect to the complete graph with the same number of vertices.
**Output:** A graph is returned.
**Argument options:** ($G_1$, $G_2$) to compute the complement of graph $G_1$ with respect to graph $G_2$.
**Additional information:** The complement of graph $G_1$ with respect to graph $G_2$ is the graph on the same vertices that contains all the edges of $G_2$ that are *not* in the edge set of $G_1$.
**See also:** networks[induce], networks[complete], networks[duplicate]

### networks[complete](n)

Generates the complete graph on n vertices. n is a non-negative integer.
**Output:** A graph is returned.
**Argument options:** ($\{v_1, ..., v_m\}$) to generate the complete graph on vertices $v_1$ through $v_m$. ♣ ($\text{int}_1, ..., \text{int}_m$) to generate the graph with complete parts containing $\text{int}_1$ through $\text{int}_m$ vertices.
**Additional information:** When a graph has more than one complete part, each vertex in part $x$ is connected to every other vertex in part $x$. ♣ Unless the vertex names are given as a set, default names for vertices and edges are automatically generated.
**See also:** networks[petersen], networks[random], networks[cycle], networks[complement], networks[edges], networks[void], networks[gunion]

### networks[components](G)

Computes the connected subgraphs of graph G.
**Output:** A set of sets containing vertices is returned.

**Additional information:** A connected subgraph of G is a set of vertices connected to other vertices in the set by either an undirected edge or an edge directed into the set.

**See also:** networks[bicomponents], networks[connect], networks[connectivity], networks[induce], networks[cyclebase]

### networks[connect]({$v_{1,1}$, ..., $v_{1,n}$}, {$v_{2,1}$, ..., $v_{2,m}$}, G)

Connects, with an undirected edge, every vertex in $v_{1,1}$ through $v_{1,n}$ with every vertex in $v_{2,1}$ through $v_{2,m}$. All vertices in both sets must be in graph G.

**Output:** An expression sequence containing the names of the added edges is returned.

**Argument options:** ({$v_{1,1}$, ..., $v_{1,n}$}, {$v_{2,1}$, ..., $v_{2,m}$}, *'directed'*, G) to specify that the edges are all directed out of the first set and into the second. ✸ ({$v_{1,1}$, ..., $v_{1,n}$}, {$v_{2,1}$, ..., $v_{2,m}$}, *names*=[$name_1$, ..., $name_{n*m}$], *weights*=[$w_1$, ..., $w_{n*m}$], G) to specify names and weights for individual edges.

**See also:** networks[connectivity], networks[bicomponents], networks[components]

### networks[connectivity](G)

Computes the edge connectivity of graph G.

**Output:** A non-negative integer is returned.

**Argument options:** (G, n) to supply a known lower-bound, n, for the connectivity of G. ✸ (G, n, $name_1$, $name_2$) to assign a set of edges that defines a minimum cut to $name_1$ and a set of super-saturated edges to $name_2$.

**See also:** networks[connect], networks[countcuts], networks[flow], networks[components], networks[bicomponents], networks[mincut]

### networks[contract](e, G)

Contracts graph G by removing edge e and collapsing its ends.

**Output:** The single vertex formed by the contraction is returned.

**Argument options:** ($v_1$, $v_2$, G) to contract the edge connecting $v_1$ to $v_2$. ✸ ({$e_1$, ..., $e_n$}, G) or ([$e_1$, ..., $e_n$], G) to contract edges $e_1$ through $e_n$. All vertices incident with these edges are collapsed into one vertex. ✸ ({{$v_{1,1}$, $v_{1,2}$}, ..., {$v_{n,1}$, $v_{n,2}$}}, G) or ([{$v_{1,1}$, $v_{1,2}$}, ..., {$v_{n,1}$, $v_{n,2}$}], G) to contract edges defined by the vertex pairs {$v_{1,1}$, $v_{1,2}$} through {$v_{n,1}$, $v_{n,2}$}.

**Additional information:** If edges are specified as vertex pairs, each vertex pair must have exactly one edge connecting it.
**See also:** networks[shrink], networks[addedge], networks[addvertex], networks[delete]

### networks[countcuts](G)

Computes the number of minimum network cuts of undirected graph G.
**Output:** A non-negative integer is returned.
**See also:** networks[counttrees], networks[connectivity], networks[djspantree], networks[mincut]

### networks[counttrees](G)

Computes the number of unique spanning trees of undirected graph G.
**Output:** A non-negative integer is returned.
**See also:** networks[spantree], networks[countcuts]

### networks[cycle](n)

Generates the cyclic graph on n vertices. n is a non-negative integer.
**Output:** A graph is returned.
**Additional information:** Default names for vertices and edges are automatically generated.
**See also:** networks[cyclebase], networks[complete], networks[path], networks[petersen], networks[girth]

### networks[cyclebase](G)

Calculates a cycle basis in undirected graph G.
**Output:** A set of sets of edges is returned.
**Additional information:** To compute the cycle basis, a spanning tree for G is found. Then, all the fundamental cycles are found with respect to this spanning tree.
**See also:** networks[cycle], networks[span], networks[components]

### networks[daughter](v, Tree)

Displays the daughter(s) of vertex v in directed tree Tree.
**Output:** A set of vertices is returned.
**Argument options:** ([$v_1$, ..., $v_n$], Tree) or ({$v_1$, ..., $v_n$}, Tree) to return the list or set of daughters of vertices $v_1$ through $v_n$.

♣ (Tree) to return the daughter table for directed tree Tree.
**Additional information:** A daughter of v is a vertex connected to it by an edge with v as its tail. ♣ Undirected edges are not considered. ♣ In a directed tree, v can have any number of daughters.
**See also:** networks[ancestor], networks[shortpathtree], networks[path], networks[neighbors], networks[departures]

### networks[degreeseq](G)

Displays a list containing the degrees of each vertex in graph G. The list is sorted in ascending order.
**Output:** A list of non-negative integers is returned.
**Additional information:** The list returned has a number of entries equal to the number of vertices in G.
**See also:** networks[maxdegree], networks[mindegree], networks[vdegree]

### networks[delete](v, G)

Deletes the vertex v from graph G. All incident edges are removed as well.
**Output:** The graph object details are returned.
**Argument options:** (e, G) to delete the edge e from graph G. ♣ ({$v_1$, ..., $v_n$}, G) or ([$v_1$, ..., $v_n$], G) to remove vertices $v_1$ through $v_n$ from G. ♣ ({$e_1$, ..., $e_n$}, G) or ([$e_1$, ..., $e_n$], G) to remove edges $e_1$ through $e_n$ from G.
**Additional information:** The new G is returned both as an updated parameter and as formal output to the command.
**See also:** networks[addedge], networks[addvertex], networks[contract], networks[shrink], networks[complement]

### networks[departures](v, G)

Displays the neighbors, found along *outgoing* edges, of vertex v in graph G.
**Output:** A set of vertices is returned.
**Argument options:** (G) to return the departures table for graph G.
**Additional information:** A departure vertex of v is defined as any vertex that is joined to it by a undirected edge or by an edge that is directed away from it (i.e., with v as its tail).
**See also:** networks[neighbors], networks[departures], networks[outdegree]

### networks[diameter](G)

Computes the maximum shortest distance between any two vertices in graph G.
**Output:** If the graph is connected, an integer is returned. If the graph is disconnected, then infinity is returned.
**Additional information:** Edge weights are assumed to represent distances. ♣ Undirected edges are assumed bidirectional.
**See also:** networks[girth], networks[allpairs], networks[shortpathtree]


### networks[dinic](G, $v_s$, $v_t$)

Computes the maximum flow from source vertex $v_s$ to sink vertex $v_t$ in graph G, using the Dinic augmenting-path flow algorithm.
**Output:** A non-negative integer is returned.
**Argument options:** (G, $v_s$, $v_t$, *'maxflow'*=n) to provide an upper bound of n for the flow in G. ♣ (G, $v_s$, $v_t$, $name_1$, $name_2$) to assign the set of saturated edges in the maximum flow to $name_1$, and to assign the vertex set of the graph induced by $name_1$ to $name_2$.
**Additional information:** During its calculations, the networks [flow] command calls the networks[dinic] command.
**See also:** networks[flow], networks[connectivity], networks[shortpathtree], networks[spantree]


### networks[djspantree](G, $v_{\text{root}}$)

Partitions the edges of graph G into the minimum number of edge-disjoint spanning trees.
**Output:** A table is returned that associates each vertex of G with a set of edges in G (possibly empty) representing a forest in G.
**Additional information:** This command uses Edmond's matroid partitioning algorithm to partition G.
**See also:** networks[spantree], networks[countcuts], networks[flow]


### networks[duplicate](G)

Creates a new graph with properties identical to those of graph G.
**Output:** The procedure that defines the new graph is returned.
**See also:** networks[new], networks[complement], networks[getlabel]

### networks[edges](G)

Displays the edges of graph G.
**Output:** A set of edges is returned.
**Argument options:** ($[v_{tail}, v_{head}]$, G) to return the names of all *directed* edges connecting $v_{tail}$ to $v_{head}$. ♣ ($\{v_1, v_2\}$, G) to return the names of all *undirected* edges connecting $v_1$ and $v_2$. ♣ ($\{v_1, v_2\}$, G, *all*) to return the names of all edges, either directed or undirected, connecting $v_1$ and $v_2$.
**See also:** networks[vertices], networks[addedge], networks[eweight]


### networks[ends](e, G)

Displays the end vertices of edge e in graph G.
**Output:** If e is *directed*, then a list of two elements, representing the tail and head, respectively, is returned. If e is *undirected*, then a set of two elements is returned.
**Argument options:** (G) to return a set containing two-element lists or sets for each edge in graph G. ♣ ($[e_1, ..., e_n]$, G) or ($\{e_1, ..., e_n\}$, G) to return a set or list containing two-element lists or sets for edges $e_1$ through $e_n$.
**Additional information:** If an edge passed to networks[ends] is not defined, then NULL is returned.
**See also:** networks[head], networks[tail], networks[edges], networks[addedge], networks[incident], networks[induce]


### networks[eweight](e, G)

Displays the weight of edge e in graph G.
**Output:** An expression is returned.
**Argument options:** (G) to return the edge weight table for graph G. ♣ ($[e_1, ..., e_n]$, G) to return a list containing the weights of edges $e_1$ through $e_n$, in that order.
**See also:** networks[vweight], networks[edges], networks[addedge], networks[allpairs]


### networks[flow](G, $v_s$, $v_t$)

Computes the maximum flow in graph G from source vertex $v_s$ to sink vertex $v_t$.
**Output:** A non-negative integer is returned.
**Argument options:** (G, $v_s$, $v_t$, *'maxflow'*=n) to provide an upper bound of n for the appropriate flow. ♣ (G, $v_s$, $v_t$, $name_1$, $name_2$) to assign the set of saturated edges in the maximum flow to $name_1$

and the vertex set of the graph induced by name$_1$ to name$_2$.
**See also:** networks[dinic], networks[flowpoly],
networks[connectivity], networks[shortpathtree],
networks[spantree], networks[djspantree]

### networks[flowpoly](G, var)

Computes the flow polynomial in variable var of undirected graph G.
**Output:** A polynomial in var is returned.
**See also:** networks[flow], networks[tuttepoly],
networks[rankpoly], networks[spanpoly],
networks[acycpoly], networks[chrompoly],
networks[charpoly]

### networks[fundcyc]({e$_1$, ..., e$_n$}, G)

Computes the fundamental cycle in G of edge set e$_1$ through e$_n$.
**Output:** A graph is returned.
**See also:** networks[span], networks[rank], networks[induce],
networks[incident], networks[ends]

### networks[getlabel](G)

Returns the internal label of graph G.
**Output:** A positive integer is returned
**Additional information:** Each graph has a unique internal label to distinguish it from possible copies.
**See also:** networks[duplicate], networks[new]

### networks[girth](G)

Computes the length of the shortest cycle in graph G.
**Output:** If the graph has a cycle, an integer is returned. If the graph has no cycle, then infinity is returned.
**Argument options:** (G, name) to assign a list of sets defining the edges of the shortest cycle to name.
**Additional information:** Edge weights are assumed to represent distances.
**See also:** networks[cycle], networks[diameter]

### networks[graphical]([int$_1$, ..., int$_n$])

Determines whether the list of non-negative integers int$_1$ through int$_n$ is graphical.
**Output:** If the integers are graphical, a list of edges defined as sets

of integers, that represents one realization of the graph is returned. Otherwise, the value FAIL is returned.

**Argument options:** ([$int_1$, ..., $int_n$], '*MULTI*') to allow multigraphs to be considered.

**See also:** networks[vdegree], networks[mindegree], networks[maxdegree], networks[degreeseq]

### networks[gsimp](G)

Creates a simple graph from the possible multigraph G.

**Output:** A graph is returned.

**Additional information:** All loops are removed. ♣ Directed edges are replaced with undirected edges. ♣ Multiple edges are replaced with single edges whose weight equals the total weight of those multiple edges.

**See also:** networks[induce], networks[gunion]

### networks[gunion]($G_1$, $G_2$)

Creates a new graph from the union of graphs $G_1$ and $G_2$.

**Output:** A graph is returned.

**Argument options:** ($G_1$, $G_2$, '*SIMPLE*') to make sure that the resulting graph is simple.

**Additional information:** The vertex set of the new graph is the union of the vertex sets of $G_1$ and $G_2$. ♣ Each edge that appears in either $G_1$ or $G_2$ is represented in the new graph.

**See also:** networks[gsimp], networks[void], networks[complete]

### networks[head](e, G)

Displays the head vertex of directed edge e in graph G.

**Output:** A vertex name is returned.

**Argument options:** (G) to return the head vertex table for graph G. ♣ ([$e_1$, ..., $e_n$], G) or (\{$e_1$, ..., $e_n$\}, G) to return a set or list containing the head vertices of edges $e_1$ through $e_n$, in that order.

**Additional information:** If an edge passed to networks[head] is undirected, then NULL is returned.

**See also:** networks[ends], networks[tail], networks[edges], networks[addedge], networks[incident]

### networks[incidence](G)

Computes the incidence matrix of graph G.

**Output:** A matrix with number of rows equal to the number of vertices of G and number of columns equal to the number of edges

of G is returned. All elements of the matrix are either 0, 1, or $-1$.
**Additional information:** If edge $e_j$ is not incident to vertex $v_i$, then the $[i, j]^{th}$ element is 0. ♣ If edge $e_j$ is a *directed* edge with $v_i$ as its tail, then the $[i, j]^{th}$ element is $-1$. ♣ In all other cases, the $[i, j]^{th}$ element is 1.
**See also:** networks[adjacency], networks[addedge], networks[incident], networks[ends]

### networks[incident](v, G)

Displays the edges of graph G that are incident to vertex v.
**Output:** A set of edges is returned.
**Argument options:** ($\{v_1, ..., v_n\}$, G) to return all edges incident with $v_1$ through $v_n$. These are, by definition, *not* cut edges. ♣ ($\{v_1, ..., v_n\}$, G, *'In'*) to return all edges incident to $v_1$ through $v_n$ that are either undirected or directed *into* the vertex set. ♣ ($\{v_1, ..., v_n\}$, G, *'Out'*) to return all edges incident to $v_1$ through $v_n$ that are either undirected or directed *out* of the vertex set.
**See also:** networks[edges], networks[ends], networks[neighbors], networks[arrivals], networks[departures], networks[induce]

### networks[indegree](v, G)

Calculates the number of edges directed into vertex v in graph G.
**Output:** A non-negative integer is returned.
**See also:** networks[vdegree], networks[outdegree], networks[arrivals], networks[mindegree], networks[maxdegree], networks[degreeseq]

### networks[induce]($\{e_1, ..., e_n\}$, G)

Computes the subgraph of G induced by edges $e_1$ through $e_n$.
**Output:** A graph is returned.
**Argument options:** ($\{v_1, ..., v_n\}$, G) to compute the subgraph of G induced by vertices $v_1$ through $v_n$.
**Additional information:** The subgraph induced by an edge set contains that edge set and all vertices adjacent to those edges. ♣ The subgraph induced by a vertex set contains that vertex set and all edges with *both* ends in that set.
**See also:** networks[span], networks[rank], networks[incident], networks[ends], networks[fundcyc], networks[gsimp]

### networks[maxdegree](G)

Calculates the maximum degree of the vertices in graph G.
**Output:** A non-negative integer is returned.
**Argument options:** (G, name) to assign the name of a vertex of maximum degree to name.
**See also:** networks[mindegree], networks[vdegree], networks[degreeseq]

### networks[mincut](G, $v_s$, $v_t$)

Calculates the minimum cut, with source vertex $v_s$ and sink vertex $v_t$, of graph G.
**Output:** A set of edges is returned.
**Argument options:** (G, $v_s$, $v_t$, name) to assign the value of the corresponding flow to name.
**See also:** networks[connectivity], networks[countcuts]

### networks[mindegree](G)

Calculates the minimum degree of the vertices in graph G.
**Output:** A non-negative integer is returned.
**Argument options:** (G, name) to assign the name of a vertex of minimum degree to name.
**See also:** networks[maxdegree], networks[vdegree], networks[degreeseq]

### networks[neighbors](v, G)

Displays the neighbors of vertex v in graph G.
**Output:** A set of vertices is returned.
**Argument options:** (G) to return the neighbor table for graph G.
**Additional information:** A neighbor of v is defined as any vertex joined to it by an edge, regardless of the direction of that edge.
**See also:** networks[arrivals], networks[departures], networks[vdegree]

### networks[new](G)

Creates a new graph G without any vertices or edges.
**Output:** The procedure that defines G is returned.
**Additional information:** Often networks[new](G) must be called before other commands can act upon G.
**See also:** networks[addvertex], networks[addedge], networks[complete], networks[duplicate], networks[getlabel]

### networks[outdegree](v, G)

Calculates the number of edges directed out of vertex v in graph G.
**Output:** A non-negative integer is returned.
**See also:** networks[vdegree], networks[indegree], networks[departures], networks[mindegree], networks[maxdegree], networks[degreeseq]

### networks[path]([$v_{start}$, $v_{end}$], Tree)

Finds a path in the directed tree Tree from vertex $v_{start}$ to vertex $v_{end}$, if possible.
**Output:** If a path exists, a list containing the vertices traversed on the path is returned. If no path exists, the value FAIL is returned.
**Additional information:** The first and last elements of the returned list are $v_{start}$ and $v_{end}$, respectively.
**See also:** networks[cycle], networks[shortpathtree], networks[spantree], networks[djspantree], networks[ancestor], networks[daughter]

### networks[petersen](n)

Generates the Petersen graph on n vertices. n is a non-negative integer.
**Output:** A graph is returned.
**Additional information:** Default names for vertices and edges are automatically generated.
**See also:** networks[cycle], networks[complete]

### networks[random](n)

Generates a random graph on n vertices. Each possible edge has a 0.5 probability of appearing.
**Output:** A graph is returned.
**Argument options:** (n, m) to generate a random graph with m edges on n vertices. ♣ (n, *prob*=expr) to use the probability of expr, a numeric value between 0 and 1, when determining the inclusion or exclusion of each possible edge. A value of 1 corresponds to the complete graph on n vertices. A value of 0 corresponds to the void graph on n vertices.
**Additional information:** When the default probability or a given probability is used, each edge is treated independently.
**See also:** networks[complete], networks[void], networks[edges]

### networks[rank]({$e_1$, ..., $e_n$}, G)

Computes the rank in G of the edge set containing $e_1$ through $e_n$.
**Output:** A non-negative integer value is returned.
**Additional information:** The rank of edge set $e_1$ through $e_n$ equals the number of vertices of G less the number of vertices in the induced graph of the edge set.
**See also:** networks[span], networks[fundcyc], networks[induce]


### networks[rankpoly](G, $var_r$, $var_c$)

Computes the *Whitney rank polynomial* of undirected graph G with rank variable $var_r$ and corank variable $var_c$.
**Output:** A polynomial in $var_r$ and $var_c$ is returned.
**Additional information:** For more information on Whitney rank polynomials, see the on-line help page.
**See also:** networks[rank], networks[flowpoly], networks[tuttepoly], networks[spanpoly], networks[acycpoly], networks[chrompoly], networks[charpoly]


### networks[shortpathtree](G, $v_{root}$)

Constructs a shortest path spanning tree of edge-weighted graph G with root vertex $v_{root}$.
**Output:** A graph is returned.
**Additional information:** Undirected edges are assumed bidirectional. ♣ All edge weights must be non-negative, numeric values. ♣ The final weights of the vertices in the spanning tree are equal to the distances from the vertices to the root. ♣ Dijkstra's algorithm is used to find the spanning tree.
**See also:** networks[spantree], networks[path], networks[ancestor], networks[daughter], networks[allpairs], networks[flow], networks[diameter]


### networks[show](G)

Returns all the information known about graph G in table form.
**Output:** A table that may contain other tables is returned. All tables are fully displayed.
**Additional information:** The different entries of the table returned by networks[show] are detailed in op(graph).
**See also:** networks[edges], networks[vertices], networks[ends], networks[neighbors], op(graph)

**networks[shrink]({$v_1$, ..., $v_n$}, G)**

Contracts graph G by collapsing vertices $v_1$ through $v_n$.
**Output:** The single vertex formed by the contraction is returned.
**Argument options:** ({$v_1$, ..., $v_n$}, G, name) to name the newly created vertex name.
**Additional information:** All edges that had both ends in $v_1$ through $v_n$ are removed.
**See also:** networks[contract], networks[delete]


**networks[span]({$e_1$, ..., $e_n$}, G)**

Computes the span in G of the edge set containing $e_1$ through $e_n$.
**Output:** A set of edges is returned.
**Additional information:** The span of edge set $e_1$ through $e_n$ is the complete set of edges with both ends in the set of vertices encompassed by that edge set.
**See also:** networks[spantree], networks[rank], networks[fundcyc], networks[induce], networks[cyclebase]


**networks[spanpoly](G, $var_p$)**

Computes the span polynomial in variable $var_p$ of undirected graph G.
**Output:** A polynomial in $var_p$ is returned.
**Argument options:** (G, num), where num is a numeric value between 0 and 1, to determine the number of connected subgraphs when each edge of G has probability num.
**Additional information:** The variable $var_p$ represents edge probability. The resulting polynomial represents the number of connected subgraphs when each edge of G has probability $var_p$.
**See also:** networks[span], networks[flowpoly], networks[tuttepoly], networks[rankpoly], networks[acycpoly], networks[chrompoly], networks[charpoly]


**networks[spantree](G, $v_{root}$)**

Constructs a minimum weight spanning tree of edge-weighted graph G with root vertex $v_{root}$.
**Output:** A graph is returned.
**Argument options:** (G) to find the minimum weight spanning tree from a random vertex in G. ♣ (G, $v_{root}$, name) to assign the sum of the weight of the edges of the spanning tree to name.
**Additional information:** Undirected edges are assumed to be bidi-

rectional. ♣ networks[spantree] returns FAIL if G is not connected.

**See also:** networks[djspantree], networks[shortpathtree], networks[counttrees], networks[path], networks[ancestor], networks[daughter], networks[allpairs], networks[flow]

### networks[tail](e, G)

Displays the tail vertex of directed edge e in graph G.
**Output:** A vertex name is returned.
**Argument options:** (G) to return the tail vertex table for graph G.
♣ ([$e_1$, ..., $e_n$], G) or ({$e_1$, ..., $e_n$}, G) to return a set or a list containing the tail vertices of edges $e_1$ through $e_n$, in that order.
**Additional information:** If an edge passed to networks[tail] is undirected, then NULL is returned.
**See also:** networks[ends], networks[head], networks[edges], networks[addedge], networks[incident]

### networks[tuttepoly](G, $var_i$, $var_e$)

Computes the Tutte polynomial of undirected graph G. The internal activity variable is $var_i$ and the external activity variable is $var_e$.
**Output:** A polynomial in $var_i$ and $var_e$ is returned.
**Additional information:** For more information on Tutte polynomials, see the on-line help page.
**See also:** networks[flowpoly], networks[rankpoly], networks[spanpoly], networks[acycpoly], networks[chrompoly], networks[charpoly]

### networks[vdegree](v, G)

Calculates the number of undirected edges in graph G incident with vertex v.
**Output:** A non-negative integer is returned.
**See also:** networks[indegree], networks[outdegree], networks[neighbors], networks[mindegree], networks[maxdegree], networks[degreeseq]

### networks[vertices](G)

Displays the vertices of graph G.
**Output:** A set of vertices is returned.
**See also:** networks[edges], networks[addvertex], networks[vweight]

### networks[void](n)

Generates the void graph on n vertices. n is a non-negative integer.
**Output:** A graph is returned.
**Argument options:** ($\{v_1, ..., v_m\}$) to generate the void graph on vertices $v_1$ through $v_m$.
**Additional information:** Unless the vertex names are given as a set, default names are automatically generated. ♣ The void graph for a set of vertices is the graph with those vertices and no edges.
**See also:** networks[complete], networks[random], networks[petersen], networks[cycle], networks[edges], networks[gunion]

### networks[vweight](v, G)

Displays the weight of vertex v in graph G.
**Output:** An expression is returned.
**Argument options:** (G) to return the vertex weight table for graph G. ♣ ([$v_1, ..., v_n$], G) to return a list containing the weights of vertices $v_1$ through $v_n$, in that order.
**See also:** networks[eweight], networks[vertices], networks[addvertex]

### op(graph)

Displays the procedure that defines the various elements of graph.
**Output:** A proc data structure, that takes one parameter, is displayed.
**Argument options:** (graph)(_Edges) or (graph)(_Vertices) to return a set containing the names of all edges or vertices in the graph. ♣ (graph)(_EdgeIndex) to return a table matching up vertex pairs with the edges they define. ♣ (graph)(_Head) or (graph)(_Tail) to return a table matching up directed edges with their heads or tails. ♣ (graph)(_Eweight) or (graph)(_Vweight) to return a table matching up edges or vertices with their weights. ♣ (graph)(_Ends) to return a table matching up edges with vertex pairs that define them. ♣ (graph)(_Ancestor) or (graph)(_Daughter) to return a table matching up vertices with their ancestors or daughters. ♣ (graph)(_Neighbors) to return a table matching up vertices with sets of vertices that are their neighbors. ♣ (graph)(_Status) to return a set containing keywords, like *DIRECTED*, *PETERSEN*, etc., that describe graph. ♣ (graph)(_Econnectivity) to return a list that has as its first element the edge connectivity of graph. ♣ (graph)(_Counttrees) or (graph)(_Countcuts) to return an integer representing the number of spanning trees or minimum network cuts of graph. ♣ (graph)(_Bicomponents) to return a list

the same as the result of networks[bicomponents](graph). ♣
(graph)(_Emaxname) to return the highest automatically generated edge index currently used.

**Additional information:** Each of the above elements have default values that are returned if no other values can be found, except for _Econnectivity, _Countcuts, _Counttrees, and _Bicomponents. These four are equal to their own names until the corresponding commands have been run upon graph.

**See also:** networks[new], networks[show], networks[edges], networks[head], networks[tail], networks[ends], networks[vertices], networks[ancestor], networks[daughter], networks[neighbors], networks[connectivity], networks[countcuts], networks[counttrees], networks[bicomponents], type(*graph*),
*table*

### type(expr, *graph*)

Tests whether expr is a graph.
**Output:** A boolean value of either true or false is returned.
**See also:** networks[new], op(*graph*)

# Number Theory

## Introduction

The commands dealing with number theory make up a significant proportion of Maple's built-in capability. Among the types of numbers for which commands are provided are integers, rationals, continued fractions, Gaussian integers, p-adic numbers, primes, and special irrational values such as $\pi$, $e$, etc. Some of these commands reside in the standard Maple library, but many more live in the three packages numtheory, GaussInt, and padic.

Most of the output from (and input to) number theory commands is in the form of numeric values (i.e., no symbolic variables). There are only a few special numeric data structures created specifically for this field that need to be explained. But first, a few words about how integers, rationals, and continued fractions are represented both internally and externally in Maple.

## Integers

Integers are perhaps the most basic number type and the most common one dealt with in number theory. Integers of great size can be represented. You are not likely to ever exceed the size limit—the largest integer that can be created has more than $500,000$ digits in it! You are more likely to run into time and memory problems trying to compute a significant value that large.

The two internal structures for positive and negative integers are INTPOS and INTNEG, respectively. Both structures contain standard information in their first word, including the data type (either INTPOS or INTNEG) and the length in words of the remainder of their representation. Then there follows that number of words, each word containing a digit in base 10000 with the digits arranged sequentially from lowest to highest order, describing the absolute value of the integer.

Externally, there is little difference between the representation of positive and negative integers. While negative integers are always preceded by a minus sign, positive numbers are *not* traditionally preceded by a plus sign. When negating an already negative integer, place the original negative integer in brackets; two minus signs in a row cause a syntax error.

```
> -3;
```

$$-3$$

```
> - -3
syntax error:
- -3
  ^

> -(-3);
```

$$3$$

The continuation character, \, is used externally in the representation of integers of greater length than can be displayed on one line of your display.

## Rationals

Rational numbers consist of integer numerators and denominators. Maple automatically simplifies rationals into their lowest, unique form. That is, the greatest common divisor of numerator and denominator is calculated and divided through before the value is stored. There is no way to turn this simplification off.

Internally, a rational number is three words long. The first word contains the data type, RATIONAL. The second word contains a pointer to either a positive or negative integer representing the numerator. The third word contains a pointer to a *positive* integer representing the denominator. The denominator is always a positive integer that is different from 0 or 1.

Externally, rational numbers are usually represented as the numerator placed directly over the denominator, separated by a horizontal line. If either numerator or denominator is too large for one line, the continuation character, \, is used to separate each line of output and the division character, /, is used to separate numerator from denominator.

```
> 32/12;
```

$$\frac{8}{3}$$

```
> -12345/3456;
```

$$\frac{-4115}{1152}$$

```
> 12345/-3456;
syntax error:
12345/-3456;
          ^
```

```
> 23^84/3^23;
```

2427382549574526520289288003118894443416642011423558\
75804873725484560136099847276449874524642689968\
8692688118772641/94143178827

The numerator and denominator of a rational number can be accessed with the numer and denom commands, respectively.

## Continued Fractions

Continued fractions do not have their own separate data type in Maple. A continued fraction representation consists of a recursive series of integers plus fractions that have as their denominators another set of an integer plus a fraction. This repeats until the the denominator is an integer without a fractional component. For example, the following is the continued fraction representation of $43/23$.

```
> numtheory[cfrac](43/23);
```

$$1 + \cfrac{1}{1 + \cfrac{1}{6 + \cfrac{1}{1 + \cfrac{1}{2}}}}$$

## P-adic Numbers

The padic package allows for dozens of different manipulations with p-adic numbers. Internally, the data structure for a p-adic number is of type function. The first operand of returns the base number for the series (the p in p-adic), the second operand returns the first power of the base with nonzero coefficient, and the third and last operand returns an ordered list of the coefficients for subsequent powers of the base. These structures are unique and cannot easily be converted into other data types (e.g., series, polynomials, etc.).

## Gaussian Integers

The GaussInt package allows for dozens of different manipulations with Gaussian integers. Gaussian integers are complex numbers of the form $a + b*I$, where both $a$ and $b$ are integers. From this, you can deduce that all *standard* integers are Gaussian integers with imaginary parts equal to 0.

Internally, the data structure for a Gaussian integer is of type +. The first operand of the data structure returns the real part of the Gaussian integer and the second operand returns the imaginary part (times $I$).

## Parameter Types Specific to This Chapter

| | |
|---|---|
| gint | a Gaussian integer |
| p | a prime number |
| $expr_p$ | a p-adic number |

## Command Listing

### bernoulli(n)

Computes the $n^{th}$ Bernoulli number. n must be a non-negative integer.
**Output:** A rational number is returned.
**Argument options:** (n, var) to compute the $n^{th}$ Bernoulli polynomial in variable var.

**Additional information:** For more information on the exponential generating function, see the on-line help page.
**See also:** numtheory[B], euler, numtheory[fermat], numtheory[mersenne]
*LA = 7 LI = 15*

### chrem([$n_1$, ..., $n_x$], [$m_1$, ..., $m_x$])

Computes the unique positive integer int such that int mod $m_i$ = $n_i$ for all $i$ from 1 to x. The $n_i$s and the $m_i$s are all integers.
**Output:** If an integer solution can be found, a positive integer is returned. Otherwise, an error message is returned.
**Argument options:** ([$poly_1$, ..., $poly_x$], [$m_1$, ..., $m_x$]) to apply the command to polynomials $poly_1$ through $poly_x$. A polynomial is returned.
**Additional information:** The Chinese Remainder Theorem is used to calculate the result. ♣ If integers $m_1$ through $m_x$ are not pairwise relatively prime, chrem will not find an answer.
**See also:** numtheory[mcombine], GaussInt[GImcmbine]
*LI = 18*

### euler(n)

Computes the $n^{th}$ Euler number. n must be a non-negative integer.
**Output:** An integer is returned.
**Argument options:** (n, var) to compute the $n^{th}$ Euler polynomial in variable var.
**Additional information:** For more information on the exponential generating function, see the on-line help page.
**See also:** numtheory[E], bernoulli, numtheory[fermat], numtheory[mersenne]
*LI = 74*

### GaussInt package

Provides commands for calculating with *Gaussian integers*. Gaussian integers are complex numbers of the form $a + b * I$ where both $a$ and $b$ are integers.
**Additional information:** To access the command fnc from the GaussInt package, use the long form of the name GaussInt[fnc], or with(GaussInt, fnc) to load in a pointer for fnc only, or with(GaussInt) to load in pointers to all the commands.

## GaussInt[GIbasis](gint$_1$, gint$_2$)

Determines whether Gaussian integers gint$_1$ and gint$_2$ form a basis for the Gaussian integer domain.
**Output:** A boolean value of either true or false is returned.
**Additional information:** gint$_1$ and gint$_2$ form a basis if every Gaussian integer can be written as n * gint$_1$ + m * gint$_2$, where n and m are integers.

## Gaussint[GIchrem]([gint$_{a1}$, ..., gint$_{an}$], [gint$_{b1}$, ..., gint$_{bn}$])

Performs the chinese remainder algorithm on Gaussian integers gint$_{a1}$ through gint$_{an}$ modulus the Gaussian integers gint$_{b1}$ through gint$_{bn}$.
**Output:** A Gaussian integer is returned.
**See also:** GaussInt[GIrem], *chrem*

## GaussInt[GIdivisor](gint)

Computes the set of divisors found in the first quadrant of Gaussian integer gint.
**Output:** A set containing all the divisors of gint is returned.
**See also:** GaussInt[GInodiv], GaussInt[GIfactor], GaussInt[GIfactors], GaussInt[GIfacset], numtheory[divisors]

## GaussInt[GIfacset](gint)

Calculates the prime factors of Gaussian integer gint.
**Output:** A set of Gaussian integers representing each prime factor of gint is returned. No multiplicities are returned.
**See also:** GaussInt[GIfactor], GaussInt[GIfactors], GaussInt[GIdivisor], numtheory[factorset]

## GaussInt[GIfactor](gint)

Computes the Gaussian integer factorization of Gaussian integer gint.
**Output:** A product of factors in Gaussian integer form is returned.
**Additional information:** The expand command can be used to return the result to its original form. ♣ For more information on the form of the result, see the on-line help page.
**See also:** GaussInt[GIfactors], GaussInt[GIfacset], GaussInt[GIdivisor], ifactor, numtheory[sq2factor]

Number Theory   271

### GaussInt[GIfactors](gint)

Calculates the Gaussian integer factors of Gaussian integer gint.
**Output:** A list with two elements is returned. The first element is either $1, -1, I, -I$, or $0$. The second element is a list of lists, where each internal list has two elements, a distinct Gaussian integer in the factorization and its multiplicity.
**Additional information:** For more information on the form of the result, see the on-line help page.
**See also:** GaussInt[GIfactor], GaussInt[GIfacset], GaussInt[GIdivisor], ifactor

### GaussInt[GIgcd]($gint_1$, ..., $gint_n$)

Computes the first quadrant associate of the greatest common divisor of Gaussian integers $gint_1$ through $gint_n$.
**Output:** A Gaussian integer is returned.
**See also:** GaussInt[GIlcm], *gcd*, *igcd*, *gcdex*

### GaussInt[GIgcdex]($gint_1$, $gint_2$, $name_1$, $name_2$)

Computes the first quadrant associate of the greatest common divisor of Gaussian integers $gint_1$ and $gint_2$, using the extended Euclidean algorithm.
**Output:** A Gaussian integer is returned. As well, $name_1$ and $name_2$ return the values of expressions such that $name_1$ * $gint_1$ + $name_2$ * $gint_2$ = GaussInt[GIgcd]($gint_1$, $gint_2$).
**See also:** GaussInt[GIgcd], *gcdex*, *igcdex*

### GaussInt[GIhermite](M)

Computes the Hermite Normal Form (reduced row echelon form) of matrix of Gaussian integers M.
**Output:** A matrix with the same dimensions as M is returned.
**Argument options:** (M, name) to assign a transformation matrix to name, such that GaussInt[GIhermite](M) = evalm(name &* M).
**Additional information:** The Hermite Normal Form is obtained through a series of row operations on M.
**See also:** *linalg[hermite]*, GaussInt[GIsmith]

### GaussInt[GIissqr](gint)

Determines if Gaussian integer gint is the square of a Gaussian integer.
**Output:** A boolean value of either true or false is returned.
**See also:** GaussInt[GIsqrt], GaussInt[GIfactor]

### GaussInt[GIlcm](gint$_1$, ..., gint$_n$)

Computes the first quadrant associate of the lowest common multiple of Gaussian integers gint$_1$ through gint$_n$.
**Output:** A Gaussian integer is returned.
**See also:** GaussInt[GIgcd], *lcm*, ilcm

### GaussInt[GImcmbine](gint$_1$, gint$_2$, gint$_3$, gint$_4$)

Computes a Gaussian integer, say gint, such that gint = gint$_2$ mod gint$_1$ and gint = gint$_4$ mod gint$_3$.
**Output:** If a solution can be found, a Gaussian integer is returned. Otherwise, the value FAIL is returned.
**Additional information:** The Chinese Remainder Theorem is used to calculate the result.
**See also:** chrem, numtheory[mcombine]

### GaussInt[GInearest](cmplx)

Computes the Gaussian integer nearest to complex value cmplx (i.e., the shortest *distance* away from cmplx).
**Output:** A Gaussian integer is returned.
**Additional information:** If cmplx is equidistant from more than one Gaussian integer, the solution of smallest norm is returned.
**See also:** GaussInt[GIsqrt], GaussInt[GInorm]

### GaussInt[GInodiv](gint)

Computes the number of non-associated divisors of Gaussian integer gint.
**Output:** A non-negative integer is returned.
**See also:** GaussInt[GIdivisor]

### GaussInt[GInorm](gint)

Computes the norm of Gaussian integer gint.
**Output:** A Gaussian integer is returned.
**Additional information:** Gaussian integers are complex numbers of the form $a + b * I$ where both $a$ and $b$ are integers. ♣ The norm of is defined as the value $a^2 + b^2$.
**See also:** GaussInt[GInearest]

### GaussInt[GIorder](gint$_1$, gint$_2$)

Computes the order (modulo gint$_2$) of Gaussian integer gint$_1$.
**Output:** If the order can be determined, a positive integer is returned. Otherwise, the value FAIL is returned.

**Additional information:** The result is the smallest positive integer $m$ such that $\text{gint}_1{}^m = 1 (mod\ \text{gint}_2)$. ♣ If $\text{gint}_1$ and $\text{gint}_2$ are not coprime, then the command fails.
**See also:** GaussInt[GIsqrt], numtheory[order]

### GaussInt[GIphi](gint)

Computes the number of Gaussian integers in the *reduced system* modulo gint.
**Output:** A positive integer is returned.

### GaussInt[GIprime](gint)

Determines if Gaussian integer gint is prime.
**Output:** A boolean value of true or false is returned.
**Additional information:** By definition, a Gaussian integer is prime if it is not *unity* or *zero* and if it is divisible only by 1 and itself.
**See also:** isprime, GaussInt[GIsieve]

### GaussInt[GIquadres](gint$_1$, gint$_2$)

Determines if Gaussian integer $\text{gint}_1$ is a *quadratic residue* modulo $\text{gint}_2$.
**Output:** If $\text{gint}_1$ is a quadratic residue, the value 1 is returned. If $\text{gint}_1$ is a quadratic nonresidue, the value $-1$ is returned.
**Additional information:** $\text{gint}_1$ is a quadratic residue modulo $\text{gint}_2$ if there exists a Gaussian integer $m$ such that $m^2 = \text{gint}_1\ mod\ \text{gint}_2$.
**See also:** GaussInt[GIissqr]

### Gaussint[GIquo](gint$_1$, gint$_2$)

Computes the quotient when Gaussian integer $\text{gint}_1$ is divided by Gaussian integer $\text{gint}_2$.
**Output:** A Gaussian integer is returned.
**Argument options:** (gint$_1$, gint$_2$, name) to assign the remainder of the division to name.
**Additional information:** If more than one Gaussian integer is valid as the quotient, the solution of smallest norm is returned.
**See also:** GaussInt[GIrem], *quo*, iquo, GaussInt[GInorm]

### Gaussint[GIrem](gint$_1$, gint$_2$)

Computes the remainder when Gaussian integer $\text{gint}_1$ is divided by Gaussian integer $\text{gint}_2$.
**Output:** A Gaussian integer is returned.

**Argument options:** (gint$_1$, gint$_2$, name) to assign the quotient of the division to name.
**See also:** GaussInt[GIquo], GaussInt[GIchrem], *rem*, irem, divide

### Gaussint[GIroots](poly)

Computes the Gaussian integer roots of univariate polynomial poly with Gaussian integer coefficients.
**Output:** A list of lists is returned. Each internal list consists of two elements, a Gaussian integer root of poly and its multiplicity.
**See also:** *isolve*

### Gaussint[GIsieve](gint)

Generates a list of all prime Gaussian integers which have a norm less than gint$^2$ and follow the rule $0 \le y < x$ (in the $x, y$-plane).
**Output:** A list with two elements is returned. The first element is the number of prime Gaussian integers that exist, and the second element is a list containing those Gaussian integers.
**Additional information:** For every prime generated (with the exception of $1 + I$), there are seven associated Gaussian primes. For more information their generation, see the on-line help page.
**See also:** GaussInt[GIprime], GaussInt[GInorm]

### GaussInt[GIsmith](M)

Computes the Smith Normal Form of matrix of Gaussian integers M.
**Output:** A square matrix of the same dimension as M is returned.
**Argument options:** (M, name$_1$, name$_2$) to assign transformation matrices to name$_1$ and name$_2$, such that GaussInt[GIsmith](M, var) = evalm(name$_1$ &* M &* name$_2$).
**See also:** *linalg[smith]*, GaussInt[GIhermite]

### GaussInt[GIsqrt](gint)

Computes the Gaussian integer that best approximates the square root of Gaussian integer gint.
**Output:** A Gaussian integer is returned.
**Additional information:** If n is a perfect square, then the result of GaussInt[GIsqrt] is exact. Otherwise, the distance between gint and the result is bounded by $\sqrt{1/2}$.
**See also:** GaussInt[GIissqr]

## ifactor(n)

Calculates the prime integer factorization of integer n and displays it as a product.
**Output:** A product of distinct prime factors to their appropriate powers is returned.
**Argument options:** (rat) to calculate the integer factorization of rational value rat. The result is a fraction of products of prime factors and their exponents. ♣ (n, *squfof*) to use a square free factorization method. ♣ (n, *pollard*) to use Pollard's rho method. ♣ (n, *lenstra*) to use Lenstra's elliptic curve method. ♣ (n, *easy*) to only compute those factors that are "easy" to compute. The remainder of the integer is represented by a placeholder of the form _cxx where xx represents its length in digits.
**Additional information:** By default, the Morrison-Brillhart algorithm is used to compute the prime factorization of n. ♣ Use the expand command to return the result to its original form.
**See also:** ifactors, GaussInt[GIfactor], numtheory[sq2factor], isprime, numtheory[factorset], *expand*
*LA = 7 LI = 107*

## ifactors(n)

Calculates the prime integer factors of integer n.
**Output:** A list with two elements is returned. The first element is either 1, −1, or 0, depending on whether n is positive, negative, or zero, respectively. The second element is a list of lists, where each internal list has two elements, a distinct prime number in the factorization and its multiplicity.
**Additional information:** This command needs to be defined by readlib(ifactors) before being invoked. ♣ Primes with a multiplicity of zero are not included in the result. ♣ The Morrison-Brillhart algorithm is used to factor n.
**See also:** isqrfree, ifactor, isprime, numtheory[factorset]
**FL = 210−211** *LI = 294*

## igcd(int$_1$, ..., int$_n$)

Computes the greatest common divisor of integers int$_1$ through int$_n$.
**Output:** A non-negative integer is returned.
**Additional information:** The integer GCD of the null sequence is equal to 0.
**See also:** ilcm, *gcd*, GaussInt[GIgcd], *gcdex*
*LI = 108*

### igcdex(int₁, int₂)

Computes the greatest common divisor of integers $int_1$ and $int_2$ using the extended Euclidian algorithm.

**Output:** A non-negative integer is returned.

**Argument options:** ($int_1$, $int_2$, $name_1$, $name_2$) to assign to $name_1$ and $name_2$ values such that $name_1 * int_1 + name_2 * int_2 =$ igcd($int_1$, $int_2$).

**See also:** igcd, *gcdex*, GaussInt[GIgcdex]

*LI = 109*


### ilcm(int₁, ..., intₙ)

Computes the lowest common multiple of integers $int_1$ through $int_n$.

**Output:** A non-negative integer is returned.

**Additional information:** The integer LCM of the null sequence is equal to 1.

**See also:** igcd, *lcm*, GaussInt[GIlcm]

*LI = 108*


### iquo(int₁, int₂)

Computes the quotient when integer $int_1$ is divided by integer $int_2$.

**Output:** An integer is returned.

**Argument options:** ($int_1$, $int_2$, name) to assign the remainder of the same division to name.

**See also:** irem, *quo*, GaussInt[GIquo]

**FL = 215−216** *LI = 117*


### iratrecon(int₁, int₂, posint₁, posint₂, name₁, name₂)

Reconstructs a signed rational number from its image $int_1$ modulo $int_2$. $name_1$ and $name_2$ are assinged integers such that $name_1/name_2 = int_1$ mod $int_2$.

**Output:** If a reconstruction can be found, a value of true is returned. Otherwise, a value of false is returned.

**Additional information:** This command needs to be defined by readlib(iratrecon) before being invoked. ✦ The bounds on the values assinged to $name_1$ and $name_2$ are such that abs($name_1$) <= $posint_1$ and abs($name_2$) <= $posint_2$.

**See also:** mod, *ratrecon*

*LI = 296*

## irem(int$_1$, int$_2$)

Computes the remainder when integer int$_1$ is divided by integer int$_2$.
**Output:** An integer is returned.
**Argument options:** (int$_1$, int$_2$, name) to assign the quotient of the same division to name.
**Additional information:** If irem is not passed two integers, it remains unevaluated.
**See also:** iquo, *Rem*, *rem*, GaussInt[GIrem]
**FL = 215−216** *LI = 117*

## iroot(int, n)

Calculates the integer approximation to the $n^{th}$ root of int.
**Output:** A non-negative integer is returned.
**Additional information:** This command needs to be defined by readlib(iroot) before being invoked. ♣ For perfect powers, iroot has no approximation error. For other integers and powers, the error is always less than 1. ♣ If int is less than 0 and n is even, then 0 is returned.
**See also:** isqrt, psqrt, numtheory[mroot]
*LI = 121*

## isprime(n)

Determines if integer n is a prime number using a probabilistic testing routine.
**Output:** A boolean value of true or false is returned.
**Argument options:** (n, posint) to use posint tests to determine if n is prime. The default value is 5.
**Additional information:** By definition, all non-positive integers are not prime. ♣ For more information on the algorithm used in isprime, see the on-line help page.
**See also:** ithprime, nextprime, prevprime, numtheory[safeprime], ifactor, ifactors, numtheory[factorset]
**FL = 119** *LA = 7*

## isqrfree(n)

Calculates the square free integer factorization of integer n.
**Output:** A list with two elements is returned. The first element is either 1, −1, or 0, depending on whether n is positive, negative, or zero, respectively. The second element is a list of lists, where each internal list has two elements, a distinct factor and its multiplicity.
**Additional information:** This command needs to be defined by

readlib(isqrfree) before being invoked.
**See also:** ifactors, *sqrfree*, numtheory[factorset]
*LI = 298*

### isqrt(n)

Calculates the integer approximation to the square root of n.
**Output:** A non-negative integer is returned.
**Additional information:** For perfect squares, isqrt has no approximation error. For other integers, the error is always less than 1.
♣ If int is less than 0, then 0 is returned.
**See also:** *sqrt*, iroot, psqrt, issqr, numtheory[msqrt], numtheory[mroot]
*LI = 121*

### issqr(n)

Determines if integer n is the square of an integer.
**Output:** A boolean value of true or false is returned.
**Additional information:** This command needs to be defined by readlib(issqr) before being invoked.
**See also:** isqrt, iroot, psqrt, *sqrt*, numtheory[msqrt]
*LI = 299*

### ithprime(n)

Calculates the $n^{th}$ prime number (beginning at 2). n must be a positive integer.
**Output:** A positive integer is returned.
**See also:** isprime, nextprime, prevprime

### nextprime(n)

Calculates the smallest prime number larger than integer n.
**Output:** A positive integer is returned.
**Additional information:** If n is itself prime, the next highest prime is returned. ♣ If n is non-positive, then the result is always 2.
**See also:** prevprime, isprime, ithprime, numtheory[safeprime]

### numtheory package

Provides commands for performing calculations in number theory. Among the topics included are prime number generation and testing, special numbers (such as Jacobi, Legendre, Mersenne, and Fermat numbers), and integer roots and divisors.
**Additional information:** To access the command fnc from the

numtheory package, use the long form of the name numtheory [fnc], or with(numtheory, fnc) to load in a pointer for fnc only, or with(numtheory) to load in pointers to all the commands. ♣ Several commands that are listed in the on-line help page for the numtheory package are also available, under the same names, in the main library. These commands are *not* listed in the following numtheory package commands.

### numtheory[B](n)

This command is identical to the bernoulli command.
**See also:** bernoulli

### numtheory[cfrac](rat)

Computes the continued fraction expansion of rational value rat.
**Output:** A continued fraction is returned.
**Argument options:** (expr) where expr is of type real or complex float, algebraic, transcendental, rational polynomial, or series, to compute the appropriate continued fraction expansion. ♣ (expr, n) to compute up to n+1 partial quotients. Ten is the default value. ♣ (expr, $name_1$, $name_2$), where expr is of type real or complex rational, float, algebraic, or transcendental, to assign a list containing convergents of the continued fraction to $name_1$ and a list containing denominators of the continued fraction to $name_2$. ♣ (expr, *'centered'*), where expr is of type real or complex rational, float, algebraic, or transcendental, to calculate the *centered* continued fraction. ♣ (expr, var), where expr is a multivariate of type rational polynomial, series, or algebraic, to perform the expansion with respect to the variable var. ♣ (ratpoly, option), where expr is a rational polynomial and option is either *simple* or *regular*, to calculate the appropriate form of the expansion. ♣ (expr, option), where expr is of type series or algebraic and option is either *simple*, *semisimple*, or *simregular*, to calculate the appropriate form of expansion.
**Additional information:** Applying numtheory[cfrac] to an existing continued fraction returns its expanded form. If it was fully expanded, the original expression is returned. ♣ There is more information on various optional arguments in the lengthy on-line help page.
**See also:** convert(cfrac), numtheory[nthdenom], numtheory[nthnumer], numtheory[nthconver], numtheory[cfracpol]

### numtheory[cfracpol](poly)

Computes the continued fraction expansions of each of the real roots of univariate rational polynomial poly.
**Output:** An expression sequence of lists, each list containing quotients for a particular real root, is returned.
**Argument options:** (poly, n) to compute up to n+1 quotients. Ten is the default value.
**See also:** convert(cfrac)

### numtheory[cyclotomic](n, var)

Calculates the $n^{th}$ cyclotomic polynomial in var.
**Output:** A univariate polynomial in var is returned. All coefficients in the polynomial are either 1 or $-1$.
**Additional information:** The cyclotomic polynomial is defined as the minimum polynomial for the roots of unity.

### numtheory[divisors](n)

Computes the positive divisors of integer n.
**Output:** A set is returned.
**Additional information:** numtheory[divisors](n) and numtheory [divisors](-n) always return the same result.
**See also:** numtheory[tau], numtheory[sigma], numtheory[mipolys], ifactor, ifactorset

### numtheory[E](n)

This command is identical to the euler command.
**See also:** euler

### numtheory[F](n)

This command is identical to the numtheory[fermat] command.
**See also:** numtheory[fermat]

### numtheory[factorset](n)

Calculates the prime factors of integer n.
**Output:** A set is returned.
**Additional information:** No multiplicities are returned.
**See also:** ifactor, ifactors, numtheory[factorEQ], isprime

### numtheory[factorEQ](n, int)

Calculates the integer factorization of n in the Euclidean ring $Z(\sqrt{int})$.

**Output:** A product of factors is returned.
**Additional information:** Use expand to multiply out the factors again. ♣ For more information about valid values for int, see the on-line help page.
**See also:** numtheory[factorset], *factor*, *expand*

### numtheory[fermat](n)

Computes the $n^{th}$ Fermat number. n must be a non-negative integer.
**Output:** A non-negative integer is returned.
**Argument options:** (n, name) to assign information about whether the $n^{th}$ Fermat number is prime or composite to name.
**Additional information:** The formula for the $n^{th}$ fermat number is $2^{2^n} + 1$.
**See also:** numtheory[F], *euler*, *bernoulli*, numtheory[mersenne]

### numtheory[GIgcd](gint$_1$, ..., gint$_n$)

This command is identical to the GaussInt[GIgcd] command.
**See also:** GaussInt[GIgcd]

### numtheory[imagunit](n)

Computes the square root of $-1 (mod\ n)$.
**Output:** If it is possible to compute the square root, a non-negative integer is returned. Otherwise, the value FAIL is returned.
**Additional information:** The result is an ''imaginary unit'' integer $m$ such that $m^2 = -1 (mod\ n)$.
**See also:** numtheory[mroot], numtheory[msqrt], *msolve*

### numtheory[index](n$_1$, n$_2$, n$_3$)

This command is identical to the numtheory[mlog] command.
**See also:** numtheory[mlog]

### numtheory[invphi](n)

Computes all individual positive integers that have exactly n positive integers that are relatively prime to and not exceeding them.
**Output:** A list of positive integer values is returned.
**See also:** numtheory[phi], numtheory[primroot], numtheory[pprimroot]

### numtheory[issqrfree](n)

Determines if integer n is square free.
**Output:** A boolean value of true or false is returned.
**Additional information:** n is square free if it is not divisible by any perfect square.
**See also:** ifactor, numtheory[mobius]

### numtheory[J]($n_1$, $n_2$)

This command is identical to the numtheory[jacobi] command.
**See also:** numtheory[jacobi]

### numtheory[jacobi]($n_1$, $n_2$)

Computes the Jacobi symbol of integers $n_1$ and $n_2$.
**Output:** A value of 1 or $-1$ is returned if $n_1$ is relatively prime to $n_2$ and $n_2$ is a positive, odd integer. Otherwise, a value of 0 is returned.
**Additional information:** If $n_2$ has integer factorization of $p_1^{k1} * ... * p_n^{kn}$, then the Jacobi symbol is defined by the product of numtheory[legendre]($n_1$, $p_1$)$^{k1}$ through numtheory[legendre] ($n_1$, $p_n$)$^{kn}$.
**See also:** numtheory[J], numtheory[legendre], numtheory[mobius]

### numtheory[kronecker]({ineq$_1$, ..., ineq$_x$}, {var$_{1,1}$, ..., var$_{1,m}$}, {var$_{2,1}$, ..., var$_{2,n}$})

Computes the Diophantine approximation, in the nonhomogeneous case, of inequations ineq$_1$ through ineq$_x$ with respect to the two sets of variables, var$_{1,1}$ through var$_{1,m}$ and var$_{2,1}$ through var$_{2,n}$.
**Output:** An expression sequence with m + n lists is returned. Each list contains an equation representing solutions for the given variables.
**Additional information:** Such Diophantine approximations are also called *Minkowski's linear forms*. ✽ For more information on these parameters and alternate calling structures, see the on-line help page.
**See also:** numtheory[minkowski], *isolve*

### numtheory[L]($n_1$, $n_2$)

This command is identical to the numtheory[legendre] command.
**See also:** numtheory[legendre]

### numtheory[lambda](n)

Computes the size of the largest cyclic group generated by $g^i \pmod{n}$.
**Output:** A positive integer is returned.
**Additional information:** This command is derived from Carmichael's theorem. ♣ For more information, see the on-line help page.
**See also:** numtheory[primroot], numtheory[pprimroot], numtheory[order]

### numtheory[legendre]($n_1$, $n_2$)

Computes the Legendre symbol of integers $n_1$ and $n_2$.
**Output:** If $n_1$ is a quadratic residue of $n_2$, a value of $1$ is returned. If $n_1$ is a quadratic nonresidue of $n_2$, a value of $-1$ is returned.
**Additional information:** $n_1$ is a quadratic residue of $n_2$ if there exists an integer m such that $m^2$ is congruent to $n_1$ $(mod\ n_2)$.
**See also:** numtheory[L], numtheory[jacobi], numtheory[mobius]

### numtheory[M](n)

This command is identical to the numtheory[mersenne] command.
**See also:** numtheory[mersenne]

### numtheory[mcombine]($n_1$, $m_1$, $n_2$, $m_2$)

Computes an integer (say $int$) such that $int = m_1\ mod\ n_1$ and $int = m_2\ mod\ n_2$. $n_1$, $m_1$, $n_2$, and $m_2$ must be integers.
**Output:** If an integer solution can be found, a positive integer is returned. Otherwise, the value FAIL is returned.
**Additional information:** The Chinese Remainder Theorem is used to calculate the result.
**See also:** chrem, GaussInt[GImcmbine]

### numtheory[mersenne](n)

Determines if $2^{n-1}$ is a prime number (i.e., a Mersenne prime).
**Output:** If $2^{n-1}$ is prime, then that value is returned. Otherwise, false is returned.
**Argument options:** ([n]) to compute the $n^{th}$ Mersenne prime.
**See also:** numtheory[M], euler, bernoulli, numtheory[fermat]

### numtheory[minkowski]({ineq$_1$, ..., ineq$_x$}, {var$_{1,1}$, ..., var$_{1,m}$}, {var$_{2,1}$, ..., var$_{2,n}$})

Computes the Diophantine approximation, in the homogeneous case, of inequations ineq$_1$ through ineq$_x$ with respect to the two sets of variables, var$_{1,1}$ through var$_{1,m}$ and var$_{2,1}$ through var$_{2,n}$.
**Output:** An expression sequence with m + n lists is returned. Each list contains an equation representing solutions for the variables given.
**Additional information:** Such Diophantine approximations are also called *Minkowski's linear forms*. For more information on the parameters and alternate calling structures, see the on-line help page.
**See also:** numtheory[kronecker], *isolve*

### numtheory[mipolys](n, p)

Computes the number of monic irreducible univariate polynomials of degree n over $Z(mod\ p)$. n must be a positive integer.
**Output:** A positive integer is returned.
**Argument options:** (n, p, m) to compute the number of monic irreducible polynomials of degree n over the Galois field defined by $p^m$.
**Additional information:** The answer is calculated using results from numtheory[mobius] and numtheory[divisors]. ♣ For more information, see the on-line help page.
**See also:** numtheory[mobius], numtheory[divisors]

### numtheory[mlog](n$_1$, n$_2$, n$_3$)

Computes the discrete logarithm of n$_1$ to the base n$_2$ modulo n$_3$.
**Output:** If it is possible to compute the discrete logarithm, a non-negative integer is returned. Otherwise, the value FAIL is returned.
**Argument options:** (n$_1$, n$_2$, n$_3$, name) to assign the characteristic of the domain of the result to name. This means that all solutions can be found from numtheory[mlog](n$_1$, n$_2$, n$_3$) + k * name, for all non-negative $k$.
**See also:** numtheory[primroot], numtheory[order]

### numtheory[mobius](n)

Computes the Mobius function of positive integer n.
**Output:** If there are an even number of *twists*, a value of 1 is returned. If there are an odd number of *twists*, a value of $-1$ is returned. A value of 0 is returned if and only if n contains a square integer factor greater than 1.

**Additional information:** A Mobius function is sometimes called a *lattice of divisors*.
**See also:** numtheory[issqrfree], numtheory[legendre], numtheory[jacobi], numtheory[mipolys]

### numtheory[mroot]($n_1$, $n_2$, $n_3$)

Computes the $n_2{}^{th}$ root of $n_1$ modulo $n_3$.
**Output:** If it is possible to compute the root, a non-negative integer is returned. Otherwise, the value FAIL is returned.
**Additional information:** The result is an integer $m$ such that $m^{n_2} = n_1 \ (mod\ n_3)$. In most cases, $n_2$ must be a prime number.
**See also:** numtheory[msqrt], *msolve*, numtheory[imagunit], numtheory[primroot], numtheory[rootsunity]

### numtheory[msrqt]($n_1$, $n_2$)

Computes the square root of $n_1$ modulo $n_2$.
**Output:** If it is possible to compute the root, a non-negative integer is returned. Otherwise, the value FAIL is returned.
**Additional information:** The result is an integer $m$ such that $m^2 = n_1 \ (mod\ n_2)$.
**See also:** numtheory[mroot], *msolve*, numtheory[imagunit], numtheory[primroot], numtheory[rootsunity]

### numtheory[nearestp]([$list_1$, ..., $list_n$], [$x_1$, ..., $x_n$])

Computes the lattice point nearest to the point defined by real numbers $x_1$ through $x_n$ if the n-element lists $list_1$ through $list_n$ represent vectors defining an n-dimensional lattice space.
**Output:** An n-element list of integers is returned.
**Additional information:** For more information on how the lattice point is generated, see the on-line help page.
**See also:** numtheory[minkowski], numtheory[kronecker]

### numtheory[nthconver](expr, n)

Computes the $n^{th}$ convergent of continued fraction expr.
**Output:** An expression is returned.
**Additional information:** numtheory[nthconver](expr, n) is equivalent to numtheory[nthnumer](expr, n) / numtheory[nthdenom](expr, n). ♣ Remember that expr must be a continued fraction in either list or fractional form. ♣ If n exceeds the number of partial quotients of expr, an error message is returned.
**See also:** convert(cfrac), numtheory[cfrac], numtheory[nthnumer], numtheory[nthdenom]

### numtheory[nthdenom](expr, n)

Computes the $n^{th}$ denominator of continued fraction expr.
**Output:** An expression is returned.
**Additional information:** Remember that expr must be a continued fraction in either list or fractional form. ♣ If n exceeds the number of partial quotients of expr, an error message is returned.
**See also:** convert(cfrac), numtheory[cfrac], numtheory[nthnumer], numtheory[nthconver]

### numtheory[nthnumer](expr, n)

Computes the $n^{th}$ numerator of continued fraction expr.
**Output:** An expression is returned.
**Additional information:** Remember that expr must be a continued fraction in either list or fractional form. ♣ If n exceeds the number of partial quotients of expr, an error message is returned.
**See also:** convert(cfrac), numtheory[cfrac], numtheory[nthdenom], numtheory[nthconver]

### numtheory[nthpow]($n_1$, $n_2$)

Determines the largest natural number $m$ such that $m^{n_2}$ divides $n_1$. $n_1$ must be a nonzero integer and $n_2$ must be a natural number.
**Output:** A natural number raised to the power $n_2$ is returned.
**See also:** convert(cfrac), ifactor

### numtheory[order]($n_1$, $n_2$)

Computes the order of integer $n_1$ in the multiplicative group modulo $n_2$.
**Output:** If the order can be determined, a positive integer is returned. Otherwise, the value FAIL is returned.
**Additional information:** The result is the smallest integer m such that $n_1{}^m = 1(mod\ n_2)$. If $n_1$ and $n_2$ are not coprime, then the command fails.
**See also:** numtheory[primroot], numtheory[pprimroot], numtheory[lambda], numtheory[mlog]

### numtheory[phi](n)

Computes the number of positive integers relatively prime to and not exceeding positive integer n.

**Output:** A positive integer is returned.
**See also:** numtheory[invphi], numtheory[primroot], numtheory[pprimroot]

### numtheory[pprimroot](n)

Computes the smallest pseudo primitive root of positive integer n, if possible.
**Output:** If a solution is possible, a positive integer is returned. Otherwise, the value FAIL is returned.
**Argument options:** $(n_1, n_2)$ to compute the smallest pseudo primitive root of $n_2$ that is greater than $n_1$, if possible.
**Additional information:** A pseudo primitive root of n is defined to be a generator of the cyclic group under multiplication modulo n containing the integers relatively prime to n. ♣ If such a generator does not exist, the smallest positive integer relatively prime to n and not exceeding n is returned.
**See also:** numtheory[primroot], numtheory[rootsunity], numtheory[order], numtheory[lambda], numtheory[mroot]

### numtheory[primroot](n)

Computes the smallest primitive root of positive integer n.
**Output:** If it is possible to compute a primitive root of n, a positive integer is returned. Otherwise, the value FAIL is returned.
**Argument options:** $(n_1, n_2)$ to compute the smallest primitive root of $n_2$ that is greater than $n_1$, if possible.
**Additional information:** A primitive root of n is defined to be a generator of the cyclic group under multiplication modulo n containing the integers relatively prime to n.
**See also:** numtheory[pprimroot], numtheory[rootsunity], numtheory[order], numtheory[lambda], numtheory[mroot]

### numtheory[rootsunity](p, n)

Computes all the $p^{th}$ roots of unity modulo n. p must be a prime number.
**Output:** An expression sequence of positive integers is returned.
**Additional information:** Note that every expression sequence returned contains at least the root 1.
**See also:** numtheory[primroot], numtheory[pprimroot], numtheory[order], numtheory[mroot]

### numtheory[safeprime](n)

Calculates the smallest "safe" prime number larger than n.
**Output:** A positive prime integer is returned.
**Additional information:** A safe prime is a prime number $p$ such that $(p-1)/2$ is also prime. ♣ If n is itself a safe prime, the next smallest safe prime is returned.
**See also:** nextprime, isprime

### numtheory[sigma](n)

Computes the sum of all positive divisors of integer n.
**Output:** A non-negative integer is returned.
**Argument options:** sigma[k](n) to compute the sum of the $k^{th}$ power of each positive divisor of n.
**Additional information:** Because of a peculiarity of Maple, when using the [k] option to numtheory[sigma], the command must first be loaded in with either with(numtheory) or with(numtheory, sigma), *not* called with the long form of the name, numtheory [sigma].
**See also:** numtheory[divisors], numtheory[tau], ifactor, ifactorset

### numtheory[sq2factor](n)

Computes the integer factorization of n, an integer in $Z(\sqrt{2})$.
**Output:** A product of factors in $Z(\sqrt{2})$ form is returned.
**Additional information:** An integer in $Z(\sqrt{2})$ form looks like $a + b\sqrt{2}$, where $a$ and $b$ are integers. ♣ The expand command can be used to return the result to its original form. ♣ For more information on the form of the result, see the on-line help page.
**See also:** ifactor, GaussInt[GIfactor]

### numtheory[tau](n)

Computes the number of positive divisors of integer n.
**Output:** A non-negative integer is returned.
**Additional information:** numtheory[tau](n) and numtheory[tau](-n) always return the same result.
**See also:** numtheory[divisors], numtheory[sigma], ifactor, ifactorset

### numtheory[thue](eqn, [var$_1$, var$_2$])

Computes the resolutions of Thue equation eqn with respect to var$_1$ and var$_2$.

**Output:** An expression sequence of lists is returned. Each list contains two equations representing a particular solution set for $var_1$ and $var_2$.
**Argument options:** (ineq, [$var_1$, $var_2$]) to compute the resolutions of Thue *inequality* ineq. ♣ (eqn, [$var_1$, $var_2$], n) to compute the resolutions of Thue equation eqn with the limiting factor of n. Default value is ten. ♣ When no solution can be found, call with (eqn, [$var_1$, $var_2$], name) to assign an interval of constant terms that might provide a solution to name.
**Additional information:** For more information on definitions and solutions of Thue equations and inequalities, see the on-line help page.
**See also:** *isolve*

### op(expr$_p$)

Displays the internal structure of p-adic number expr$_p$ and allows you access to its operands.
**Output:** A list containing three further data components is returned.
**Argument options:** (0, op(expr$_p$)) to return the special data handle PADIC. ♣ (1, op(expr$_p$)) to return the integer base p. ♣ (2, op(expr$_p$)) to return the integer exponent of the first lowest nonzero term. ♣ (3, op(expr$_p$)) to return a list of the coefficients of the powers of expr$_p$, starting with the power found with op(2, op(expr$_p$)).
**Additional information:** Terms with a zero coefficient that are not of lower exponent than the first nonzero term are represented in the internal data structure. ♣ Examining the $0^{th}$ operand of an expression is the best method of determining whether it is in p-adic form. ♣ If there is an O() order term in the p-adic number, it is represented by a 1 in the last element of the coefficient list.
**See also:** padic[ratvaluep], *subsop*

### padic package

Provides commands for performing calculations with p-adic numbers.
**Additional information:** To access the command fnc from the padic package, use the long form of the name padic[fnc], or with(padic, fnc) to load in a pointer for fnc only, or with(padic) to load in pointers to all the commands. ♣ If any of the parameters provided to commands in the padic package are of an unexpected type, then an unevaluated command is typically returned. ♣ Most commands in the padic package take an expression and a prime number as parameters. If the expression is already in p-adic form, it may not be necessary to supply the prime number. ♣ Commands

in this package that are of the form padic[fnc*p*](expr, p, n), where fnc is a standard Maple function such as sin, exp, etc., can also be accessed with the padic[evalp](fnc(expr), p, n) syntax. ♣ If an integer value denoting the order is not given as the last parameter to a command in the padic package, the value of the global variable Digitsp ($= 10$ if not defined) is used. ♣ If the numeric values supplied to a command do not lead to a p-adic valuation, then a value of FAIL is typically returned.

### padic[arccoshp](expr, p , n)

Evaluates to size n the p-adic arc-hyperbolic cosine of expr, an expression in rational and/or p-adic numbers.
**Output:** A p-adic expression is returned.
**See also:** padic[evalp], padic[arcsinhp], padic[arctanhp], padic[coshp], *arccosh*

### padic[arccosp](expr, p , n)

Evaluates to size n the p-adic arc-cosine of expr, an expression in rational and/or p-adic numbers.
**Output:** A p-adic expression is returned.
**See also:** padic[evalp], padic[arcsinp], padic[arctanp], padic[cosp], *arccos*

### padic[arccothp](expr, p , n)

Evaluates to size n the p-adic arc-hyperbolic cotangent of expr, an expression in rational and/or p-adic numbers.
**Output:** A p-adic expression is returned.
**See also:** padic[evalp], padic[arcsechp], padic[arccschp], padic[cothp], *arccoth*

### padic[arccotp](expr, p , n)

Evaluates to size n the p-adic arc-cotangent of expr, an expression in rational and/or p-adic numbers.
**Output:** A p-adic expression is returned.
**See also:** padic[evalp], padic[arcsecp], padic[arccscp], padic[cotp], *arccot*

### padic[arccschp](expr, p , n)

Evaluates to size n the p-adic arc-hyperbolic cosecant of expr, an expression in rational and/or p-adic numbers.

**Output:** A p-adic expression is returned.
**See also:** padic[evalp], padic[arcsechp], padic[arccothp], padic[cschp], *arccsch*

### padic[arccscp](expr, p , n)

Evaluates to size n the p-adic arc-cosecant of expr, an expression in rational and/or p-adic numbers.
**Output:** A p-adic expression is returned.
**See also:** padic[evalp], padic[arcsecp], padic[arccotp], padic[cscp], *arccsc*

### padic[arcsechp](expr, p , n)

Evaluates to size n the p-adic arc-hyperbolic secant of expr, an expression in rational and/or p-adic numbers.
**Output:** A p-adic expression is returned.
**See also:** padic[evalp], padic[arccschp], padic[arccothp], padic[sechp], *arcsech*

### padic[arcsecp](expr, p , n)

Evaluates to size n the p-adic arc-secant of expr, an expression in rational and/or p-adic numbers.
**Output:** A p-adic expression is returned.
**See also:** padic[evalp], padic[arccscp], padic[arccotp], padic[secp], *arcsec*

### padic[arcsinhp](expr, p , n)

Evaluates to size n the p-adic arc-hyperbolic sine of expr, an expression in rational and/or p-adic numbers.
**Output:** A p-adic expression is returned.
**See also:** padic[evalp], padic[arcsinhp], padic[arctanhp], padic[sinhp], *arcsinh*

### padic[arcsinp](expr, p , n)

Evaluates to size n the p-adic arc-sine of expr, an expression in rational and/or p-adic numbers.
**Output:** A p-adic expression is returned.
**See also:** padic[evalp], padic[arcsinp], padic[arctanp], padic[sinp], *arcsin*

### padic[arctanhp](expr, p , n)

Evaluates to size n the p-adic arc-hyperbolic tangent of expr, an expression in rational and/or p-adic numbers.
**Output:** A p-adic expression is returned.
**See also:** padic[evalp], padic[arccoshp], padic[arcsinhp], padic[tanhp], *arctanh*

### padic[arctanp](expr, p , n)

Evaluates to size n the p-adic arc-tangent of expr, an expression in rational and/or p-adic numbers.
**Output:** A p-adic expression is returned.
**See also:** padic[evalp], padic[arccosp], padic[arcsinp], padic[tanp], *arctan*

### padic[coshp](expr, p , n)

Evaluates to size n the p-adic hyperbolic cosine of expr, an expression in rational and/or p-adic numbers.
**Output:** A p-adic expression is returned.
**See also:** padic[evalp], padic[sinhp], padic[tanhp], padic[arccoshp], *cosh*

### padic[cosp](expr, p , n)

Evaluates to size n the p-adic cosine of expr, an expression in rational and/or p-adic numbers.
**Output:** A p-adic expression is returned.
**See also:** padic[evalp], padic[sinp], padic[tanp], padic[arccosp], *cos*

### padic[cothp](expr, p , n)

Evaluates to size n the p-adic hyperbolic cotangent of expr, an expression in rational and/or p-adic numbers.
**Output:** A p-adic expression is returned.
**See also:** padic[evalp], padic[sechp], padic[cschp], padic[arccothp], *coth*

### padic[cotp](expr, p , n)

Evaluates to size n the p-adic cotangent of expr, an expression in rational and/or p-adic numbers.
**Output:** A p-adic expression is returned.

Number Theory    293

**Argument options:**
**See also:** padic[evalp], padic[secp], padic[cscp], padic[arccotp], *cot*

### padic[cschp](expr, p , n)

Evaluates to size n the p-adic hyperbolic cosecant of expr, an expression in rational and/or p-adic numbers.
**Output:** A p-adic expression is returned.
**See also:** padic[evalp], padic[sechp], padic[cothp], padic[arccschp], *csch*

### padic[cscp](expr, p , n)

Evaluates to size n the p-adic cosecant of expr, an expression in rational and/or p-adic numbers.
**Output:** A p-adic expression is returned.
**See also:** padic[evalp], padic[secp], padic[cotp], padic[arccscp], *csc*

### padic[evalp](expr, p, n)

Determines to size n the p-adic value of expr, an expression or polynomial in rational and/or p-adic numbers.
**Output:** A p-adic expression is returned.
**Additional information:** expr usually involves a previously known mathematical function (e.g., cos, sqrt, log, etc.). ♣ Most of these functions have a direct counterpart command in the padic package, typically named padic[namep] where name is the name of the function. ♣ See the on-line help page for padic[evalp] for a complete listing of the available functions.
**See also:** padic[valuep]

### padic[expp](expr, p , n)

Evaluates to size n the p-adic exponential of expr, an expression in rational and/or p-adic numbers.
**Output:** A p-adic expression is returned.
**See also:** padic[evalp], padic[logp], padic[sqrtp], *exp*

### padic[logp](expr, p, n)

Evaluates to size n the p-adic logarithm of expr, an expression in rational and/or p-adic numbers.
**Output:** A p-adic expression is returned.
**Additional information:** If p is not a prime integer or n is equal

to zero, a value of FAIL is returned.
**See also:** padic[evalp], padic[expp], padic[sqrtp], *log*

### padic[ordp](expr, p)

Determines the p-adic order of expr, an expression or polynomial in rational and/or p-adic numbers.
**Output:** A non-negative integer is returned.
**Additional information:** If p is not a prime integer, a value of 0 is returned.
**See also:** padic[evalp], padic[valuep]

### padic[ratvaluep]($expr_p$, n)

Calculates the sum of the first n terms of the p-adic number $expr_p$.
**Output:** A rational number is returned.
**Argument options:** ($expr_p$) to calculate the sum of the first Digitsp terms of $expr_p$. Digitsp is a globally defined variable with a default value of ten.
**Additional information:** If $expr_p$ is a rational number, then $expr_p$ itself is returned. ♣ Warning: providing ratvaluep with a first parameter that is *not* a p-adic number may cause Maple to crash.
**See also:** padic[evalp], padic[sqrtp], padic[expp], padic[logp], *sqrt*

### padic[rootp](expr, p , n)

Calculates to size n all the roots in the p-adic number field for expr, an expression or polynomial in rational and/or p-adic numbers.
**Output:** If it can be computed, an expression sequence of p-adic expressions is returned.
**Additional information:** The same result can be computed by using the command padic[evalp](RootOf(expr), p, n).
**See also:** padic[evalp], padic[sqrtp], padic[expp], padic[logp], *sqrt*

### padic[sechp](expr, p , n)

Evaluates to size n the p-adic hyperbolic secant of expr, an expression in rational and/or p-adic numbers.
**Output:** A p-adic expression is returned.
**See also:** padic[evalp], padic[cschp], padic[cothp], padic[arcsechp], *sech*

### padic[secp](expr, p , n)

Evaluates to size n the p-adic secant of expr, an expression in rational and/or p-adic numbers.
**Output:** A p-adic expression is returned.
**See also:** padic[evalp], padic[cscp], padic[cotp], padic[arcsecp], *sec*

### padic[sinhp](expr, p , n)

Evaluates to size n the p-adic hyperbolic sine of expr, an expression in rational and/or p-adic numbers.
**Output:** A p-adic expression is returned.
**See also:** padic[evalp], padic[sinhp], padic[tanhp], padic[arcsinhp], *sinh*

### padic[sinp](expr, p , n)

Evaluates to size n the p-adic sine of expr, an expression in rational and/or p-adic numbers.
**Output:** A p-adic expression is returned.
**See also:** padic[evalp], padic[sinp], padic[tanp], padic[arcsinp], *sin*

### padic[sqrtp](expr, p , n)

Evaluates to size n one p-adic square root of expr, an expression or polynomial in rational and/or p-adic numbers.
**Output:** A p-adic expression is returned.
**Additional information:** The other square root is calculated by $-1$ times the result. ♣ The same result can be computed by using the command padic[evalp](sqrt(expr), p, n) or padic[evalp]((expr)^(1/2), p, n). ♣ If p is not a prime integer or n is equal to zero, a value of FAIL is returned.
**See also:** padic[evalp], padic[rootp], padic[expp], padic[logp], *sqrt*

### padic[tanhp](expr, p , n)

Evaluates to size n the p-adic hyperbolic tangent of expr, an expression in rational and/or p-adic numbers.
**Output:** A p-adic expression is returned.
**See also:** padic[evalp], padic[coshp], padic[sinhp], padic[arctanhp], *tanh*

### padic[tanp](expr, p , n)

Evaluates to size n the p-adic tangent of expr, an expression in rational and/or p-adic numbers.
**Output:** A p-adic expression is returned.
**See also:** padic[evalp], padic[cosp], padic[sinp], padic[arctanp], *tan*

### padic[valuep](expr, p)

Evaluates the p-adic valuation of expr, an expression or polynomial in rational and/or p-adic numbers.
**Output:** A positive integer raised to an integer exponent is returned.
**See also:** padic[evalp], padic[ordp]

### prevprime(n)

Calculates the largest prime number smaller than integer n.
**Output:** A prime integer is returned.
**Additional information:** If n is itself prime, the next smallest prime is returned (if it exists). If n is less than 3, an error message is returned.
**See also:** nextprime, isprime, ithprime

### Prime(int)

Represents the inert isprime command for integer int.
**Output:** An unevaluated Prime call is returned.
**Argument options:** Prime(int) *mod* $int_p$ to evaluate the inert command call modulo $int_p$. $int_p$ must be a prime.
**See also:** isprime, *mod*

### type(expr, *complex(integer)*)

Tests whether expr is a Gaussian integer (i.e., a complex number with integer components).
**Output:** A boolean value of either true or false is returned.
**Additional information:** For the result to be true, both type(*Re*(expr), *integer*) and type(*Im*(expr), *integer*) must return true.
**See also:** GaussInt[GIbasis], GaussInt[GIgcd], *Re*, *Im*, *type*, *op(cmplx)*, *subsop*

### type(expr, *facint*)

Tests whether expr is a factored integer form.
**Output:** A boolean value of either true or false is returned.

**Additional information:** The integers 0, 1, and −1 are automatically factored integers. All other integers must be passed through the ifactor command first.
**See also:** type(*integer*), ifactor
*LI = 228*

**type(expr,** *primeint***)**

Tests whether expr is a prime integer.
**Output:** A boolean value of either true or false is returned.
**See also:** type(*integer*), isprime

# Statistics

## Introduction

The stats package in Maple has been redesigned and rewritten for Maple V Release 3. So great is the change in usage and functionality that the package bears little resemblance to its former self. What's more, the improvements are so sweeping that the stats package now rates its own separate chapter. You may notice that there are no Manual Cross References in this chapter. This is because none of this code existed when the most recent editions of the manuals were written.

## The New Package Structure

Not only is this stats package new to Maple, but the command hierarchy, the internal structure that allows you access to the various commands, is substantially different from any other package in Maple. Currently, the stats package is the only Maple package that contains *multiple levels* of commands. To demonstrate what this means, let's load in the package the way we would any other.

> with(stats);

$$[describe, fit, importdata, random, statevalf, statplots, transform]$$

All of these "commands" that you see here (with the exception of importdata) are actually *subpackages* of the stats package. For example, the describe subpackage contains commands for descriptive statistical calculations. From, this point, there are two ways to access the individual commands within describe.

You can use the long form of the command.

```
> describe[mean]([92,34,55,67,80,92,50,44,87,71]);
```

$$\frac{336}{5}$$

Or you can load in all the commands in describe using the with command again.

```
> with(describe);
```

$[coefficientofvariation, count, countmissing, covariance,$
$\quad decile, geometricmean, harmonicmean, kurtosis,$
$\quad linearcorrelation, mean, meandeviation, median, mode,$
$\quad moment, percentile, quadraticmean, quantile, quartile,$
$\quad range, skewness, standarddeviation, variance]$

The above could also have been accomplished by initially calling with(stats, describe).

## Statistical Lists

As you can see in the previous call to describe[mean], statistical data is typically enclosed within a list structure. To deal with more complicated statistical data, the concept of a *statistical list* is introduced. Apart from simple expressions (typically numerical values), statistical lists can also contain values in ranges (e.g., 27..50 represents a single value in the range 27 through 50), weighted elements (e.g., Weight(60, 4) represents four elements with value 60), and *missing* elements represented by the variable name missing. Weights and ranges can be combined, as well.

As an example, here is a more complicated statistical list.

```
> slist := [25, 97, 50..55, Weight(60, 5), 44,
> Weight(80..89, 3)];
```

$slist := [\,25, 97, 50..55, \text{Weight}(\,60, 5\,), 44, \text{Weight}(\,80..89, 3\,)\,]$

And here are some more complicated commands working upon that statistical list.

```
> variance(slist);
```

$$\frac{13445}{36}$$

```
> evalf(");
```

$$373.4722222$$

```
> standarddeviation(slist);
```

$$\frac{1}{6}\sqrt{13445}$$

```
> evalf(");
```

$$19.32542942$$

```
> harmonicmean(slist);
```

$$\frac{2272069800}{39935501}$$

```
> evalf(");
```

$$56.89348432$$

# Command Indexing

Another interesting feature of many of the statistics commands is that they can be indexed by specific values to alter their operation. For example, in the command decile, the particular decile desired (1 through 9) is specified as an index, not as a traditional parameter.

```
> decile[5](slist);
```

$$60$$

```
> evalf(describe[decile[2]](slist));
```

$$51.45000000$$

In the last example, you see how to place the index value when using the long form of the command name.

Try some commands from the other subpackages (fit, random, statevalf, transform) to get a feel for their capabilities.

## Statistical Plots

The statplots subpackage contains many useful plotting routines for analyzing statistical data. Box plots, histograms, and one- and two-dimensional scatter plots are among those available.

```
> with(statplots);
Warning: new definition for     quantile
```

$[boxplot, histogram, notchedbox, quantile, quantile2, scatter1d,$
$\quad scatter2d, symmetry, xscale, xshift, xyexchange]$

Following is a large set of data for test scores out of 10, and two examples of the available plotting commands acting on that data.

```
> testdata := [1,4,8,5,6,9,3,6,5,0,3,10,10,5,6,5,8,7,9,
> 9,9,5,6,2,7,1,10,8,8,6,7,5,7,4,4,3,6,7,10,9,8,7,4,6,
> 6,5,4,3,6,7,5,8,6,9,2,10,10,9,8,9]:
> scatter1d[stacked](testdata);
```

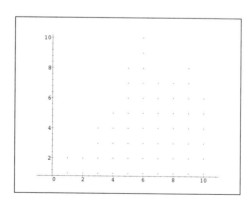

```
> boxplot(testdata);
```

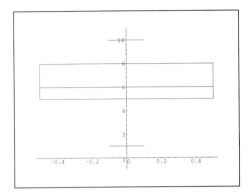

# Parameter Types Specific to This Chapter

| | |
|---|---|
| 2DPlot | a two-dimensional plot structure |
| distrib | a statistical distribution function |
| Ms | a statistical matrix |
| Slist | a statistical list |
| subpckg | a package name |

# Command Listing

### describe subpackage

Provides descriptive commands for statistical data analysis.
**Additional information:** To access the command fnc directly from the describe subpackage, use the long form of the name, describe[fnc], after loading in a pointer to the subpackage with with(stats). ♣ To use the short form of a command within describe (e.g., count), you must first load the pointers in its subpackage with with(describe).
**See also:** stats package

### describe[coefficientofvariation]([$expr_1$, ..., $expr_n$])

Computes the coefficient of variation of expressions $expr_1$ through $expr_n$
**Output:** An expression is returned.

**Argument options:** (Slist) to compute the coefficient of variation of Slist, a statistical list. Classes are represented by their class mark and missing data are ignored entirely. For more information on statistical lists, see the introduction to this chapter. ♣
describe[coefficientofvariation[0]](Slist) to treat Slist as the entire population. This is the default. ♣
describe[coefficientofvariation[1]](Slist) to treat Slist as a sample of the population.
**Additional information:** The coefficient of variation is a measure of the relative dispersion of the data and is equal to the *standard deviation* divided by the *mean*.
**See also:** describe[standarddeviation], describe[mean]

### describe[count]([expr$_1$, ..., expr$_n$])

Counts the number of expressions in expr$_1$ through expr$_n$.
**Output:** A positive integer is returned.
**Argument options:** (Slist) to count the number a statistical elements represented in Slist, a statistical list. Any element that equals the string missing is ignored by this command. For more information on statistical lists, see the introduction to this chapter.
**Additional information:** op can be used to find the number of elements in a list of expressions. Keep in mind that with a statistical list including occurrences of the Weight command, the number of elements of the list may differ from the number of statistical elements.
**See also:** Weight, *op(list)*, transform[classmark], transform[scaleweight], describe[countmissing], describe[range], transform[deletemissing]

### describe[countmissing](Slist)

Counts the number of missing elements defined in statistical list Slist.
**Output:** A non-negative integer is returned.
**Additional information:** For more information on statistical lists, see the introduction to this chapter.
**See also:** describe[count], transform[deletemissing]

### describe[covariance](list$_1$, list$_2$)

Computes the covariance between list$_1$ and list$_2$, two lists of expressions with equal numbers of elements.
**Output:** An expression is returned.
**Argument options:** (Slist$_1$, Slist$_2$) to compute the covariance between Slist$_1$ and Slist$_2$, two statistical lists. Classes are represented

by their class mark and missing data are ignored entirely. For more information on statistical lists, see the introduction to this chapter.
**Additional information:** When dealing with two statistical lists, there must be an equal number of observations in each list and corresponding elements must have equal weight. ♣ For a definition of the covariance, see the on-line help page.
**See also:** transform[classmark], describe[variance], describe[linearcorrelation]

### describe[decile[n]]([num$_1$, ..., num$_n$])

Computes the $n^{th}$ decile of numeric values num$_1$ through num$_n$
**Output:** An expression is returned.
**Argument options:** (Slist) to compute the decile of Slist, a statistical list. For more information on statistical lists, see the introduction to this chapter. ♣ (Slist, num) to specify that the size of all gaps between classes in Slist is num. For more information about the role of gaps, see the on-line help page for describe[gaps].
**Additional information:** All values passed to describe[decile[n]] must be numeric and n must be an integer from 1 to 9, inclusive. ♣ Deciles falling between entries are interpolated.
**See also:** describe[quartile], describe[quantile], describe[percentile], describe[mean], transform[classmark]

### describe[geometricmean]([expr$_1$, ..., expr$_n$])

Computes the geometric mean of expressions expr$_1$ through expr$_n$.
**Output:** An expression is returned.
**Argument options:** (Slist) to compute the geometric mean of Slist, a statistical list. Classes are represented by their class mark and missing data are ignored entirely. For more information on statistical lists, see the introduction to this chapter.
**Additional information:** Remember that the geometric mean is not necessarily equal to any of the individual elements being averaged.
**See also:** transform[classmark], describe[mean], describe[harmonicmean], describe[quadraticmean]

### describe[harmonicmean]([expr$_1$, ..., expr$_n$])

Computes the harmonic mean of expressions expr$_1$ through expr$_n$.
**Output:** An expression is returned.
**Argument options:** (Slist) to compute the harmonic mean of Slist, a statistical list. Classes are represented by their class mark and missing data are ignored entirely. For more information on statistical lists, see the introduction to this chapter.

**Additional information:** Remember that the harmonic mean is not necessarily equal to any of the individual elements being averaged.
**See also:** transform[classmark], describe[mean], describe[geometricmean], describe[quadraticmean]

### describe[kurtosis]([expr$_1$, ..., expr$_n$])

Computes the moment coefficient of kurtosis of expressions expr$_1$ through expr$_n$
**Output:** An expression is returned.
**Argument options:** (Slist) to compute the moment coefficient of kurtosis of Slist, a statistical list. Classes are represented by their class mark and missing data are ignored entirely. For more information on statistical lists, see the introduction to this chapter. ♣ describe[kurtosis[0]](Slist) to treat Slist as the entire population. This is the default. ♣ describe[kurtosis[1]](Slist) to treat Slist as a sample of the population.
**Additional information:** For more information on what the moment coefficient of kurtosis is, see the on-line help page.
**See also:** transform[classmark], describe[variance], describe[standarddeviation], describe[moment], describe[skewness]

### describe[linearcorrelation](list$_1$, list$_2$)

Computes the coefficient of linear correlation between list$_1$ and list$_2$, two lists of expressions with equal numbers of elements.
**Output:** An expression is returned.
**Argument options:** (Slist$_1$, Slist$_2$) to compute the coefficient of linear correlation between Slist$_1$ and Slist$_2$, two statistical lists. Classes are represented by their class mark and missing data are ignored entirely. For more information on statistical lists, see the introduction to this chapter.
**Additional information:** When dealing with two statistical lists, there must be an equal number of observations in each list and corresponding elements must have equal weight. ♣ The coefficient of linear correlation equals the covariance of list$_1$ and list$_2$ divided by the square root of the product of the two individual variances.
**See also:** transform[classmark], describe[variance], describe[covariance]

### describe[mean]([expr$_1$, ..., expr$_n$])

Computes the arithmetic mean of expressions expr$_1$ through expr$_n$.
**Output:** An expression is returned.

**Argument options:** (Slist) to compute the mean of Slist, a statistical list. Classes are represented by their class mark and missing data are ignored entirely. For more information on statistical lists, see the introduction to this chapter.
**Additional information:** Remember that the mean is not necessarily equal to any of the individual elements being averaged.
**See also:** transform[classmark], describe[geometricmean], describe[harmonicmean], describe[quadraticmean], describe[meandeviation], describe[mode], describe[median]

### describe[meandeviation]([expr$_1$, ..., expr$_n$])

Computes the mean deviation of expressions expr$_1$ through expr$_n$.
**Output:** An expression is returned.
**Argument options:** (Slist) to compute the mean deviation of Slist, a statistical list. Classes are represented by their class mark and missing data are ignored entirely. For more information on statistical lists, see the introduction to this chapter.
**Additional information:** The mean deviation is found by subtracting the mean from the data and then computing the mean of the absolute values of the resulting elements.
**See also:** transform[classmark], describe[mean], describe[variance], describe[standarddeviation]

### describe[median]([num$_1$, ..., num$_n$])

Computes the median of numeric values num$_1$ through num$_n$.
**Output:** An expression is returned.
**Argument options:** (Slist) to compute the median of Slist, a statistical list. For more information on statistical lists, see the introduction to this chapter. ♣ (Slist, num) to specify that the size of all gaps between classes in Slist is num. For more information about the role of gaps, see the on-line help page for describe[gaps].
**Additional information:** All values passed to describe[median] must be numeric. ♣ If a sorted list has an odd number of elements, the median is the middle element. Otherwise, the median is the arithmetic mean of the two middle elements. ♣ Medians falling between classes are interpolated by a formula which depend on the particular circumstances.
**See also:** describe[quartile], describe[quantile], describe[decile], describe[percentile], describe[mode], describe[mean], transform[classmark]

### describe[mode]([num$_1$, ..., num$_n$])

Computes the mode of numeric values num$_1$ through num$_n$.
**Output:** If the mode is a single value, that value is returned. Otherwise, an expression sequence of values is returned.
**Argument options:** (Slist) to compute the mode of Slist, a statistical list. For more information on statistical lists, see the introduction to this chapter.
**Additional information:** All values passed to describe[mode] must be numeric. ♣ The mode of a list is the element with the highest frequency. ♣ Modes falling within classes are interpolated.
**See also:** describe[mean], describe[median], transform[classmark], describe[count]


### describe[moment[n]]([expr$_1$, ..., expr$_n$])

Computes the n$^{th}$ moment of expressions expr$_1$ through expr$_n$
**Output:** An expression is returned.
**Argument options:** (Slist) to compute the moment of Slist, a statistical list. Classes are represented by their class mark and missing data are ignored entirely. For more information on statistical lists, see the introduction to this chapter. ♣ describe[moment[n, option]](Slist) to compute the moment about a particular point. option can be a numeric value or the name of any command from stats[describe] that returns a single value. Most commonly used values are *mean*, *median*, or *mode*. ♣ describe[moment[n, option, 0]](Slist) to treat Slist as the entire population. This is the default. ♣ describe[moment[n, option, 1]](Slist) to treat Slist as a sample of the population.
**Additional information:** The moment can be a measure of dispersion, asymmetry, or flattening, depending on the parameters passed to it.
**See also:** transform[classmark], describe[variance], describe[standarddeviation], describe[kurtosis]


### describe[percentile[n]]([num$_1$, ..., num$_n$])

Computes the n$^{th}$ percentile of numeric values num$_1$ through num$_n$
**Output:** An expression is returned.
**Argument options:** (Slist) to compute the percentile of Slist, a statistical list. For more information on statistical lists, see the introduction to this chapter. ♣ (Slist, num) to specify that the size of all gaps between classes in Slist is num. For more information about the role of gaps, see the on-line help page for describe[gaps].
**Additional information:** All values passed to describe[percent-

Statistics  309

ile[n]] must be numeric, and n must be an integer from 1 to 99, inclusive. ♣ Percentiles falling between entries are interpolated.
**See also:** describe[quartile], describe[quantile], describe[decile], describe[mean], transform[classmark]

### describe[quadraticmean]([expr$_1$, ..., expr$_n$])

Computes the quadratic mean of expressions expr$_1$ through expr$_n$.
**Output:** An expression is returned.
**Argument options:** (Slist) to compute the quadratic mean of Slist, a statistical list. Classes are represented by their class mark and missing data are ignored entirely. For more information on statistical lists, see the introduction to this chapter.
**Additional information:** The quadratic mean is the square root of the mean of the squares of the elements.
**See also:** transform[classmark], describe[mean], describe[geometricmean], describe[harmonicmean], describe[mean]

### describe[quantile[a]]([num$_1$, ..., num$_n$])

Computes the $a^{th}$ quantile of numeric values num$_1$ through num$_n$
**Output:** An expression is returned.
**Argument options:** (Slist) to compute the quantile of Slist, a statistical list. For more information on statistical lists, see the introduction to this chapter. ♣ (Slist, num) to specify that the size of all gaps between classes in Slist is num. For more information about the role of gaps, see the on-line help page for describe[gaps].
**Additional information:** All values passed to describe[quantile[a]] must be numeric and a must be a numeric value between 0 and 1, inclusive. ♣ Quantiles falling between entries are interpolated.
**See also:** describe[decile], describe[quartile], describe[percentile], describe[mean], transform[classmark]

### describe[quartile[n]]([num$_1$, ..., num$_n$])

Computes the $n^{th}$ quartile of numeric values num$_1$ through num$_n$
**Output:** An expression is returned.
**Argument options:** (Slist) to compute the quartile of Slist, a statistical list. For more information on statistical lists, see the introduction to this chapter. ♣ (Slist, num) to specify that the size of all gaps between classes in Slist is num. For more information about the role of gaps, see the on-line help page for describe[gaps].
**Additional information:** All values passed to describe[quartile[n]] must be numeric and n must be an integer from 1 to 3, inclusive.

♣ Quartiles falling between entries are interpolated.
**See also:** describe[decile], describe[quantile], describe[percentile], describe[mean], transform[classmark]

### describe[range]([num$_1$, ..., num$_n$])

Computes the range of the numeric values num$_1$ through num$_n$.
**Output:** A numeric range is returned.
**Argument options:** (Slist) to compute the range of elements in Slist, a statistical list. Any element that equals the string missing is ignored and classes are assumed to cover their entire range. For more information on statistical lists, see the introduction to this chapter.
**Additional information:** All values passed to describe[range] must be numeric.
**See also:** describe[count]

### describe[skewness]([expr$_1$, ..., expr$_n$])

Computes the moment coefficient of skewness of expressions expr$_1$ through expr$_n$.
**Output:** An expression is returned.
**Argument options:** (Slist) to compute the moment coefficient of skewness of Slist, a statistical list. Classes are represented by their class mark and missing data are ignored entirely. For more information on statistical lists, see the introduction to this chapter.
♣ describe[skewness[0]](Slist) to treat Slist as the entire population. This is the default. ♣ describe[skewness[1]](Slist) to treat Slist as a sample of the population.
**Additional information:** For more information on what the moment coefficient of skewness is, see the on-line help page.
**See also:** transform[classmark], describe[variance], describe[standarddeviation], describe[moment], describe[kurtosis]

### describe[standarddeviation]([expr$_1$, ...,expr$_n$])

Computes the standard deviation of expressions expr$_1$ through expr$_n$.
**Output:** An expression is returned.
**Argument options:** (Slist) to compute the standard deviation of Slist, a statistical list. Classes are represented by their class mark and missing data are ignored entirely. For more information on statistical lists, see the introduction to this chapter. ♣ describe[standarddeviation[0]](Slist) to treat Slist as the entire population. This is the default. ♣ describe[standarddeviation[1]]

Statistics 311

(Slist) to treat Slist as a sample of the population.
**Additional information:** The standard deviation is a measure of spread of the data and is equal to the square root of the *variation*.
**See also:** transform[classmark], describe[variance], describe[meandeviation]

### describe[variance]([expr$_1$, ..., expr$_n$])

Computes the variance of expressions expr$_1$ through expr$_n$
**Output:** An expression is returned.
**Argument options:** (Slist) to compute the variance of Slist, a statistical list. Classes are represented by their class mark and missing data are ignored entirely. For more information on statistical lists, see the introduction to this chapter. ♣ describe[variance[0]](Slist) to treat Slist as the entire population. This is the default. ♣ describe[variance[1]](Slist) to treat Slist as a sample of the population.
**Additional information:** The variance is a measure of spread of the data and is equal to the square of the *standard deviation*.
**See also:** transform[classmark], describe[covariance], describe[meandeviation], describe[standarddeviation], describe[coefficientofvariation], describe[mean]

### fit subpackage

Provides a command for fitting a linear curve to statistical data.
**Additional information:** There is currently only one command in the fit subpackage, leastsquare. ♣ To access the command leastsquare directly from the fit subpackage, use the long form of the name, fit[leastsquare], after loading in a pointer to the subpackage with with(stats). ♣ To use the short form of a command within fit, you must first load the pointers in its subpackage with with(fit).
**See also:** fit[leastsquare], stats package

### fit[leastsquare[[var$_1$, ..., var$_n$]]]([Slist$_1$, ..., Slist$_n$])

Fits, using a weighted least-squares method, a linear curve to the data in statistical lists Slist$_1$ through Slist$_n$, which represent variables var$_1$ through var$_n$, respectively.
**Argument options:** fit[leastsquare[[var$_1$, ..., var$_n$], eqn, name$_1$, ..., name$_m$]]([Slist$_1$, ..., Slist$_n$]) to fit the data to a linear equation of the form specified in eqn. The unknowns in eqn that represent the coefficients must be specified with name$_1$ through name$_m$.
**Additional information:** var$_n$ is always seen to be the dependent variable, while var$_1$ through var$_{n-1}$ are the independent variables.

♣ For more information and examples, see the on-line help page for fit[leastsquare].
**See also:** describe[correlation]

### random subpackage

Provides random generators for several different statistical distributions.

**Additional information:** All of the random distribution commands take similar forms of parameter sequences. If no parameters are specified (i.e., ()), then a single random value is generated. If the single parameter *'generator'* is supplied, a random number generator for the distribution is supplied. Use (n, option$_1$, option$_2$) to control optional methods of the random number generation. option$_1$ controls what random generator is used and can be either *'default'* for the existing generator, or a user supplied command name representing a random number generator returning a value between 0 and 1. option$_2$ can be *'auto'* (the default value), *'builtin'* to use a special method to generate the values, or *'inverse'* to use the appropriate inverse density function method. ♣ To access the command fnc directly from the random subpackage, use the long form of the name, random[fnc], after loading in a pointer to the subpackage with with(stats). ♣ To use the short form of a command within random (e.g., beta), you must first load the pointers in its subpackage with with(random).
**See also:** stats package

### random[beta[int$_1$, int$_2$]](n)

Generates n random numbers from the Beta distribution with degrees of freedom int$_1$ and int$_2$.
**Output:** An expression sequence of floating-point numbers is returned.
**Argument options:** See the entry for random package for more details on argument options.
**Additional information:** int$_1$ and int$_2$ must be *positive* integer values. For more information on their effect, see the on-line help page for stats[distributions]. ♣ For examples of the uses of the optional parameters, see the on-line help page for stats[random]. In most cases, though, the default values for these options should suffice.
**See also:** random package, random[chisquare], random[fratio], random[gamma], random[normald], random[studentst]

### random[binomiald[posint, num]](n)

Generates n random numbers from the binomial distribution with contributing factors posint and num.
**Output:** An expression sequence of floating-point numbers is returned.
**Argument options:** See the entry for random package for more details on argument options.
**Additional information:** posint must be a positive integer and num must be a numeric value between 0 and 1, inclusive. For more information on their effect, see the on-line help page for stats[distributions]. ♣ If this command doesn't work for you, there should be a workaround available from your technical support representative. ♣ For examples of the uses of the optional parameters, see the on-line help page for stats[random]. In most cases, though, the default values for these options should suffice.
**See also:** random package, random[negativebinomial], random[normald], random[uniform]


### random[cauchy[$num_1$, $num_2$]](n)

Generates n random numbers from the Cauchy distribution with contributing factors $num_1$ and $num_2$.
**Output:** An expression sequence of floating-point numbers is returned.
**Argument options:** See the entry for random package for more details on argument options.
**Additional information:** $num_1$ and $num_2$ must be numeric values. For more information on their effect, see the on-line help page for stats[distributions]. ♣ For examples of the uses of the optional parameters, see the on-line help page for stats[random]. In most cases, though, the default values for these options should suffice.
**See also:** random package, random[laplaced], random[weibull], random[gamma], random[uniform]


### random[chisquare[int]](n)

Generates n random numbers from the chi-square distribution dependent on the value int.
**Output:** An expression sequence of floating-point numbers is returned.
**Argument options:** See the entry for random package for more details on argument options.
**Additional information:** int must be a *positive* integer value. For more information on its effect, see the on-line help page for

stats[distributions]. ♣ For examples of the uses of the optional parameters, see the on-line help page for stats[random]. In most cases, though, the default values for these options should suffice.
**See also:** random package, random[beta], random[fratio], random[gamma], random[normald], random[studentst]

### random[discreteuniform[a, b]](n)

Generates n random numbers from the discrete uniform distribution between a and b.
**Output:** An expression sequence of floating-point numbers is returned.
**Argument options:** See the entry for random package for more details on argument options.
**Additional information:** a and b must be numeric values, such that a < b. For more information on this distribution, see the on-line help page for stats[distributions]. ♣ For examples of the uses of the optional parameters, see the on-line help page for stats[random]. In most cases, though, the default values for these options should suffice.
**See also:** random package, random[uniform]

### random[empirical[$num_1$, ..., $num_m$]](n)

Generates n random natural numbers from 1 to m, depending on the related probabilities $num_1$ through $num_m$.
**Output:** An expression sequence of natural numbers is returned.
**Argument options:** See the entry for random package for more details on argument options.
**Additional information:** $num_1$ through $num_m$ must be numeric values and must sum to 1. For more information on this distribution, see the on-line help page for stats[distributions]. ♣ For examples of the uses of the optional parameters, see the on-line help page for stats[random]. In most cases, though, the default values for these options should suffice.
**See also:** random package

### random[exponential[$num_1$, $num_2$]](n)

Generates n random numbers from the exponential distribution with contributing factors $num_1$ and $num_2$.
**Output:** An expression sequence of floating-point numbers is returned.
**Argument options:** See the entry for random package for more details on argument options.
**Additional information:** $num_1$ and $num_2$ must be numeric values

and $num_1$ must be positive, as well. For more information on their effect, see the on-line help page for stats[distributions]. ✸ For examples of the uses of the optional parameters, see the on-line help page for stats[random]. In most cases, though, the default values for these options should suffice.

**See also:** random package, random[cauchy], random[gamma], random[uniform]

### random[fratio[$int_1$, $int_2$]](n)

Generates n random numbers from the *Fisher f* distribution with degrees of freedom $int_1$ and $int_2$.

**Output:** An expression sequence of floating-point numbers is returned.

**Argument options:** See the entry for random package for more details on argument options.

**Additional information:** $int_1$ and $int_2$ must be *positive* integer values. For more information on their effect, see the on-line help page for stats[distributions]. ✸ For examples of the uses of the optional parameters, see the on-line help page for stats[random]. In most cases, though, the default values for these options should suffice.

**See also:** random package, random[beta], random[chisquare], random[gamma], random[normald], random[studentst]

### random[gamma[$num_1$, $num_2$]](n)

Generates n random numbers from the Gamma distribution with contributing factors $num_1$ and $num_2$.

**Output:** An expression sequence of floating-point numbers is returned.

**Argument options:** See the entry for random package for more details on argument options.

**Additional information:** $num_1$ and $num_2$ must be numeric values. For more information on their effect, see the on-line help page for stats[distributions]. ✸ For examples of the uses of the optional parameters, see the on-line help page for stats[random]. In most cases, though, the default values for these options should suffice.

**See also:** random package, random[beta], random[chisquare], random[fratio], random[normald], random[studentst]

### random[laplaced[$num_1$, $num_2$]](n)

Generates n random numbers from the Laplace distribution with contributing factors $num_1$ and $num_2$.

**Output:** An expression sequence of floating-point numbers is returned.

**Argument options:** See the entry for random package for more details on argument options.

**Additional information:** $num_1$ and $num_2$ must be numeric values. For more information on their effect, see the on-line help page for stats[distributions]. ♣ For examples of the uses of the optional parameters, see the on-line help page for stats[random]. In most cases, though, the default values for these options should suffice.

**See also:** random package, random[gamma], random[weibull], random[uniform]

### random[logistic[$num_1$, $num_2$]](n)

Generates n random numbers from the logistic distribution with contributing factors $num_1$ and $num_2$.

**Output:** An expression sequence of floating-point numbers is returned.

**Argument options:** See the entry for random package for more details on argument options.

**Additional information:** $num_1$ and $num_2$ must be numeric values. For more information on their effect, see the on-line help page for stats[distributions]. ♣ For examples of the uses of the optional parameters, see the on-line help page for stats[random]. In most cases, though, the default values for these options should suffice.

**See also:** random package, random[gamma], random[uniform]

### random[lognormal](n)

Generates n random numbers from the logarithmic normal distribution.

**Output:** An expression sequence of floating-point numbers is returned.

**Argument options:** random[lognormal[$num_1$, $num_2$]](n) to use the values of $num_1$ and $num_2$ as the $\mu$ and $\sigma$ values for the normal distribution, respectively. ♣ See the entry for random package for more details on argument options.

**Additional information:** The default values of $\mu$ and $\sigma$ are 0 and 1, respectively. ♣ For examples of the uses of the optional parameters, see the on-line help page for stats[random]. In most cases, though, the default values for these options should suffice.

**See also:** random package, random[normald]

### random[negativebinomial[posint, num]](n)

Generates n random numbers from the negative binomial distribution with contributing factors posint and num.
**Output:** An expression sequence of floating-point numbers is returned.
**Argument options:** See the entry for random package for more details on argument options.
**Additional information:** posint must be a positive integer and num must be a numeric value between 0 and 1, inclusive. For more information on their effect, see the on-line help page for stats[distributions]. ♣ If this command doesn't work for you, there should be a workaround available from your technical support representative. ♣ For examples of the uses of the optional parameters, see the on-line help page for stats[random]. In most cases, though, the default values for these options should suffice.
**See also:** random package, random[binomiald]

### random[normald](n)

Generates n random numbers from the normal distribution.
**Output:** An expression sequence of floating-point numbers is returned.
**Argument options:** random[lognormal[$num_1$, $num_2$]](n) to use the values of $num_1$ and $num_2$ as the $\mu$ and $\sigma$ values for the normal distribution, respectively. ♣ See the entry for random package for more details on argument options.
**Additional information:** The default values of $\mu$ and $\sigma$ are 0 and 1, respectively. ♣ For examples of the uses of the optional parameters, see the on-line help page for stats[random]. In most cases, though, the default values for these options should suffice.
**See also:** random package, random[beta], random[chisquare], random[fratio], random[gamma], random[lognormal]

### random[poisson[num]](n)

Generates n random numbers from the Poisson distribution dependent on the value num.
**Output:** An expression sequence of floating-point numbers is returned.
**Argument options:** See the entry for random package for more details on argument options.
**Additional information:** num must be a *positive* numeric value. For more information on its effect, see the on-line help page for stats[distributions]. ♣ For examples of the uses of the optional parameters, see the on-line help page for stats[random]. In most

cases, though, the default values for these options should suffice.
**See also:** random package, random[chisquare], random[studentst]

### random[studentst[int]](n)

Generates n random numbers from the *Students' T distribution* dependent on the value num.
**Output:** An expression sequence of floating-point numbers is returned.
**Argument options:** See the entry for random package for more details on argument options.
**Additional information:** int must be a *positive* integer value. For more information on its effect, see the on-line help page for stats[distributions]. ♣ For examples of the uses of the optional parameters, see the on-line help page for stats[random]. In most cases, though, the default values for these options should suffice.
**See also:** random package, random[chisquare], random[poisson]

### random[uniform[a, b]](n)

Generates n random numbers from the uniform distribution between the values a and b.
**Output:** An expression sequence of floating-point numbers is returned.
**Argument options:** See the entry for random package for more details on argument options.
**Additional information:** a and b must be numeric values, such that a < b. ♣ For examples of the uses of the optional parameters, see the on-line help page for stats[random]. In most cases, though, the default values for these options should suffice.
**See also:** random package, random[gamma], random[normald]

### random[weibull[$num_1$, $num_2$]](n)

Generates n random numbers from the Weibull distribution with contributing factors $num_1$ and $num_2$.
**Output:** An expression sequence of floating-point numbers is returned.
**Argument options:** See the entry for random package for more details on argument options.

**Additional information:** num$_1$ and num$_2$ must be numeric values. For more information on their effect, see the on-line help page for stats[distributions]. ♣ For examples of the uses of the optional parameters, see the on-line help page for stats[random]. In most cases, though, the default values for these options should suffice.

**See also:** random package, random[laplaced], random[cauchy], random[gamma], random[uniform]

### statevalf subpackage

Provides commands for numerical evaluation of statistical functions.

**Additional information:** To access the command fnc directly from the statevalf subpackage, use the long form of the name, statevalf[fnc], after loading in a pointer to the subpackage with with(stats). ♣ To use the short form of a command within statevalf (e.g., cdf), you must first load the pointers in its subpackage with with(statevalf).

**See also:** stats package

### statevalf[cdf, distrib](num)

Computes the continuous cumulative density for the distrib function at the value num.

**Output:** A positive numeric value is returned.

**Additional information:** Sometimes there are restrictions on the value of num. ♣ If distrib needs extra values enclosed in square brackets, they must be supplied here. For more information on the continuous distributions available, see the entries under the random subpackage and the help page for stats[distributions].

**See also:** statevalf[icdf], statevalf[pdf]

### statevalf[dcdf, distrib](int)

Computes the discrete cumulative probability for the distrib function at the value num.

**Output:** A positive numeric value is returned.

**Additional information:** Sometimes there are restrictions on the value of int. ♣ If distrib needs extra values enclosed in square brackets, they must be supplied here. For more information on the discrete distributions available, see the entries under the random subpackage and the help page for stats[distributions].

**See also:** statevalf[idcdf], statevalf[pf]

### statevalf[icdf, distrib](num)

Computes the inverse continuous cumulative density for the distrib function at the value num.
**Output:** A numeric value is returned.
**Additional information:** num must be between 0 and 1, or an error message is returned. ♣ If distrib needs extra values enclosed in square brackets, they must be supplied here. For more information on the continuous distributions available, see the entries under the random subpackage and the help page for stats[distributions].
**See also:** statevalf[cdf], statevalf[pdf]

### statevalf[idcdf, distrib](num)

Computes the inverse discrete cumulative probability for the distrib function at the value num.
**Output:** A numeric value is returned.
**Additional information:** num must be between 0 and 1, or an error message is returned. ♣ If distrib needs extra values enclosed in square brackets, they must be supplied here. For more information on the discrete distributions available, see the entries under the random subpackage and the help page for stats[distributions].
**See also:** statevalf[dcdf], statevalf[pf]

### statevalf[pdf, distrib](num)

Computes the continuous probability density for the distrib function at the value num.
**Output:** A positive numeric value is returned.
**Additional information:** Sometimes there are restrictions on the value of num. ♣ If distrib needs extra values enclosed in square brackets, they must be supplied here. For more information on the continuous distributions available, see the entries under the random subpackage and the help page for stats[distributions].
**See also:** statevalf[cdf], statevalf[icdf]

### statevalf[pf, distrib](int)

Computes the discrete probability for the distrib function at the value num.
**Output:** A positive numeric value is returned.
**Additional information:** Sometimes there are restrictions on the value of int. ♣ If distrib needs extra values enclosed in square brackets, they must be supplied here. For more information on the discrete distributions available, see the entries under the random subpackage and the help page for stats[distributions].
**See also:** statevalf[dcdf], statevalf[idcdf]

## stats package

Provides commands for many calculations in statistics.
**Additional information:** Almost all of the commands in the stats package are found in one of its *subpackages*; describe, fit, random, statevalf, statplots, and transform. ♣ To load in pointers to all the subpackages (but *not* to the the functions contained therein), use with(stats). To load in the pointer to just one subpackage, use with(stats, subpckg). To access the command fnc directly from a subpackage within stats, use the long form of the name stats[subpckg, fnc].
**See also:** describe subpackage, fit subpackage, random subpackage, statevalf subpackage, statplots subpackage, transform subpackage

## stats[importdata](filename, n)

Reads in raw data in n streams from the external file filename.
**Output:** An expression sequence of n statistical lists is returned.
**Additional information:** The numerical data in filename must be separated with spaces and/or end-of-line characters. ♣ As the data is read, each of the n statistical lists gets one value, in order, then a second value is added to each list, and so on until all the values are read. ♣ The character * is interpreted as a missing value and is stored as *missing* in its respective statistical list. ♣ For more information on how formatted input/output is handled, see the entry for sscanf.
**See also:** *sscanf*

## statplots subpackage

Provides commands for plotting statistical data two-dimensionally.
**Additional information:** An unfortunate drawback of this subpackage is that you cannot add additional options to the command (e.g., *title*=str) to control standard options to two-dimensional plots. ♣ To access the command fnc directly from the statplots subpackage, use the long form of the name, statplots[fnc], after loading in a pointer to the subpackage with with(stats). ♣ To use the short form of a command within statplots (e.g., histogram), you must first load the pointers in its subpackage with with(statplots).
**See also:** stats package

## statplots[boxplot]([$num_1$, ..., $num_n$])

Creates a two-dimensional box plot of the numerical data $num_1$ through $num_n$.

**Output:** A two-dimensional plot is displayed.
**Argument options:** (Slist) to plot statistical list Slist. ♣ stats[boxplot[$num_c$]](Slist) to center the box plot about the horizontal value $num_c$. Default is 0. ♣ stats[boxplot[$num_c$, $num_w$]](Slist) to specify the width of the box plot as $num_w$. Default is 1.
**Additional information:** The box plot consists of horizontal lines representing the first quartile, median, and third quartile, two lines extending from the center, and any data elements lying outside these structures. ♣ Classes are represented by their class mark and missing data are ignored. ♣ Use plots[display] to combine multiple box plots.
**See also:** statplots[notchedbox], statplots[xyexchange], describe[quartile], describe[median], *plots[display]*

### statplots[histogram]([$num_1$, ..., $num_n$])

Creates a two-dimensional histogram of the numerical data $num_1$ through $num_n$.
**Output:** A two-dimensional plot is displayed.
**Argument options:** (Slist) to plot statistical list Slist.
**Additional information:** After the data is tallied, individual points are represented with lines of length proportionate to their frequency, and classes are represented by rectangles with areas proportionate to their weight. Conflicting lines and rectangles overlap. ♣ Missing data are ignored.
**See also:** statplots[notchedbox], statplots[xyexchange], transform[tally], transform[tallyinto]

### statplots[notchedbox]([$num_1$, ..., $num_n$])

Creates a two-dimensional notched box plot of the numerical data $num_1$ through $num_n$.
**Output:** A two-dimensional plot is displayed.
**Argument options:** (Slist) to plot statistical list Slist. ♣ stats[notchedbox[$num_c$]](Slist) to center the plot about the horizontal value $num_c$. Default is 0. ♣ stats[notchedbox[$num_c$, $num_w$]](Slist) to specify the width of the plot as $num_w$. Default is 1.
**Additional information:** The notched box plot consists of horizontal lines representing the first quartile, median, and third quartile, two lines extending from the center, and any data elements lying outside these structures. ♣ Classes are represented by their class mark and missing data are ignored. ♣ Use plots[display] to combine multiple notched box plots.
**See also:** statplots[boxplot], statplots[xyexchange], describe[quartile], describe[median], *plots[display]*

## statplots[quantile]([num₁, ..., numₙ])

Creates a two-dimensional quantile plot of the numerical data $num_1$ through $num_n$.

**Output:** A two-dimensional plot is displayed.

**Argument options:** (Slist) to plot the quantile plot of statistical list Slist.

**Additional information:** A quantile plot sorts the data from the two lists and then plots points corresponding to equal quantiles from the two lists. ♣ Values from Slist are plotted along the vertical axis. ♣ Ranges are plotted as rectangles and weighted values are plotted as horizontal lines.

**See also:** statplots[quantile2], statplots[scatter1d], describe[quantile], transform[statsort]

## statplots[quantile2](list$_x$, list$_y$)

Creates a two-dimensional quantile-quantile plot of the numerical data in list$_x$ and list$_y$, two lists of equal length.

**Output:** A two-dimensional plot is displayed.

**Argument options:** (Slist$_x$, Slist$_y$) to plot the quantile-quantile plot of statistical lists Slist$_x$ and Slist$_y$.

**Additional information:** A quantile-quantile plot sorts the data from the two lists and then plots points corresponding to equal quantiles from the two lists. ♣ Values from list$_x$ are plotted along the horizontal axis and values from list$_y$ along the vertical axis. ♣ Classes plotted against classes are represented as rectangles.

**See also:** statplots[quantile], statplots[scatter2d], describe[quantile], transform[statsort]

## statplots[scatter1d]([num₁, ..., numₙ])

Creates a one-dimensional scatter plot of the numerical data $num_1$ through $num_n$.

**Output:** A one-dimensional plot is displayed within the structure of a two-dimensional plot.

**Argument options:** (Slist) to plot the one-dimensional scatter plot of statistical list Slist. ♣ statplots[scatter1d[*projected*]](Slist) to plot the data points at their actual $x$-axis values. This is the default behavior. ♣ statplots[scatter1d[*stacked*]](Slist) to plot the data points stacked at their $x$-axis values, much as in a histogram. ♣ statplots[scatter1d[*jittered*]](Slist) to plot equal $x$-axis values clustered vertically about that value.

**Additional information:** Missing data are ignored and classes are

plotted as line segments. ♣ To place these plots along the axes of another two-dimensional plot, use the xshift and xyexchange commands.
**See also:** statplots[scatter2d], statplots[histogram], statplots[symmetry], statplots[xshift], statplots[xyexchange]

### statplots[scatter2d](list$_x$, list$_y$)

Creates a two-dimensional scatter plot of the numerical data in list$_x$ versus the numerical data in list$_y$.
**Output:** A two-dimensional plot is displayed.
**Argument options:** (Slist$_x$, Slist$_y$) to plot the two-dimensional scatter plot of statistical lists Slist$_x$ and Slist$_y$.
**Additional information:** The first list always supplies the horizontal elements, while the second list always supplies the vertical.
♣ Classes versus classes are plotted as rectangles, classes versus points are plotted as lines, and points versus points are plotted as points.
**See also:** statplots[xshift], statplots[xyexchange], statplots[scatter1d]

### statplots[symmetry]([num$_1$, ..., num$_n$])

Creates a two-dimensional symmetry plot of the numerical data num$_1$ through num$_n$.
**Output:** A two-dimensional plot is displayed.
**Argument options:** (Slist) to plot the symmetry plot of statistical list Slist.
**Additional information:** A symmetry plot includes a plot of the line $y = x$ and measures the symmetry of the data about the median value. ♣ For more information on the definition of symmetry plots, see the on-line help page. ♣ Class data are not handled by this command.
**See also:** statplots[scatter1d], describe[median]

### statplots[xscale[num]](2DPlot)

Scales the $x$ (horizontal) values in two-dimensional plot 2DPlot by a factor of num.
**Output:** A two-dimensional plot is displayed.
**See also:** statplots[xshift], statplots[xyexchange], *plots[display]*

### statplots[xshift[num]](2DPlot)

Shifts the $x$ (horizontal) values in two-dimensional plot 2DPlot by num.
**Output:** A two-dimensional plot is displayed.
**See also:** statplots[xscale], statplots[xyexchange], *plots[display]*

### statplots[xyexchange](2DPlot)

Exchanges the $x$ (horizontal) values and $y$ (vertical) values in two-dimensional plot 2DPlot, thereby effecting a rotation of 90° clockwise.
**Output:** A two-dimensional plot is displayed.
**See also:** statplots[xscale], statplots[xshift], *plots[display]*

### transform subpackage

Provides commands for transforming lists of statistical data.
**Additional information:** To access the command fnc directly from the transform subpackage, use the long form of the name, transform[fnc], after loading in a pointer to the subpackage with with(stats). ♣ To use the short form of a command within transform (e.g., frequency), you must first load the pointers in its subpackage with with(transform).
**See also:** stats package

### transform[apply[fnc]]([expr$_1$, ..., expr$_n$])

Applies the function fnc to each expression expr$_1$ through expr$_n$.
**Output:** A list of expressions is returned.
**Argument options:** (Slist) to apply fnc to statistical list Slist.
**Additional information:** Missing items are not changed.
**See also:** Weight, transform[divideby], transform[multiapply], transform[remove]

### transform[classmark](Slist)

Replaces any class ranges present in statistical list Slist by their respective class marks.
**Output:** A new statistical list is returned.
**Additional information:** The class mark of a particular class range is its midpoint. ♣ Many of the computations in the describe subpackage call classmark on their data. ♣ For examples of the uses of class ranges, see the introduction to this chapter.

### transform[cumulativefrequency](Slist)

Replaces each element of statistical list Slist with the cumulative frequency of itself and the elements before it.
**Output:** A list of positive integers is returned.
**Additional information:** If there are no Weight elements, then the list returned contains [1, 2, 3, ...]. ♣ The order of the elements is preserved. ♣ Missing data have a frequency of one.
**See also:** transform[frequency], transform[scaleweight], transform[statvalue], transform[tally]

### transform[deletemissing](Slist)

Deletes any missing elements found in statistical list Slist.
**Output:** A new statistical list is returned.
**Additional information:** For more information on statistical lists, see the introduction to this chapter.
**See also:** describe[count], describe[countmissing]

### transform[divideby[num]]([expr$_1$, ..., expr$_n$])

Divides each expression expr$_1$ through expr$_n$ by the numeric value num.
**Output:** A list of expressions is returned.
**Argument options:** (Slist) to divide each element of statistical list Slist by num. ♣ transform[divide[option]]([expr$_1$, ..., expr$_n$]) or transform[divide[option]](Slist) to divide by the value of the data represented by option. option can be any one of the descriptive names from the describe subpackage, such as *standarddeviation*, *meandeviation*, or *variance*.
**Additional information:** Missing items are not changed.
**See also:** transform[apply], transform[remove]

### transform[frequency](Slist)

Replaces each element of statistical list Slist with its weight.
**Output:** A list of positive integers is returned.
**Additional information:** If there are no Weight elements, then the list returned contains all ones. ♣ The order of the elements is preserved. ♣ Missing data have a frequency of one. ♣ This command does not tally up identical elements.
**See also:** transform[cumulativefrequency], transform[scaleweight], transform[statvalue], transform[tally]

**transform[moving[posint]]([expr$_1$, ..., expr$_n$])**

Replaces each expression expr$_1$ through expr$_n$ with the average of its posint neighbors (including itself).
**Output:** A list of expressions is returned.
**Argument options:** (Slist) to perform the operation on statistical list Slist. ♣ transform[moving[posint, fnc]]([expr$_1$, ..., expr$_n$]) to use the statistical function fnc to act upon the neighboring values. The default is *mean*. ♣ transform[moving[posint, fnc, list]] ([expr$_1$, ..., expr$_n$]), where list is a list with posint elements, to weigh the respective members of each neighborhood according to those elements when calculating the average. Default is a list of ones.
**Additional information:** This computation is sometimes called a *moving average* and is useful for smoothing the data. If an element of the list does not have sufficient neighbors on both sides of it, then it is not replaced with an average value. ♣ Missing data are included in sizing neighborhoods.
**See also:** Weight, describe[mean]


**transform[multiapply[fnc]]([list$_1$, ..., list$_n$])**

Applies the n-parameter function fnc to the corresponding elements from each of the m-element lists list$_1$ through list$_n$.
**Output:** An m-element list of expressions is returned.
**Argument options:** ([Slist$_1$, ..., Slist$_n$]) to apply a function to the elements of statistical lists Slist$_1$ through Slist$_n$.
**Additional information:** This command resembles the map command.
**See also:** transform[apply], *map*


**transform[remove[num]]([expr$_1$, ..., expr$_n$])**

Subtracts the numeric value num from each expression expr$_1$ through expr$_n$.
**Output:** A list of expressions is returned.
**Argument options:** (Slist) to subtract num from each element of statistical list Slist. ♣ transform[remove[option]]([expr$_1$, ..., expr$_n$]) or transform[remove[option]](Slist) to remove the value of the data represented by option. option can be any one of the descriptive names from the describe subpackage, such as *mean*, *median*, or *mode*.
**Additional information:** If there are any classes being transformed, num is subtracted from both boundaries of the class. ♣ Missing items are not changed.
**See also:** transform[apply], transform[divideby]

### transform[scaleweight[num]](Slist)

Scales the frequency of each value in Slist by a factor of num.
**Output:** A statistical list is returned.
**Additional information:** If, after scaling, the frequency of any value is not one, the scaled frequency is represented with a Weight command. If the scaled frequency is one, the value is not represented with a Weight command.
**See also:** Weight, transform[frequency], transform[tally]

### transform[split[n]]([expr$_1$, ..., expr$_n$])

Splits the expressions expr$_1$ through expr$_n$ into n statistical lists all with the same weight (number of elements).
**Output:** A list of statistical lists is returned.
**Argument options:** (Slist) to split up statistical list Slist.
**Additional information:** If necessary, the resulting lists will have non-integer weights. This is accomplished with adept use of the Weight command. ♣ Missing items are included in the weights. ♣ For examples of splitting lists of unwieldy lengths, see the on-line help page.
**See also:** Weight

### transform[standardscore]([expr$_1$, ..., expr$_n$])

Replaces each expression expr$_1$ through expr$_n$ with its related standard score.
**Output:** A list of expressions is returned.
**Argument options:** (Slist) to determine standard scores in statistical list Slist. ♣ transform[standardscore[0]](Slist) to treat Slist as the entire population. This is the default. ♣ transform[standardscore[1]](Slist) to treat Slist as a sample of the population.
**Additional information:** The standard score of element expr$_i$ is (expr$_i$ - mean(data))/standarddeviation(data). ♣ The order of the elements is preserved. ♣ Missing elements are unchanged.
**See also:** transform[moving], describe[mean], describe[standarddeviation]

### transform[statsort](Slist)

Sorts the elements is statistical list Slist in ascending order.
**Output:** A statistical list is returned.
**Additional information:** If there are overlapping classes or the command is unable to order two or more quantities, an error message is returned. ♣ To sort a list with purely numeric values, use sort. ♣ The class mark of a particular class range is its midpoint.

♣ All missing data is sorted to the end of the list.
**See also:** transform[tally], transform[frequency]

### transform[statvalue](Slist)

Replaces each Weight command in statistical list Slist with the corresponding unweighted value or range.
**Output:** A statistical list is returned.
**Additional information:** Basically, all Weight commands are eliminated.
**See also:** Weight, transform[cumulativefrequency], transform[frequency]

### transform[tally]([expr$_1$, ..., expr$_n$])

Replaces any elements with identical values with one grouped Weight command.
**Output:** A statistical list is returned.
**Argument options:** (Slist) to group elements in statistical list Slist with identical values. Ranges must be *exactly* identical to be grouped together.
**Additional information:** To tally the data into *specific* ranges, use the transform[tallyinto] command.
**See also:** transform[tallyinto], transform[frequency], transform[statsort]

### transform[tallyinto]([expr$_1$, ..., expr$_n$, [a$_1$..b$_1$, ..., a$_m$..b$_m$]])

Groups elements from expr$_1$ through expr$_n$ into weighted entries with ranges a$_1$..b$_1$ through a$_m$..b$_m$.
**Output:** A statistical list is returned.
**Argument options:** (Slist, [a$_1$..b$_1$, ..., a$_m$..b$_m$]) to group elements in statistical list Slist. ♣ transform[tallyinto[name]](Slist, [a$_1$..b$_1$, ..., a$_m$..b$_m$]) to place all the elements that do not fall into the specified ranges into a list assigned to name.
**Additional information:** A value expr falls in a range a..b only if a$\leq$expr$<$b. ♣ If you do not specify the name parameter and there are values that do not fall into the specified ranges, an error message is produced.
**See also:** Weight, transform[tally], transform[statsort]

# Standard Functions and Constants

## Introduction

### Standard Functions

Maple contains dozens of predefined commands for representing values of well-known (and lesser-known) mathematical functions at particular points. Among these are included such functions as $\sin x$, $\arctan x$, $\exp x$, the error function, and the sine integral.

For the most part, if you provide one of these commands with a real (non-decimal) numeric value, the function remains unevaluated. This does not mean that Maple is unable to compute a floating-point value, but merely that Maple is allowing you to carry the *exact* value through subsequent calculations. In a few cases where exact results, which differ from the unevaluated representation, can be obtained, real values are returned. A few examples are worth a thousand words.

```
> sin(6*Pi/19);
```

$$sin\left(\frac{6}{19}\pi\right)$$

```
> arctan(2/3);
```

$$arctan\left(\frac{2}{3}\right)$$

```
> exp(3);
```

$$e^3$$

```
> erf(12);
```

$$erf(12)$$

## 332 Functions and Constants

```
> Si(-3);
```

$$-Si(3)$$

```
> sin(Pi/3);
```

$$\frac{1}{2}\sqrt{3}$$

```
> exp(0);
```

$$1$$

Of course, given Maple's *symbolic* nature, you can also provide standard functions with expressions containing unassigned variable names. In these cases, the command is almost always left unevaluated. If, at a later time, you assign values to the variables, these unevaluated expressions may automatically evaluate.

```
> cosh(a+b);
```

$$cosh(a+b)$$

```
> mycall := sin(expand((x+3)^4));
```

$$mycall := sin(x^4 + 12x^3 + 54x^2 + 108x + 81)$$

```
> x := .0123:
> mycall;
```

$$.6093043741$$

```
> x := 'x':
```

One way to get standard functions to evaluate automatically to a floating-point value is to provide them with floating-point input in the first place. Another way to achieve the same results is to apply evalf to the unevaluated command call.

```
> cos(3.141);
```

$$-.9999998244$$

> BesselK(1., 2.45);

$$.07864292861$$

> ln(3/2);

$$ln\left(\frac{3}{2}\right)$$

> evalf(");

$$.4054651081$$

Many standard functions also accept complex values as input. As with real-valued input, if the components (i.e. the real and imaginary parts) of the complex value are nondecimal, then the functions tend to remain unevaluated.

> temp := ln(1+2*I);

$$temp := ln(1+2\,I\,)$$

At this point, there are a couple of options for handling the unevaluated command call. Firstly, you can call evalc (evaluate over the complex field) to express the command call in terms of a complex value whose components are exact values (but may contain unevaluated calls to other functions).

> evalc(temp);

$$\frac{1}{2}ln(\,5\,) + I\,arctan(\,2\,)$$

Similarly, if you want the components to be floating-point values, you can wrap evalf around the unevaluated command call.

> evalf(temp);

$$.8047189562 + 1.107148718\,I$$

And, of course, placing floating-point values in the original input achieves the same goal.

> sin(3.4 - 2.112*I);

$$-1.071451899 + 3.936685520\,I$$

## Predefined Constants

Mathematics is full of specific values with special meanings. As outlined in the *Getting Started With Maple* chapter, many of these constants have special representations in Maple. Several are irrational in nature (i.e., with no exact rational representation) and like standard functions with exact parameters, they remain unevaluated unless otherwise requested.

> 3*Pi;

$$3\pi$$

> gamma/Catalan;

$$\frac{\gamma}{Catalan}$$

> evalf(");

$$.6301717756$$

There are a few other "constants" that don't quite fit this mold; they don't evaluate to anything other than themselves. Those values are true, false, I, and infinity.

## Command Listing

### abs(expr)

Represents the absolute value of expr.
**Output:** If expr is a real value, its absolute value is returned. If expr is a complex value, the *norm* of the value is returned. Otherwise, an unevaluated abs command is returned.
**See also:** *evalc*, signum
**FL = 12  *LI = 5***

### Ai(expr)

Represents the Airy wave function Ai, one of the linearly independent solutions for $w$ in the equation $w'' - wz = 0$.
**Output:** An unevaluated Ai command is returned.
**Additional information:** To evaluate to a floating-point value, use evalf. Unlike other standard functions, using a floating-point value

for input is not enough to cause evaluation to occur. ♣ For more information, see the on-line help page.
**See also:** Bi, *evalf*
***LI* = 5**

### arccos(expr)

Represents the inverse cosine of expr.
**Output:** If expr is a floating-point value, a floating-point value (in radians) is returned. If expr is a complex value with floating-point components, a complex value is returned. If expr is a special rational or irrational value (e.g., 0, $1/\sqrt{2}$), a multiple of $\pi$ is returned. Otherwise, an unevaluated arccos command is returned.
**Additional information:** To convert from radians to degrees, use convert(*degrees*). ♣ The accuracy to which the inverse cosine is calculated depends on global variable Digits.
**See also:** arcsin, arctan, arccsc, arccosh, cos

### arccosh(expr)

Represents the inverse hyperbolic cosine of expr.
**Output:** If expr is a floating-point value, a floating-point value (in radians) is returned. If expr is a complex value with floating-point components, a complex value is returned. If expr is a special rational or irrational value (e.g., 0, 1), a multiple of $\pi$ is returned. Otherwise, an unevaluated arccosh command is returned.
**Additional information:** To convert from radians to degrees, use convert(*degrees*). ♣ The accuracy to which the inverse hyperbolic cosine is calculated depends on global variable Digits.
**See also:** arcsinh, arctanh, arccsch, arccos, cosh

### arccot(expr)

Represents the inverse cotangent of expr.
**Output:** If expr is a floating-point value, a floating-point value (in radians) is returned. If expr is a complex value with floating-point components, a complex value is returned. If expr is a special rational or irrational value (e.g., 0, 1), a multiple of $\pi$ is returned. Otherwise, an unevaluated arccot command is returned.
**Additional information:** To convert from radians to degrees, use convert(*degrees*). ♣ The accuracy to which the inverse cotangent is calculated depends on global variable Digits.
**See also:** arcsec, arccsc, arctan, arctanh, tan

### arccoth(expr)

Represents the inverse hyperbolic cotangent of expr.
**Output:** If expr is a floating-point value, a floating-point value (in radians) is returned. If expr is a complex value with floating-point components, a complex value is returned. If expr is a special rational or irrational value (e.g., $0, \infty$), a multiple of $\pi$ is returned. Otherwise, an unevaluated arccoth command is returned.
**Additional information:** To convert from radians to degrees, use convert(*degrees*). ♣ The accuracy to which the inverse hyperbolic cotangent is calculated depends on global variable Digits.
**See also:** arcsech, arccsch, arctanh, arctan, tanh

### arccsc(expr)

Represents the inverse cosecant of expr.
**Output:** If expr is a floating-point value, a floating-point value (in radians) is returned. If expr is a complex value with floating-point components, a complex value is returned. If expr is a special rational or irrational value (e.g., $1, \infty$), a multiple of $\pi$ is returned. Otherwise, an unevaluated arccsc command is returned.
**Additional information:** To convert from radians to degrees, use convert(*degrees*). ♣ The accuracy to which the inverse cosecant is calculated depends on global variable Digits.
**See also:** arcsec, arccot, arccos, arccsch, csc

### arccsch(expr)

Represents the inverse hyperbolic cosecant of expr.
**Output:** If expr is a floating-point value, a floating-point value (in radians) is returned. If expr is a complex value with floating-point components, a complex value is returned. If expr is a special rational or irrational value (e.g., $\infty, -\infty$), a multiple of $\pi$ is returned. Otherwise, an unevaluated arccsch command is returned.
**Additional information:** To convert from radians to degrees, use convert(*degrees*). ♣ The accuracy to which the inverse hyperbolic cosecant is calculated depends on global variable Digits.
**See also:** arcsech, arccoth, arccosh, arccsc, csch

### arcsec(expr)

Represents the inverse secant of expr.
**Output:** If expr is a floating-point value, a floating-point value (in radians) is returned. If expr is a complex value with floating-point components, a complex value is returned. If expr is a special rational or irrational value (e.g., $1, \infty$), a multiple of $\pi$ is returned.

Otherwise, an unevaluated arcsec command is returned.
**Additional information:** To convert from radians to degrees, use convert(*degrees*). ♣ The accuracy to which the inverse secant is calculated depends on global variable Digits.
**See also:** arccsc, arccot, arcsin, arcsech, sec

### arcsech(expr)

Represents the inverse hyperbolic secant of expr.
**Output:** If expr is a floating-point value, a floating-point value (in radians) is returned. If expr is a complex value with floating-point components, a complex value is returned. If expr is a special rational or irrational value (e.g., $1, \infty$), a multiple of $\pi$ is returned. Otherwise, an unevaluated arcsech command is returned.
**Additional information:** To convert from radians to degrees, use convert(*degrees*). ♣ The accuracy to which the inverse hyperbolic secant is calculated depends on global variable Digits.
**See also:** arccsch, arccoth, arcsinh, arcsec, sech

### arcsin(expr)

Represents the inverse sine of expr.
**Output:** If expr is a floating-point value, a floating-point value (in radians) is returned. If expr is a complex value with floating-point components, a complex value is returned. If expr is a special rational or irrational value (e.g., $0, 1/\sqrt{2}$), a multiple of $\pi$ is returned. Otherwise, an unevaluated arcsin command is returned.
**Additional information:** To convert from radians to degrees, use convert(*degrees*). ♣ The accuracy to which the inverse sine is calculated depends on global variable Digits.
**See also:** arccos, arctan, arcsinh, sin

### arcsinh(expr)

Represents the inverse hyperbolic sine of expr.
**Output:** If expr is a floating-point value, a floating-point value (in radians) is returned. If expr is a complex value with floating-point components, a complex value is returned. If expr is a special rational or irrational value (e.g., 0), a multiple of $\pi$ is returned. Otherwise, an unevaluated arcsinh command is returned.
**Additional information:** To convert from radians to degrees, use convert(*degrees*). ♣ The accuracy to which the inverse hyperbolic sine is calculated depends on global variable Digits.
**See also:** arccosh, arctanh, arcsin, sinh

### arctan(expr)

Represents the inverse tangent of expr.

**Output:** If expr is a floating-point value, a floating-point value (in radians) is returned. If expr is a complex value with floating-point components, a complex value is returned. If expr is a special rational or irrational value (e.g., $0, \infty$), a multiple of $\pi$ is returned. Otherwise, an unevaluated arctan command is returned.

**Additional information:** To convert from radians to degrees, use convert(*degrees*). ♣ The accuracy to which the inverse tangent is calculated depends on global variable Digits.

**See also:** arcsin, arccos, arctanh, tan

### arctanh(expr)

Represents the inverse hyperbolic tangent of expr.

**Output:** If expr is a floating-point value, a floating-point value (in radians) is returned. If expr is a complex value with floating-point components, a complex value is returned. If expr is a special rational or irrational value (e.g., $0, \infty$), a multiple of $\pi$ is returned. Otherwise, an unevaluated arctanh command is returned.

**Additional information:** To convert from radians to degrees, use convert(*degrees*). ♣ The accuracy to which the inverse hyperbolic tangent is calculated depends on global variable Digits.

**See also:** arcsinh, arccosh, arctan, tanh

### BesselI(expr$_1$, expr$_2$)

Represents the *modified* Bessel function of the first kind with order expr$_1$ and argument expr$_2$.

**Output:** An unevaluated BesselI command is returned.

**Additional information:** If expr$_1$ and expr$_2$ are real values, calling evalf on the unevaluated BesselI function provides a numerical result. ♣ BesselI is known to many other Maple commands, such as diff, int, etc.

**See also:** BesselJ, BesselK, BesselY

*LI = 15*

### BesselJ(expr$_1$, expr$_2$)

Represents the Bessel function of the first kind with order expr$_1$ and argument expr$_2$.

**Output:** An unevaluated BesselJ command is returned.

**Additional information:** If expr$_1$ and expr$_2$ are real values, calling evalf on the unevaluated BesselJ function provides a numerical result. ♣ BesselJ is known to many other Maple commands, such as diff, int, etc.

**See also:** BesselI, BesselK, BesselY

**FL = 206** *LI = 16*

Functions and Constants 339

### BesselK(expr$_1$, expr$_2$)

Represents the *modified* Bessel function of the second kind with order expr$_1$ and argument expr$_2$.
**Output:** An unevaluated BesselK command is returned.
**Additional information:** If expr$_1$ and expr$_2$ are real values, calling evalf on the unevaluated BesselK function provides a numerical result. ♣ BesselK is known to many other Maple commands, such as diff, int, etc.
**See also:** BesselI, BesselJ, BesselY
*LI = 15*


### BesselY(expr$_1$, expr$_2$)

Represents the Bessel function of the second kind with order expr$_1$ and argument expr$_2$.
**Output:** An unevaluated BesselY command is returned.
**Additional information:** If expr$_1$ and expr$_2$ are real values, calling evalf on the unevaluated BesselY function provides a numerical result. ♣ BesselY is known to many other Maple commands, such as diff, int, etc.
**See also:** BesselI, BesselJ, BesselK
*LI = 16*


### Beta(expr$_1$, expr$_2$)

Represents the Beta function value at expr$_1$ and expr$_2$.
**Output:** If expr$_1$ and expr$_2$ are rational values, an exact value may be returned. If either expr$_1$ or expr$_2$ is a floating-point value, a floating-point value is returned. If either expr$_1$ or expr$_2$ is a complex value with floating-point components, a complex value is returned. Otherwise, an unevaluated Beta command is returned.
**Additional information:** Beta(expr$_1$, expr$_2$) is defined as (GAMMA(expr$_1$) * GAMMA(expr$_2$)) / GAMMA(expr$_1$ + expr$_2$).
♣ Certain values cause GAMMA to run into a singularity and fail.
♣ The accuracy to which Beta is calculated depends on global variable Digits.
**See also:** GAMMA, lnGAMMA, Psi, W


### Bi(expr)

Represents the Airy wave function Bi, one of the linearly independent solutions for $w$ in the equation $w'' - wz = 0$.
**Output:** An unevaluated Bi command is returned.
**Additional information:** To evaluate the Bi command to a floating-point value, use evalf. Unlike other standard functions, using a

floating-point value for input is not enough to cause evaluation to occur. For more information on this Airy wave functions, see the on-line help page.
**See also:** Ai, *evalf*
*LI = 5*

### Catalan

Represents Catalan's constant.
**Output:** This constant is left unevaluated unless acted upon by a command such as evalf.
**Additional information:** Catalan is defined as sum((-1)^i/(2*i+1)^2,i=0..infinity) and has a value of approximately .9159655942.

### Chi(expr)

Represents the hyperbolic cosine integral of expr.
**Output:** If expr is a floating-point value, a floating-point value is returned. If expr is a complex value with floating-point components, a complex value is returned. Otherwise, an unevaluated Chi command is returned.
**Additional information:** Chi must be defined with readlib(Chi) before it can be used. ♣ Chi(expr) is defined as gamma + ln(expr) + int((cosh(t)-1)/t, t=0..expr). ♣ The accuracy to which the hyperbolic cosine integral is calculated depends on global variable Digits.
**See also:** Si, Ci, Ssi, Shi, Ei, Li

### Ci(expr)

Represents the cosine integral of expr.
**Output:** If expr is a floating-point value, a floating-point value is returned. If expr is a complex value with floating-point components, a complex value is returned. Otherwise, an unevaluated Ci command is returned.
**Additional information:** Ci(expr) is defined as gamma + ln(I*expr) - I*Pi/2 + int((cos(t)-1)/t, t=0..expr). ♣ The accuracy to which the cosine integral is calculated depends on global variable Digits.
**See also:** Si, Ssi, Shi, Chi, FresnelC, Ei, Li
*LI = 19*

### constants

Displays all currently defined global constants.
**Output:** An expression sequence is returned.

Functions and Constants 341

**Argument options:** constants := name$_1$ ..., name$_n$ to assign name$_1$ through name$_n$ as the newly defined global constants. ♣ constants := constants, name to add name as a new global constant.

**Additional information:** The initially defined global constants include Catalan, E, false, gamma, I, infinity, Pi, and true.

**See also:** Catalan, E, false, gamma, I, infinity, Pi, true

*LA = 45*

### convert(expr, *Ei*)

Converts expr, an expression containing trigonometric, hyperbolic, or logarithmic integrals, to an expression containing exponential integrals.

**Output:** An expression is returned.

**Argument options:** ([expr$_1$, ..., expr$_n$], *Ei*) or ({expr$_1$, ..., expr$_n$}, *Ei*) to create a list or set of converted expressions, respectively.

**See also:** simplify(*Ei*), Ei, Li, Si, Ci, SSi, Chi, Shi

### convert(expr, *erf*)

Converts expr, an expression containing the complementary error function erfc, into an equivalent expression in the error function erf.

**Output:** An expression is returned.

**Argument options:** ([expr$_1$, ..., expr$_n$], *erf*) or ({expr$_1$, ..., expr$_n$}, *erf*) to create a list or set of converted expressions, respectively. ♣ (expr$_{daw}$, *erf*) to convert expr$_{daw}$, an expression containing Dawson's integral command dawson, into an equivalent expression in terms of erf.

**See also:** convert(*erfc*), erf, erfc, dawson

### convert(expr, *erfc*)

Converts expr, an expression containing the error function erf, into an equivalent expression in the complementary error function erfc.

**Output:** An expression is returned.

**Argument options:** ([expr$_1$, ..., expr$_n$], *erfc*) or ({expr$_1$, ..., expr$_n$}, *erfc*) to create a list or set of converted expressions, respectively.

**See also:** convert(*erf*), erf, erfc, dawson

### convert(expr, *exp*)

Converts expr, an expression containing trigonometric functions, into an equivalent expression in exponential form.

**Output:** An expression is returned.

**Argument options:** ([expr$_1$, ..., expr$_n$], exp) or ({expr$_1$, ..., expr$_n$}, exp) to create a list or set of converted expressions, respectively.
**See also:** convert(*expsincos*), convert(*sincos*), convert(*ln*), convert(*expln*), convert(*trig*), exp, sin, cos, tan
**FL = 65**  *LI = 37*

### convert(expr, *expln*)

Converts all trigonometric functions in expr into exponential form and all inverse trigonometric functions into logarithmic form.
**Output:** An expression is returned.
**Argument options:** ([expr$_1$, ..., expr$_n$], expln) or ({expr$_1$, ..., expr$_n$}, expln) to create a list or set of converted expressions, respectively.
**See also:** convert(*exp*), convert(*ln*), exp, ln
*LI = 38*

### convert(expr, *expsincos*)

Converts all trigonometric functions in expr into terms of sin and cos, and all hyperbolic trigonometric functions into exponential form.
**Output:** An expression is returned.
**Argument options:** ([expr$_1$, ..., expr$_n$], expsincos) or ({expr$_1$, ..., expr$_n$}, expsincos) to create a list or set of converted expressions, respectively.
**See also:** convert(*exp*), convert(*sincos*), convert(*ln*), exp, sin, cos, tan, sinh, cosh, tanh
*LI = 38*

### convert(expr, *GAMMA*)

Converts all factorials, binomial coefficients, and multinomial coefficients in expr into terms of the GAMMA function.
**Output:** An expression is returned.
**Argument options:** ([expr$_1$, ..., expr$_n$], GAMMA) or ({expr$_1$, ..., expr$_n$}, GAMMA) to create a list or set of converted expressions, respectively.
**See also:** convert(*factorial*), binomial, GAMMA
*LI = 39*

### convert(expr, *ln*)

Converts all inverse trigonometric functions in expr into terms of the logarithmic function ln.
**Output:** An expression is returned.

**Argument options:** ([expr$_1$, ..., expr$_n$], *ln*) or ({expr$_1$, ..., expr$_n$}, *ln*) to create a list or set of converted expressions, respectively.
**See also:** convert(*exp*), convert(*expsincos*), convert(*sincos*), convert(*expln*), ln, arcsin, arccos, arctan
*LI = 43*

### convert(expr, *sincos*)

Converts all trigonometric functions in expr into terms of sin and cos, and all hyperbolic trigonometric functions into terms of sinh and cosh.
**Output:** An expression is returned.
**Argument options:** ([expr$_1$, ..., expr$_n$], *sincos*) or ({expr$_1$, ..., expr$_n$}, *sincos*) to create a list or set of converted expressions, respectively.
**See also:** convert(*exp*), convert(*expsincos*), convert(*ln*), convert(*tan*), exp, sin, cos, sinh, cosh

### convert(expr, *tan*)

Converts all trigonometric functions in expr into terms of tan.
**Output:** An expression is returned.
**Argument options:** ([expr$_1$, ..., expr$_n$], *tan*) or ({expr$_1$, ..., expr$_n$}, *tan*) to create a list or set of converted expressions, respectively.
**See also:** convert(*sincos*), sin, cos, tan
*FL = 64  LI = 54*

### convert(expr, *trig*)

Converts expr, an expression in exponential functions, into an equivalent expression in terms of trigonometric and hyperbolic trigonometric functions.
**Output:** An expression is returned.
**Additional information:** Euler's formula is used for the conversion. ♣ The resulting expression is passed through combine(*trig*).
**See also:** convert(*exp*), *combine*, exp, sin, cos, sinh, cosh
*LI = 55*

### cos(expr)

Represents the cosine of expr.
**Output:** If expr is a floating-point value or a simple fractional multiple of Pi (e.g., $2\pi$, $\pi/2$, $2\pi/3$), a numerical value is returned. If expr is a complex value with floating-point components, a complex value is returned. Otherwise, an unevaluated cos command is returned.

**Additional information:** expr is assumed to be given in radians. To convert from degrees to radians, use the convert(*radians*) command. ♣ The accuracy to which the cosine is calculated depends on global variable Digits.
**See also:** sin, tan, csc, cosh, arccos, Ci Chi

### cosh(expr)

Represents the hyperbolic cosine of expr.
**Output:** If expr is a floating-point value, a numerical value is returned. If expr is a complex value with floating-point components, a complex value is returned. Otherwise, an unevaluated cosh command is returned.
**Additional information:** expr is assumed to be given in radians. To convert from degrees to radians, use the convert(*radians*) command. ♣ The accuracy to which the hyperbolic cosine is calculated depends on global variable Digits.
**See also:** sinh, tanh, csch, cos, arccosh

### cot(expr)

Represents the cotangent of expr.
**Output:** If expr is a floating-point value or a simple fractional multiple of Pi (e.g., $\pi/2$, $2\pi/3$), a numerical value is returned. If expr is a complex value with floating-point components, a complex value is returned. Otherwise, an unevaluated cot command is returned.
**Additional information:** expr is assumed to be given in radians. To convert from degrees to radians, use the convert(*radians*) command. ♣ The accuracy to which the cotangent is calculated depends on global variable Digits.
**See also:** sec, csc, tan, tanh, arctan

### coth(expr)

Represents the hyperbolic cotangent of expr.
**Output:** If expr is a floating-point value, a numerical value is returned. If expr is a complex value with floating-point components, a complex value is returned. Otherwise, an unevaluated coth command is returned.
**Additional information:** expr is assumed to be given in radians. To convert from degrees to radians, use the convert(*radians*) command. ♣ The accuracy to which the hyperbolic cotangent is calculated depends on global variable Digits.
**See also:** sech, csch, tanh, tan, arctanh

### csc(expr)

Represents the cosecant of expr.

**Output:** If expr is a floating-point value or a simple fractional multiple of Pi (e.g., $\pi/2$, $2\pi/3$), a numerical value is returned. If expr is a complex value with floating-point components, a complex value is returned. Otherwise, an unevaluated csc command is returned.

**Additional information:** expr is assumed to be given in radians. To convert from degrees to radians, use the convert(*radians*) command. ♣ The accuracy to which the cosecant is calculated depends on global variable Digits.

**See also:** sec, cot, cos, csch, arccsc

### csch(expr)

Represents the hyperbolic cosecant of expr.

**Output:** If expr is a floating-point value, a numerical value is returned. If expr is a complex value with floating-point components, a complex value is returned. Otherwise, an unevaluated csch command is returned.

**Additional information:** expr is assumed to be given in radians. To convert from degrees to radians, use the convert(*radians*) command. ♣ The accuracy to which the hyperbolic cosecant is calculated depends on global variable Digits.

**See also:** sech, coth, cosh, csc, arccsch

### csgn(expr)

Represents the sign (i.e., *positive* or *negative*) of real or complex expression expr.

**Output:** If expr = 0, then 1 is returned. If Re(expr) > 0 (or Re(expr) = 0 and Im(expr) > 0), then 1 is returned. If Re(expr) < 0 (or Re(expr) = 0 and Im(expr) < 0), then −1 is returned. Otherwise, an unevaluated signum command is returned.

**Argument options:** (*1*, expr) to compute the derivative of csgn (expr). This is 0 for all complex values with non-zero real components and undefined along the imaginary axis.

**Additional information:** The value of csgn(expr) is used to help determine if certain symmetry transformations are automatically performed. ♣ If csgn is unable to correctly determine the sign of a complex value component, increasing the value of global variable Digits may help.

**See also:** *Digits*, signum, *sign*

### dawson(expr)

Represents Dawson's integral function at expr.
**Output:** If expr is a floating-point value, a floating-point value is returned. Otherwise, an unevaluated dawson command is returned.
**Additional information:** This command must be defined with readlib(dawson) before it can be used. ♣ dawson(expr) is defined by exp(-expr^2) * int(exp(t^2), t=0..expr).
**See also:** erf, erfc, FresnelC, FresnelS, Fresnelf, Fresnelg

### dilog(expr)

Represents the dilogarithm function at expr.
**Output:** If expr is a floating-point value, a floating-point value is returned. For certain rational values of expr, a value in terms of Pi and/or ln is returned. Otherwise, an unevaluated dilog command is returned.
**Additional information:** dilog(expr) is defined by int(ln(t)/(1-t), t=1..expr).
**See also:** ln, Pi

### Dirac(expr)

Represents the Dirac delta function of expr, an algebraic expression.
**Output:** If expr is a numeric value other than 0, then a value of 0 is returned. Otherwise, an unevaluated Dirac command is returned.
**Argument options:** (n, expr) to represent the $n^{th}$ derivative of the Dirac delta function at expr.
**Additional information:** The Dirac command is most often used in results of integral transforms commands, such as fourier, mellin, and laplace. ♣ The Heaviside command is similar to Dirac.
**See also:** Heaviside, fourier, mellin, laplace

### E

Represents the base of the natural logarithm.
**Output:** This constant is left unevaluated unless acted upon by a command such as evalf.
**Additional information:** It is recommended that you use the exp command instead of E whenever possible. ♣ E has a value of approximately 2.718281828.

### Ei(n, expr)

Represents the exponential integral, where expr is an expression and n is a non-negative integer.

**Output:** If expr is a floating-point value, a floating-point value is returned. If expr is a complex value with floating-point components, a complex value is returned. Otherwise, an unevaluated Ei command is returned.

**Argument options:** (expr) to find the one-argument exponential integral, which is a *Cauchy Principal Value* integral and is defined only for real expr.

**Additional information:** Ei(n, expr) is defined as int(exp(-expr*t)/t^n, t=1..infinity). ♣ The accuracy to which the exponential integral is calculated depends on the global variable Digits.

**See also:** Si, Ci, Ssi, Shi, Chi, Li, MeijerG

*LI = 71*


### erf(expr)

Represents the error function at expr.

**Output:** If expr is a floating-point value, a floating-point value is returned. If expr is a complex value with floating-point components, a complex value is returned. Otherwise, an unevaluated erf command is returned.

**Additional information:** The erf command is known to many manipulation commands, such as factor and simplify. ♣ erf(expr) is defined by 2/sqrt(Pi) * int(exp(-t^2), t=0..expr).

**See also:** convert(*erf*), convert(*erfc*), erfc, dawson

*LA = 6*


### erfc(expr)

Represents the complementary error function at expr.

**Output:** If expr is a floating-point value, a floating-point value is returned. If expr is a complex value with floating-point components, a complex value is returned. Otherwise, an unevaluated erfc command is returned.

**Argument options:** (int, expr), where int $\geq -1$, to find the iterated integral of the complementary error function at expr.

**Additional information:** The erfc command is known to many manipulation commands, such as expand and simplify. ♣ erfc(expr) is defined by 1 - 2/sqrt(Pi) * int(exp(-t^2), t=0..expr) or, equivalently, 1 - erf(expr). ♣ For more information, see the on-line help page.

**See also:** convert(*erf*), convert(*erfc*), erf, dawson

### exp(expr)

Represents the exponential value $e^{\text{expr}}$.

**Output:** If expr is a floating-point value, a floating-point value is returned. If expr is a complex value with floating-point components, a complex value is returned. Otherwise, an unevaluated exp command is returned.

**Additional information:** There is a special value E that stands for exp(1), but it is recommended that you use the exp command whenever possible. ♣ The accuracy to which the exponential function is calculated depends on global variable Digits.

**See also:** ln, log[expr], E
**FL = 13**


### false

Represents the *false* value in boolean contexts.

**Output:** This constant is left unevaluated.

**See also:** false, evalb, *and*, *or*, *not*, logic[&and], logic[&or], logic[&not]


### FresnelC(expr)

Represents the Fresnel cosine integral of expr.

**Output:** If expr is a floating-point value, a floating-point value is returned. If expr is infinity or -infinity, then a rational value is returned. Otherwise, an unevaluated FresnelC command is returned.

**Additional information:** FresnelC(expr) is defined as int(cos (Pi/2*t^2), t=0..expr). ♣ The accuracy to which the Fresnel cosine integral is calculated depends on global variable Digits.

**See also:** FresnelS, Fresnelf, Fresnelg, dawson, Ci
*LI = 93*


### Fresnelf(expr)

Represents one of two auxiliary Fresnel functions at expr.

**Output:** If expr is a floating-point value, a floating-point value is returned. If expr is infinity, then a value of 0 is returned. Otherwise, an unevaluated Fresnelf command is returned.

**Additional information:** Fresnelf must be defined with readlib (Fresnelf) before it can be used. ♣ The other auxiliary Fresnel function is Fresnelg. ♣ For the definition of the Fresnelf command, see the on-line help page.

**See also:** Fresnelg, FresnelC, FresnelS, dawson
*LI = 94*

### Fresnelg(expr)

Represents one of two auxiliary Fresnel functions at expr.

**Output:** If expr is a floating-point value, a floating-point value is returned. If expr is infinity, then a value of 0 is returned. Otherwise, an unevaluated Fresnelg command is returned.

**Additional information:** Fresnelg must be defined with readlib (Fresnelg) before it can be used. ✽ The other auxiliary Fresnel function is Fresnelf. ✽ For the definition of the Fresnelg command, see the on-line help page.

**See also:** Fresnelf, FresnelC, FresnelS, dawson

*LI = 94*

### FresnelS(expr)

Represents the Fresnel sine integral of expr.

**Output:** If expr is a floating-point value, a floating-point value is returned. If expr is infinity or -infinity, then a rational value is returned. Otherwise, an unevaluated FresnelS command is returned.

**Additional information:** FresnelS(expr) is defined as int(sin(Pi/2*t^2), t=0..expr). ✽ The accuracy to which the Fresnel sine integral is calculated depends on global variable Digits.

**See also:** FresnelC, Fresnelf, Fresnelg, dawson, Si

*LI = 94*

### gamma

Represents Euler's constant.

**Output:** This constant is left unevaluated unless acted upon by a command such as evalf.

**Argument options:** (n), where n is a non-negative integer, to represent the special gamma constant defined by limit(sum(ln(k)^n/k, k=1..n) - ln(m)^(n+1)/(n+1), m=infinity).

**Additional information:** gamma is defined by limit(sum(1/i, i=1..n) - ln(n), n=infinity). ✽ gamma has a value of approximately .5772156649.

**See also:** GAMMA

### GAMMA(expr)

Represents the Gamma function value at expr.

**Output:** If expr is a rational value, an exact value may be returned. If expr is a floating-point value, a floating-point value is returned. If expr is a complex value with floating-point components, a complex value is returned. Otherwise, an unevaluated GAMMA command is returned.

**Argument options:** ($expr_1$, $expr_2$) to represent the incomplete

Gamma function.
**Additional information:** GAMMA(expr) is defined as int(exp(-t)*t^(expr-1), t=0..infinity). ♣ Certain values of expr cause GAMMA to run into a singularity and fail. ♣ GAMMA(expr$_1$, expr$_2$) is defined as int(exp(-t)*t^(expr$_1$-1), t=expr$_2$..infinity). ♣ The accuracy to which the Gamma function is calculated depends on global variable Digits.
**See also:** lnGAMMA, Beta, Psi, MeijerG, hypergeom
**FL = 13, 47**


### harmonic(expr)

Represents the Harmonic function value at expr.
**Output:** An unevaluated harmonic command is returned.
**Additional information:** harmonic(expr) is defined as sum(1/i, i=1..n). ♣ To evaluate the harmonic command, use evalf.
**See also:** *evalf*
*LI = 103*


### Heaviside(expr)

Represents the Heaviside step function of expr, an algebraic expression.
**Output:** If expr is a numeric value less than 0, a value of 0 is returned. If expr is a numeric value greater than or equal to 0, a value of 1 is returned. Otherwise, an unevaluated Heaviside command is returned.
**Additional information:** The Heaviside command is most often used in results of integral transforms commands, such as fourier, mellin, and laplace. ♣ The Heaviside command is related to the Dirac command in that diff(Heaviside(t), t) = Dirac(t).
**See also:** Dirac, fourier, mellin, laplace


### hypergeom([exprn$_1$, ..., exprn$_m$], [exprd$_1$, ..., exprd$_n$], expr)

Represents the generalized hypergeometric function at expr with numerator and denominator coefficients represented by exprn$_1$ through exprn$_m$ and exprd$_1$ through exprd$_n$, respectively.
**Output:** In certain special cases, a rational expression in expr is returned. Otherwise, an unevaluated hypergeom command is returned.
**Additional information:** Another name for hypergeom is *Barnes' extended hypergeometric function*. ♣ The generalized hypergeometric function is defined as an infinite sum of products of GAMMA

**See also:** *convert(hypergeom)*, GAMMA
*LI = 293*

## I

Represents the square root of $-1$.
**Output:** This constant is left unevaluated.
**Additional information:** I is an alias for (-1)^(1/2). It is used primarily for representing complex values.
**See also:** evalc

## ilog(expr)

Represents the natural integer logarithm (i.e., to base $e$) of expr.
**Output:** If expr is a numeric value or a complex value with numeric components, an integer is returned. Otherwise, an unevaluated ilog command is returned.
**Additional information:** For integer logarithms to other bases, use ilog[expr] or ilog10.
**See also:** ln, ilog[expr], ilog10

## ilog[expr](expr$_x$)

Represents the integer logarithm to base expr of expr$_x$.
**Output:** If expr and expr$_x$ are numeric values, an integer is returned. Otherwise, an unevaluated ilog[expr] command is returned.
**Additional information:** ilog[expr] must be defined with readlib(ilog) before it can be used. ♣ For integer logarithms to other bases, use ilog or ilog10.
**See also:** log[expr], ilog, ilog10

## ilog10(expr)

Represents the integer logarithm to base 10 of expr.
**Output:** If expr is a numeric value or a complex value with numeric components, an integer is returned. Otherwise, an unevaluated ilog10 command is returned.
**Additional information:** For integer logarithms to other bases, use ilog[expr] or ilog.
**See also:** log10, ilog[expr], ilog

## infinity

Represents the infinite value, $\infty$.
**Output:** This constant is left unevaluated.

**Argument options:** -infinity to represent $-\infty$.
**Additional information:** infinity is most commonly used in specifying ranges to commands such as int, sum, and product or in specifying limiting values in commands such as limit.
**See also:** *int, limit, sum, product*

### Li(n, expr)

Represents the logarithmic integral of expr. If expr is a numeric value, it must be real and greater than 1.
**Output:** If expr is a floating-point value, a floating-point value is returned. If expr is a complex value with floating-point components, a complex value is returned. Otherwise, an unevaluated Li command is returned.
**Additional information:** Li must be defined with readlib(Li) before it can be used.
**See also:** Si, Ci, Ssi, Shi, Chi, Ei

### ln(expr)

Represents the natural logarithm (i.e., to base $e$) of expr.
**Output:** If expr is a floating-point value, a floating-point value is returned. If expr is a complex value with floating-point components, a complex value is returned. Otherwise, an unevaluated ln command is returned.
**Additional information:** For logarithms to other bases, use log[expr] or log10.
**See also:** exp, log[expr], log10, ilog
**FL = 13**

### lnGAMMA(expr)

Represents the logarithm of the Gamma function at expr.
**Output:** If expr is a floating-point value, a floating-point value is returned. If expr is a complex value with floating-point components, a complex value is returned. Otherwise, an unevaluated call to ln or an unevaluated lnGAMMA command is returned.
**Additional information:** lnGAMMA(expr) is defined as ln(GAMMA(expr)). ♣ Certain values of expr cause lnGAMMA to run into a singularity and fail. ♣ The accuracy to which the logarithm of the Gamma function is calculated depends on global variable Digits.
**See also:** GAMMA, Beta, Psi

### log(expr)

This command is identical to ln.
**See also:** ln

### log[expr](expr$_x$)

Represents the general logarithm to base expr of expr$_x$.
**Output:** If expr is a floating-point value, a floating-point value is returned. Otherwise, an unevaluated ratio ln(expr$_x$)/ln(expr) is returned.
**Additional information:** If no value is provided for expr, the default is exp(1). ♣ For natural and general logarithms, use ln and log10, respectively.
See also: exp, log10, ln, ilog[expr]

### log10(expr)

Represents the logarithm to base 10 of expr.
**Output:** If expr is a floating-point value, a floating-point value is returned. If expr is a complex value with floating-point components, a complex value is returned. Otherwise, an unevaluated ratio ln(expr)/ln(10) is returned.
**Additional information:** For logarithms to other bases, use log[expr] or ln.
See also: exp, log[expr], ln, ilog10

### MeijerG(n, expr$_1$, expr$_2$)

Represents a special case of the Meijer G function with respect to integer n ($> 2$), real expression expr$_1$, and real or complex expression expr$_2$.
**Output:** Typically, an unevaluated MeijerG command is returned.
**Additional information:** On occasion, when its parameters are in a special range, the MeijerG command simplifies to an expression in terms of GAMMA or Ei. ♣ For more information on these occasions and the general definition, see the on-line help page.
See also: GAMMA, Ei
*LI = 135*

### Pi

Represents the constant $\pi$.
**Output:** This constant is left unevaluated unless acted upon by a command such as evalf.
**Additional information:** Pi equals approximately 3.141592654.

### Psi(expr)

Represents the Psi (digamma) function value at expr.
**Output:** If expr is a rational value, an exact value may be returned. If expr is a floating-point value, a floating-point value is returned.

354  Functions and Constants

If expr is a complex value with floating-point components, a complex value is returned. Otherwise, an unevaluated Psi command is returned.

**Argument options:** (n, expr) to represent the $n^{th}$ polygamma function, which equals the $n^{th}$ derivative of the digamma function.

**Additional information:** Psi(expr) is defined as diff(GAMMA(expr), expr)/GAMMA(expr). ♣ Certain values of expr cause GAMMA to run into a singularity and fail. ♣ The accuracy to which the Psi function is calculated depends on global variable Digits.

**See also:** GAMMA, lnGAMMA, Beta, W, Zeta


### piecewise(expr$_{b1}$, expr$_1$, ..., expr$_{bn}$, expr$_n$)

Represents the piecewise function: if expr$_{b1}$ then expr$_1$ else if ... else if expr$_{bn}$ then expr$_n$. expr$_{b1}$ through expr$_{bn}$ are boolean expressions that evaluate to either true of false.

**Output:** An unevaluated piecewise command is returned.

**Argument options:** (int, expr$_{b1}$, expr$_1$, ..., expr$_{bn}$, expr$_n$) to specify whether the function is continuous at the joints. A value of *-1* for int specifies that the function is *not* continuous, while a non-negative value specifies that the function is continuous and remains continuous over int derivations. ♣ (expr$_{b1}$, expr$_1$, ..., expr$_{bn}$, expr$_n$, expr$_{n+1}$) to provide a final *else* expression in case all the expr$_{bi}$ fail.

**Additional information:** piecewise is similar to the *case* statement found in many other programming languages. ♣ Unevalauted piecewise commands can be manipulated by many other commands, such as diff and plot.

**See also:** *diff, plot, if/then/elif/then/else/fi*


### sec(expr)

Represents the secant of expr.

**Output:** If expr is a floating-point value or a simple fractional multiple of Pi (e.g., $\pi/2$, $2\pi/3$), then a numerical value is returned. If expr is a complex value with floating-point components, a complex value is returned. Otherwise, an unevaluated sec command is returned.

**Additional information:** expr is assumed to be given in radians. To convert from degrees to radians, use the convert(*radians*) command. ♣ The accuracy to which the secant is calculated depends on global variable Digits.

**See also:** csc, cot, sin, sech, arcsec

### sech(expr)

Represents the hyperbolic secant of expr.
**Output:** If expr is a floating-point value, a numerical value is returned. If expr is a complex value with floating-point components, a complex value is returned. Otherwise, an unevaluated sech command is returned.
**Additional information:** expr is assumed to be given in radians. To convert from degrees to radians, use the convert(*radians*) command. ♣ The accuracy to which the hyperbolic secant is calculated depends on global variable Digits.
**See also:** csch, coth, sinh, sec, arcsech

### Shi(expr)

Represents the hyperbolic sine integral of expr.
**Output:** If expr is a floating-point value, a floating-point value is returned. If expr is a complex value with floating-point components, a complex value is returned. Otherwise, an unevaluated Shi command is returned.
**Additional information:** Shi must be defined with readlib(Shi) before it can be used. ♣ Shi(expr) is defined as int(sinh(t)/t, t=0..expr). ♣ The accuracy to which the hyperbolic sine integral is calculated depends on global variable Digits.
**See also:** Si, Ci, Ssi, Chi, Ei, Li

### Si(expr)

Represents the sine integral of expr.
**Output:** If expr is a floating-point value, a floating-point value is returned. If expr is a complex value with floating-point components, a complex value is returned. Otherwise, an unevaluated Si command is returned.
**Additional information:** Si(expr) is defined as int(sin(t)/t, t=0..expr). ♣ The accuracy to which the sine integral is calculated depends on global variable Digits.
**See also:** Ci, Ssi, Shi, Chi, Ei, Li
*LI = 188*

### signum(expr)

Represents the sign (i.e., *positive* or *negative*) of expr.
**Output:** If expr is a positive real value, 1 is returned. If expr is a negative real value, $-1$ is returned. If expr is 0, 1 is returned. If expr is a complex value, the result of expr/abs(expr) is returned. Otherwise, an unevaluated signum command is returned.

**Argument options:** (1, expr) to compute the derivative of signum (expr). This is 0 for all nonzero real values. ♣ (0, expr, num) to set the value of signum(0) to num.
**Additional information:** The value of signum at (0) can also be set by assigning a new value to the global variable _Envsignum0. See the help page for signum to learn the importance of signum(0).
♣ If signum is unable to determine the sign of a real value, increasing the value of global variable Digits may help.
**See also:** *Digits*, csgn, *sign*
*LI = 189*

### simplify(expr, *Ei*)

Simplifies expr, an expression containing exponential integrals.
**Output:** An expression is returned.
**Additional information:** This command attempts to simplify calls to Ei into calls to the sine and the integrals, Si and Ci, respectively.
♣ For more information on the identities used in these simplifications, see the on-line help page.
**See also:** *simplify*, convert(*Ei*), Ei, Si, Ci

### simplify(expr, *GAMMA*)

Simplifies expr, an expression containing occurrences of GAMMA.
**Output:** An expression is returned.
**See also:** *simplify*, GAMMA
*LI = 192*

### simplify(expr, *hypergeom*)

Simplifies expr, an expression containing hypergeometric expressions.
**Output:** An expression is returned.
**Additional information:** Once you use simplify(*hypergeom*), all future instances of hypergeom are automatically simplified.
**See also:** *simplify*, hypergeom
*LI = 192*

### simplify(expr, *ln*)

Simplifies expr, an expression containing logarithms.
**Output:** An expression is returned.
**Additional information:** The simplifications performed are ln(a^r) → r*ln(a) and ln(a+b) → ln(a)+ln(b). ♣ Be warned that this command does not check whether or not these simplifications are valid

for particular values of a, b, and r; it just does them.
**See also:** *simplify, simplify(power), combine,* ln

### simplify(expr, *sqrt*)

Simplifies expr, an expression containing square roots or powers of square roots.
**Output:** An expression is returned.
**Additional information:** All square factors are removed from within the base(s). ♣ A square root can be specified with either the sqrt command or the appropriate use of the ^ operator.
**See also:** *simplify, simplify(RootOf),* sqrt
*LI = 195*

### simplify(expr, *trig*)

Simplifies expr, an expression containing trigonometric values.
**Output:** An expression is returned.
**Additional information:** The simplifications performed include sin(x)^2 + cos(x)^2 → 1 and cosh(x)^2 - sinh(x)^2 → 1. As well, all occurrences of tan(x) are simplified to sin(x)/cos(x).
**See also:** *simplify, combine,* sin, cos, sinh, cosh, tan
**FL = 39** *LI = 196*

### sin(expr)

Represents the sine of expr.
**Output:** If expr is a floating-point value or a simple fractional multiple of Pi (e.g., $2\pi$, $\pi/2$, $2\pi/3$), a numerical value is returned. If expr is a complex value with floating-point components, a complex value is returned. Otherwise, an unevaluated sin command is returned.
**Additional information:** expr is assumed to be given in radians. To convert from degrees to radians, use the convert(*radians*) command. ♣ The accuracy to which the sine is calculated depends on global variable Digits.
**See also:** cos, tan, sinh, arcsin, Si, Ssi, Shi

### sinh(expr)

Represents the hyperbolic sine of expr.
**Output:** If expr is a floating-point value, a numerical value is returned. If expr is a complex value with floating-point components, a complex value is returned. Otherwise, an unevaluated sinh command is returned.
**Additional information:** expr is assumed to be given in radians.

To convert from degrees to radians, use the convert(*radians*) command. ♣ The accuracy to which the hyperbolic sine is calculated depends on global variable Digits.
**See also:** cosh, tanh, sin, arcsinh

### sqrt(expr)

Represents the square root of algebraic expression expr.
**Output:** If expr contains a perfect square or a floating-point value as a term in a product, the square of that term is taken. If expr is a complex value with exact components, a complex value with components in square root notation is returned. If expr is a complex value with floating-point components, a complex value with floating-point components is returned. Otherwise, an unevaluated sqrt command is returned.
**Argument options:** (expr, *symbolic*) to force simplifications based on assumptions about the domain of expr. See the help page for sqrt for more information.
**See also:** *issqr, issqrfree*
*LI = 206*

### Ssi(expr)

Represents the shifted sine integral of expr.
**Output:** If expr is a floating-point value, a floating-point value is returned. If expr is a complex value with floating-point components, a complex value is returned. Otherwise, an unevaluated Ssi command is returned.
**Additional information:** Ssi must be defined with readlib(Ssi) before it can be used. ♣ Ssi(expr) is defined as Si(expr) - Pi/2. ♣ The accuracy to which the shifted sine integral is calculated depends on global variable Digits.
**See also:** Ci, Si, Shi, Chi, Ei, Li

### tanh(expr)

Represents the hyperbolic tangent of expr.
**Output:** If expr is a floating-point value, a numerical value is returned. If expr is a complex value with floating-point components, a complex value is returned. Otherwise, an unevaluated tanh command is returned.
**Additional information:** expr is assumed to be given in radians. To convert from degrees to radians, use the convert(*radians*) command. ♣ The accuracy to which the hyperbolic tangent is calculated depends on global variable Digits.
**See also:** sinh, cosh, tan, arctanh

## tan(expr)

Represents the tangent of expr.

**Output:** If expr is a floating-point value or a simple fractional multiple of Pi (e.g., $2\pi$, $\pi/2$, $2\pi/3$), a numerical value is returned. If expr is a complex value with floating-point components, a complex value is returned. Otherwise, an unevaluated tan command is returned.

**Additional information:** expr is assumed to be given in radians. To convert from degrees to radians, use the convert(*radians*) command. ♣ The accuracy to which the tangent is calculated depends on global variable Digits.

**See also:** sin, cos, tanh, arctan

*LA = 6*

## true

Represents the *true* value in boolean contexts.

**Output:** This constant is left unevaluated.

**See also:** false, evalb, *and*, *or*, *not*, *logic[&and]*, *logic[&or]*, *logic[&not]*

## type(expr, *arctrig*)

Determines whether expr is an unevaluated inverse trigonometric function.

**Output:** A boolean value of either true or false is returned.

**Argument options:** (expr, *arctrig*(var)) to determine if expr is an inverse trigonometric function in var.

**Additional information:** An expression is trigonometric if it is a function whose name is one of arcsin, arccos, arctan, arcsec, arccsc, arccot, arcsinh, arccosh, arctanh, arcsech, arccsch, or arccot.

**See also:** type(*trig*), type(*function*)

## type(expr, *evenfunc*(var))

Determines whether the parity of expr with respect to var is even.

**Output:** A boolean value of either true or false is returned.

**See also:** type(*function*), type(*oddfunc*)

## type(expr, *function*)

Determines whether expr is an unevaluated function (command) call.

**Output:** A boolean value of either true or false is returned.
**See also:** type(*evenfunc*), type(*oddfunc*)
*LA = 77−78*

### type(name, *mathfunc*)

Determines whether name represents one of Maple's initially known mathematical functions.
**Output:** A boolean value of either true or false is returned.
**Additional information:** name must not have a parameter sequence or parentheses following it for this command to evaluate to true. As well, composition of function names and expressions of function names are not allowed. ♣ The list of initially known functions includes Ai, Beta, exp, etc. For a complete listing, see the on-line help page for type,mathfunc.
*LI = 232*

### type(expr, *oddfunc*(var))

Determines whether the parity of expr with respect to var is odd.
**Output:** A boolean value of either true or false is returned.
**See also:** type(*function*), type(*evenfunc*)

### type(expr, *trig*)

Determines whether expr is an unevaluated trigonometric function.
**Output:** A boolean value of either true or false is returned.
**Argument options:** (expr, *trig*(var)) to determine if expr is a trigonometric function in var.
**Additional information:** An expression is trigonometric if it is a function whose name is one of sin, cos, tan, sec, csc, cot, sinh, cosh, tanh, sech, csch, or coth.
**See also:** type(*arctrig*), type(*function*)
*LI = 255*

### W(expr)

Represents the $W$ or omega function value at expr.
**Output:** If expr is a rational value, an exact value may be returned. If expr is a floating-point value, a floating-point value is returned. If expr is a complex value with floating-point components, a complex value is returned. Otherwise, an unevaluated W command is returned.
**Argument options:** (n, expr) to represent $W$ function at the $n^{th}$ branch. The default is the *principal* branch.
**Additional information:** The $W$ function satisfies the equation

W(expr)*exp(W(expr))=expr at all values of expr. ♣ The accuracy to which W is calculated depends on global variable Digits. ♣ For more information on the various branches, see the on-line help page.
**See also:** exp, Beta, Psi

## Zeta(expr)

Represents the Riemann Zeta function value at expr.
**Output:** If expr is a rational value, an exact value may be returned. If expr is a floating-point value, a floating-point value is returned. If expr is a complex value with floating-point components, a complex value is returned. Otherwise, an unevaluated Zeta command is returned.
**Argument options:** (n, expr) to represent $n^{th}$ derivative of the Zeta function. ♣ (n, expr$_1$, expr$_2$) to represent the special summation sum(1/((i+expr$_2$)^expr$_1$), i=0..infinity).
**Additional information:** The Zeta function is defined by sum(1/(i^expr), i=1..infinity). Because of this definition, an unevaluated Zeta command is sometimes returned in output to sum. ♣ The accuracy to which Zeta is calculated depends on global variable Digits.
**See also:** *sum*, Psi
*LA = 6 LI = 262*

# Expression Manipulation

## Introduction

Maple is capable of creating a wide variety of data structures. Because any of these structures could be returned to you as output to a command, it is essential you have a good grasp of the tools for working with them. In many cases, you need to transform or dissect a structure before it is valid to pass to another command.

An understanding of manipulation routines gives you a better feel for how Maple works, how it performs. With this knowledge, programming your own commands is infinitely easier. For more information on Maple programming, see the chapter *Programming and System Commands*.

## Dissecting Data Structures

Every Maple object has an internal structure which you can examine and from which you can extract the individual elements (operands) that make it up. There are several commands in Maple that allow manipulation of selected parts of objects. Being able to do this is particularly handy when dealing with large objects or objects that contain *one* crucial piece of information.

Internally, Maple breaks down each object into subobjects, which are then again broken down into even smaller subobjects, until basic elements are reached. Pictorially, this creates a branching, tree-like structure. The nops command tells you how many first-level operands there are in an object, and the op command displays these operands in the form of an expression sequence.

```
> object := 3*x^2 + 2*x - 3;
```

$$object := 3\,x^2 + 2\,x - 3$$

```
> nops(object);
```

$$3$$

```
> op(object);
```

$$3x^2, 2x, -3$$

op can also be used to select individual operands out of an object and, used recursively, can delve even deeper into an object.

```
> object := x^3*exp(1) - 34/Pi;
```

$$object := x^3 e - 34\frac{1}{\pi}$$

```
> op(1, object);
```

$$x^3 e$$

```
> op(1, op(1, object));
```

$$x^3$$

The subsop command allows you to change a specified operand of an expression for a new operand. The following example doubles the current numerator of a fraction (i.e., its first operand).

```
> myfrac := (a + b)/(c + d);
```

$$myfrac := \frac{a+b}{c+d}$$

```
> subsop(1=2*op(1,myfrac), myfrac);
```

$$\frac{2a+2b}{c+d}$$

# The Same, Yet Different—Transforming Expressions

It often occurs that Maple returns a result that is in a form other than you would prefer. For example, the result may be in terms

Manipulation 365

of exponentials when you would prefer trigonometric functions, or it may be unnecessarily complex in structure and need some simplification. At these times, you want to bring Maple's expression transformation routines to the job.

Most of these commands (e.g., factor, convert, simplify, etc.) do not change the basic value of an expression, they merely rearrange how the value is expressed.

> factor(x^3 - 2*x^2 - 5*x + 6);

$$(x-1)(x-3)(x+2)$$

> expand(");

$$x^3 - 2x^2 - 5x + 6$$

> simplify(exp(ln(b*sin(c))));

$$b\,sin(c)$$

> convert(Pi/12, degrees);

$$15\,degrees$$

> convert([[1,2,3], [4,5,6], [7,8,9]], matrix);

$$\begin{bmatrix} 1 & 2 & 3 \\ 4 & 5 & 6 \\ 7 & 8 & 9 \end{bmatrix}$$

Other commands *do* change the value of the expression (e.g., subs and certain convert commands). You must be very careful that you know beforehand whether a particular command is going to change the value of the expression you pass it.

> subs(x^2=y^4, x^2*z - y^3 + x^2 - 5);

$$y^4 z - y^3 + y^4 - 5$$

> convert(7*x^5 + x^3*y^4, mod2);

$$x + x\,y$$

A common characteristic of most transformation commands is that they automatically search the input for different types of expressions and contain unique routines for dealing with each type. This allows you to extend their functionality by programming subroutines to deal with your own data types. For more information on extending existing commands, see the introduction to the *Programming* chapter.

## Parameter Types Specific to This Chapter

bin          a binary value
oct          an octal value

## Command Listing

### asubs(expr$_{old}$ = expr$_{new}$, expr)

Substitutes expr$_{new}$ for expr$_{old}$ in expr, a sum of terms.
**Output:** An expression is returned.
**Argument options:** (expr$_{old}$ = expr$_{new}$, expr, *always*) to force asubs to occur on each and every sum in expr.
**Additional information:** Whereas subs depends solely on syntactic substitution (i.e., the existence of an *exact* replica of expr$_{old}$ in the internal data structure of expr$_{new}$), asubs has the ability to match up terms of a sum (and only a sum) that do not exist side by side in the internal representation. ♣ Using the *always* option forces substitution to occur even in places that may not seem apparent. ♣ Substitution is *not* followed by evaluation. To ensure evaluation, use eval.
**See also:** subs, op, subsop, trigsubs, eval

### collect(expr, var)

Collects like terms in expr, an expression in var.
**Output:** An expression is returned.
**Argument options:** (expr, [var$_1$, ..., var$_n$]) or (expr, {var$_1$, ..., var$_n$}) to collect like terms with respect to variables var$_1$ through var$_n$. ♣ (expr$_1$, expr$_2$) to collect like terms from expr$_1$ with respect to expr$_2$. expr$_2$ is typically a known function like sin(x) or exp(2*x). ♣ (expr, [var$_1$, ..., var$_n$], *distributed*) to collect terms in distributed fashion. ♣ (expr, [var$_1$, ..., var$_n$], *recursive*) to collect terms in recursive fashion. This is the default. ♣ (expr, [var$_1$,

..., $var_n$], option, fnc), where option is *distributed* or *recursive*, to apply fnc to each coefficient in the collected result.

**Additional information:** collect can also be applied to the components of a list, set, equation, series, range, relation, or function. ♣ Recursive collecting is done by first collecting the terms with respect to $var_1$, then collecting the coefficient produced with respect to $var_2$, and so on. ♣ To sort the resulting expression, use the sort command.

**See also:** factor, expand, combine, simplify, normal, sort

**FL = 59−60** *LI = 21*

### combine(expr)

Combines terms in expr that contain occurrences of certain known functions (e.g., sin, exp).

**Output:** An expression is returned.

**Argument options:** (expr, option) to combine terms with respect to certain areas only. The available options are *exp*, *ln*, *power*, *trig*, *radical*, and *Psi*. For more information on these transformations, see the on-line help page for combine[option]. ♣ (expr, *radical*, *symbolic*) to force combinations based on assumptions about the domain of expr. See the help pages for combine[radical] and sqrt for more information. ♣ (expr, *ln*, type) to specify logarithmic combinations are performed only if a coefficient is of type type.

**Additional information:** In many cases, combine is the inverse command to expand. For example, where expand takes $\sin(a+b)$ and transforms it to $\sin(a)\cos(b) + \cos(a)\sin(b)$, combine transforms $\sin(a)\cos(b) + \cos(a)\sin(b)$ to $\sin(a+b)$. ♣ combine also works on expressions containing unevaluated Diff, Int, Sum, Limit, and Product commands. ♣ collect can be applied to the components of a list, set, equation, series, range, relation, or function.

**See also:** factor, expand, collect, simplify, *Diff*, *Int*, *Sum*, *Limit*, *Product*

**FL = 56−58** *LI = 22−26*

### convert(expr, '+')

Converts expr to an expression equaling the sum of it operands.

**Output:** An expression of type '+' is returned.

**Additional information:** Regardless of what type of structure expr is, its first-level operands are determined and added together.

**See also:** convert('*'), *type('+')*, op

**FL = 91** *LI = 30*

### convert(expr, '*')

Converts expr to an expression equaling the product of its operands.
**Output:** An expression of type '*' is returned.
**Additional information:** Regardless of what type of structure expr is, its first-level operands are determined and multiplied together.
**See also:** convert('+'), *type('\*')*, op
**FL = 212** *LI = 30*

### convert(int, *base*, posint)

Converts decimal integer int into a list of integers representing its valuation in base posint.
**Output:** A list of integers is returned.
**Argument options:** ([int$_1$, ..., int$_n$], *base*, posint$_1$, posint$_2$) to convert the integer in base posint$_1$ represented by [int$_1$, ..., int$_n$] into a list of integers representing its valuation in base posint$_2$.
**Additional information:** If the resulting list is of length n, then it is interpreted as int = list[1]*posint^0 + list[2]*posint^1 + ... + list[n]*posint^(n-1).
**See also:** convert(*binary*), convert(*octal*), convert(*hex*), convert(*decimal*)
*LI = 31*

### convert(int, *binary*)

Converts decimal integer int into its binary equivalent.
**Output:** A integer in binary base is returned.
**Argument options:** (float, *binary*) to convert decimal floating-point value float into its binary equivalent. ♣ (float, *binary*, posint) to convert decimal floating-point value float into its binary equivalent with posint digits of precision. Default is the value of Digits.
**Additional information:** Binary numbers are returned as a decimal values using only the digits 0 and 1.
**See also:** convert(*base*), convert(*decimal*), convert(*octal*), convert(*hex*)
*LI = 32*

### convert(expr, *confrac*)

Converts algebraic expression expr to a continued fraction approximation.
**Output:** Typically, a list of partial quotients is returned.
**Argument options:** (num, *confrac*, n) to convert numeric value num to a continued fraction approximation with at most n partial

quotients. ♣ (num, *confrac*, name) or (num, *confrac*, n, name) to assign the list containing convergents of the continued fraction to name. ♣ (ratpoly, *confrac*, var) to compute the continued fraction approximation of ratpoly, a rational polynomial in var. ♣ (s, *confrac*) to compute the continued fraction approximation of series s.

**Additional information:** For more information on the many options to this command see the on-line help page for convert,confrac.
**See also:** *convergs*
*LI = 33*

### convert(bin, *decimal*, *binary*)

Converts binary value bin into its decimal equivalent.
**Output:** An integer in decimal base is returned.
**Argument options:** (oct, *decimal*, *octal*) to convert octal value oct into its decimal equivalent. ♣ (str, *decimal*, *hex*) to convert str, a string representing a hexadecimal value, into its decimal equivalent.
**Additional information:** Binary numbers must be represented as decimal values using only the digits 0 and 1. ♣ Octal numbers must be represented as decimal values using only the digits 0 through 7. ♣ Hexadecimal numbers must be represented as strings using only the characters 0 through 9 and $A$ through $F$.
**See also:** convert(*base*), convert(*binary*), convert(*octal*), convert(*hex*)
*LI = 34*

### convert(expr, *degrees*)

Converts expr, an expression in radians, to an expression in degrees.
**Output:** A product of an expression and the string degrees is returned.
**Additional information:** To compute the number of degrees, expr is multiplied by $180/\pi$.
**See also:** convert(*radians*)
*LI = 35*

### convert(float, *double*, option)

Converts floating-point value float into a double precision floating-point hexadecimal number in the format specified by option. option can be *ibm*, *mips*, or *vax*.
**Output:** A string representing the hexadecimal value is returned.
**Argument options:** (str, *double*, *maple*, option) to convert the

hexadecimal value represented by str from the option format, where option can be one of *ibm*, *mips*, or *vax*, to Maple floating-point format.

**Additional information:** Hexadecimal numbers must be represented as strings using only the characters 0 through 9 and $A$ through $F$.

**See also:** convert(*hex*)

*LI = 35*


### convert(expr, *factorial*)

Converts all GAMMA functions, binomial coefficients, and multinomial coefficients in expr into terms of factorials.

**Output:** An expression in factorials is returned.

**Argument options:** ([$expr_1$, ..., $expr_n$], *factorial*) or ({$expr_1$, ..., $expr_n$}, *factorial*) to create a list or set of converted expressions, respectively.

**See also:** *convert(binomial)*, *convert(GAMMA)*, *binomial*, *GAMMA*

**FL = 65** *LI = 38*


### convert(expr, *float*)

Converts all non-floating-point numeric values in expr into floating-point values.

**Output:** An expression is returned.

**Argument options:** ([$expr_1$, ..., $expr_n$], *float*) or ({$expr_1$, ..., $expr_n$}, *float*) to create a list or set of converted expressions, respectively.

**Additional information:** Basically, convert(*float*) is just a call to the evalf command. ♣ Floating-point values are calculated to an accuracy set by global variable Digits.

**See also:** *convert(rational)*, *evalf*, *Digits*

*LI = 39*


### convert(expr, *fraction*)

This command is identical to convert(*rational*).

**See also:** *convert(rational)*


### convert(int, *hex*)

Converts decimal integer value int into its hexadecimal equivalent.

**Output:** A string representing the value in hexadecimal base is returned.

**Additional information:** Hexadecimal numbers are returned as

Manipulation 371

strings using only the characters 0 through 9 and $A$ through $F$.
**See also:** convert(*base*), convert(*decimal*), convert(*binary*), convert(*octal*), convert(*double*)
*LI = 40*


### convert(A, *list*)

Converts one-dimensional array A into a list with identical elements.
**Output:** A list is returned.
**Argument options:** (V, *list*) to convert vector V to a list. ♣ (expr, *list*) to convert the operands of expression expr into a list. ♣ (M, *list*, *list*) to convert matrix M into a list of lists. See the listing for convert(*listlist*). ♣ (T, *list*, '=') to convert table T into a list of equations of form $index = element$. ♣ ({expr$_1$, ..., expr$_n$}, *list*) to create a list from expr$_1$ through expr$_n$.
**See also:** convert(*listlist*), convert(*set*)
**FL = 91** *LA = 17* *LI = 42*


### convert(A, *listlist*)

Converts the two-dimensional array A into a list of lists with identical elements.
**Output:** A list of lists is returned.
**Argument options:** ([eqn$_1$, ..., eqn$_n$], *listlist*) to convert the list of equations in the form $index = element$ into a list containing the elements sorted appropriately.
**Additional information:** The list of lists returned from an array is identical to the one that would have been used in defining the array.
**See also:** convert(*list*)
**FL = 91** *LA =* *LI = 42*


### convert(expr, *metric*)

Converts expr, an expression with terms modelled on num*unit (e.g., 34*feet), into terms of the metric system.
**Output:** An expression containing products of numeric values and metric units is returned.
**Argument options:** (expr, *metric*, *US*) to specify that expr is in terms of U.S. units. ♣ (expr, *metric*, *imp*) to specify that expr is in terms of imperial units. This is the default. ♣ ([expr$_1$, ..., expr$_n$], *metric*) or ({expr$_1$, ..., expr$_n$}, *metric*) to create a list or set of converted expressions, respectively.
**Additional information:** For a complete list of the types of units known to convert(*metric*), see the on-line help page.
*LI = 44*

### convert(expr, mod2)

Converts expr, an expression that may contain the boolean operators and, or, and not, into its modulo 2 form.
**Output:** An expression whose only numerical values are 0 or 1 is returned.
**See also:** logic[convert]
*LI = 44*


### convert(list, multiset)

Converts list into a list of two-elements lists. For each unique value in list there exists in the result a list with that value as the first element and the multiplicity of that value in list as the second element.
**Output:** A list of lists is returned.
**Argument options:** (T, multiset) to create internal lists where the indices of table T are the first elements and the values at those indices are the second elements. ♣ (expr, multiset) to create internal lists where the unique factors are the first elements and the multiplicities of those factors are the second elements.
**See also:** convert(set)
*LI = 45*


### convert(expr, name)

This command is identical to convert(string).
**See also:** convert(string)


### convert(int, octal)

Converts decimal integer value int into its octal equivalent.
**Output:** A integer in octal base is returned.
**Argument options:** (float, octal) to convert decimal floating-point value float into its octal equivalent. ♣ (float, octal, posint) to convert the decimal floating-point value float into its octal equivalent with posint digits of precision. Default is the value of Digits.
**Additional information:** Octal numbers are returned as a decimal values using only the digits 0 through 7.
**See also:** convert(base), convert(decimal), convert(binary), convert(hex)
*LI = 46*


### convert(expr, parfrac, var)

Converts rational expression expr to a partial fraction in var.
**Output:** An expression with fractional components is returned.

**Argument options:** (expr, *parfrac*, var, *true*) if the denominator of expr is already in the desired form, that is, normal has been applied to expr.
**Additional information:** Unless *true* is provided, factor is applied to the denominator of expr before partial fraction decomposition.
**See also:** normal, factor
*LI = 47*


### convert(cmplx, *polar*)

Converts complex expression cmplx into an expression in polar coordinates.
**Output:** An unevaluated polar command is returned.
**Additional information:** The polar command has two arguments, a modulus and an argument of the complex value of the expression, respectively.
**See also:** polar, *evalc*
*LI = 48*


### convert(expr, *radians*)

Converts expr, an expression in degrees (i.e., a product of an expression and the string degrees), to an expression in radians.
**Output:** An expression is returned.
**Additional information:** To compute the number of radians, expr is multiplied by $\pi/180$.
**See also:** convert(*degrees*)
*LI = 49*


### convert(expr, *radical*)

Converts all RootOf expressions in expr into their corresponding radical notations, if possible.
**Output:** An expression is returned.
**Additional information:** RootOf(_Z^2 + 1) is converted to I.
**See also:** convert(*RootOf*)


### convert(expr, *rational*)

Converts all floating-point values in expr into rational values.
**Output:** An expression is returned.
**Argument options:** (expr, *rational*, n) to calculate the rational values to n digits of accuracy. ♣ (expr, *rational*, *exact*) to calculate the rational values to infinite precision, if possible. ♣ ([expr$_1$, ..., expr$_n$], *rational*) or ({expr$_1$, ..., expr$_n$}, *rational*) to create a list or set of converted expressions, respectively.

**Additional information:** With just two parameters, rational values are calculated to an accuracy set by global variable Digits.
**See also:** convert(*float*), *Digits*
*LI = 50*

### convert(expr, RootOf)

Converts all radical expressions and occurrences of I in expr into their corresponding RootOf notations, if possible.
**Output:** An expression in RootOf notation is returned.
**Additional information:** I is converted to RootOf(_Z^2 + 1).
**See also:** convert(*radical*)
*LI = 52*

### convert(A, set)

Converts array A into a set with identical elements. A can be of any dimension.
**Output:** A set is returned.
**Argument options:** (list, *set*) to convert the operands of list into a set. ♣ (expr, *set*) to convert the operands of expression expr into a set. ♣ (T, *set*) to convert table T into a set containing the elements of but not the indices of T.
**See also:** convert(*list*), convert(*multiset*)
*LI = 53*

### convert(expr, string)

Converts expression expr into a string.
**Output:** A string is returned.
**Argument options:** ([$expr_1$, ..., $expr_n$], *string*) or ({$expr_1$, ..., $expr_n$}, *string*) to create a list or set of converted expressions, respectively.
**Additional information:** To convert a string back to an expression, use the parse command.
**See also:** *parse*
*LI = 46*

### denom(rat)

Returns the denominator of rational expression rat.
**Output:** An expression is returned.
**Additional information:** If rat is not in normal form, it is first normalized.
**See also:** numer, normal
*LI = 127*

## expand(expr)

Expands the expression expr.
**Output:** An expanded expression is returned.
**Argument options:** (expr, subexpr$_1$, subexpr$_2$, ..., subexpr$_n$) to expand expr but stipulate that subexpressions subexpr$_1$ through subexpr$_n$ are *not* to be expanded. ♣ (rat) to expand rat, a rational polynomial or expression. Only the numerator is expanded. ♣ (eqn) to expand both sides of an equation.♣ ([expr$_1$, ..., expr$_n$]) or ({expr$_1$, ..., expr$_n$}) to create a list or set of expanded expressions, respectively.
**Additional information:** Most standard mathematical functions are known to expand. For example, sin, exp, Psi, etc. all may be affected by expand. If you do not wish them to be expanded, add them to the excepted subexpressions, as detailed above. ♣ For a more permanent solution, see expandon and expandoff. ♣ To prevent *all* expansions from happening, use the frontend command.
**See also:** factor, collect, combine, simplify, normal, sort, expandon, expandoff, frontend, *Expand*
**FL = 54−55** *LA = 3* **LI = 90**

## Expand(expr)

Represents the inert expansion command for expr.
**Output:** An unevaluated Expand call is returned.
**Argument options:** Expand(expr) *mod* posint to evaluate the inert command call modulo posint.
**Additional information:** For more information on valid parameter sequences for Expand, see the command listing for expand. ♣ Three special methods can be called upon when Factor is combined with evala. See the listing for evala for more information.
**See also:** expand, *evala*
*LI = 89*

## expandoff(fnc)

Suppresses expansion of the command fnc in future calculations.
**Output:** A NULL expression is returned.
**Argument options:** (fnc$_1$, ..., fnc$_n$) to suppress expansion of commands fnc$_1$ through fnc$_n$. ♣ () to suppress expansion of *all* commands and expressions.
**Additional information:** Before expandoff can be used, it must be defined by the call expand(expandoff()). ♣ expand uses *remember tables*, so if expansion of a particular expression containing fnc is calculated prior to calling expandoff(fnc), then that ex-

pression will still expand after expandoff(fnc) is called. ♣ To reinstate expansion for a command, use the expandon command.
**See also:** expand, expandon, frontend, *remember*

### expandon(fnc)

Checks whether command fnc has had expansion suppressed, and reinstates expansion if necessary.
**Output:** A NULL expression is returned.
**Argument options:** ($fnc_1$, ..., $fnc_n$) to reinstate expansion for commands $fnc_1$ through $fnc_n$. ♣ () to reinstate expansion for *all* commands and expressions.
**Additional information:** Before expandon can be used, it must first be defined by the call expand(expandon()). ♣ To suppress expansion for a command, use expandoff.
**See also:** expand, expandoff, frontend

### isolate(eqn, expr)

Isolates expr, a subexpression contained in equation eqn, to the lefthand side of an equivalent equation, if possible.
**Output:** An equation is returned.
**Argument options:** ($expr_1$, $expr_2$) to isolate $expr_2$ in the equation $expr_1$ = 0. ♣ (eqn, expr, n) to specify that isolate performs at most n transformation steps in its calculation.
**Additional information:** This command needs to be defined by readlib(isolate) before being invoked.
**See also:** solve, collect, combine, simplify, normal, sort, expandon, expandoff, frontend, *Expand*
*LI = 298, 336*

### lhs(eqn)

Returns the lefthand side of the equation eqn.
**Output:** An expression is returned.
**Argument options:** (ineq) to return the lefthand side of an inequality. ♣ (range) to return the lefthand side of a range.
**Additional information:** In all the above cases, the result is equivalent to the first operand of the input.
**See also:** rhs, op
*LI = 127*

### nops(expr)

Determines the number of first-level operands (components) in expr.
**Output:** An integer value is returned.
**Argument options:** (*op*(n, expr)) to extract the number of suboperands in the $n^{th}$ operand of expr.

Manipulation 377

**Additional information:** expr can be *any* type of Maple object—everything has an internal form. ♣ To produce a sequence of the actual operands themselves, use the op command.
**See also:** op, subsop
**FL = 76** *LA = 17* **LI = 147**


### normal(rat)

Normalizes rational expression rat.
**Output:** An expression is returned.
**Argument options:** (rat, *expanded*) to fully expand the numerator and denominator of the result. ♣ ([rat$_1$, ..., rat$_n$]) or ({rat$_1$, ..., rat$_n$}) to create a list or set of normalized expressions, respectively.
**Additional information:** normal also works on each component of a range, series, or function. ♣ rat is converted into *factored normal form*, where numerator and denominator are relatively prime polynomials with integer coefficients. ♣ If the input contains subexpressions that are not rational (e.g., square roots, functions, or series), these subexpressions are ''frozen'' to unique names so normalization can proceed. This may cause some expressions that are equivalent to 0 to pass by unsimplified.
**See also:** factor, expand, simplify, frontend, denom, numer, *Normal*
**FL = 58**−**59** *LA = 3* **LI = 144**


### Normal(rat)

Represents the inert normalization command for rat, a rational expression.
**Output:** An unevaluated Normal call is returned.
**Argument options:** Normal(rat) *mod* posint to evaluate the inert command call modulo posint.
**Additional information:** For more information on valid parameter sequences for Normal, see the command listing for normal.
**See also:** normal, *evala*, *evalgf*
*LI = 143*


### numer(rat)

Returns the numerator of rational expression rat.
**Output:** An expression is returned.
**Additional information:** If rat is not in normal form, it is first normalized.
**See also:** denom, normal
*LI = 127*

## op(expr)

Displays the first-level operands (components) of expr.
**Output:** An expression sequence is returned.
**Argument options:** (n, expr) to extract the $n^{th}$ operand of expr.
♣ (a..b, expr) to extract operands a through b into an expression sequence. a and b must be valid integer values. ♣ (op(n, expr)) to extract the operands of the $n^{th}$ operand of expr.
**Additional information:** expr can be *any* type of Maple object—everything has an internal form. ♣ The expression sequence returned lists operands in the order that they appear in expr. ♣ Some data types (e.g., integers) have only one operand, while others have many operands. ♣ For most data types, the $0^{th}$ operand holds type information. For a few data types (e.g., series), the $0^{th}$ operand holds other special information.
**See also:** nops, subsop, *type*
**FL = 76–77** *LA = 17* **LI = *147***

## radnormal(expr)

Puts expr, an expression containing radicals, in normal form.
**Output:** An expression is returned.
**Additional information:** This command needs to be defined by readlib(radnormal) before being invoked.
**See also:** radsimp, simplify, rationalize

## radsimp(expr)

Simplifies expr, an expression containing radicals.
**Output:** An expression is returned.
**Argument options:** (expr, name) to assign the simplified expression with rationalized denominator to name.
**Additional information:** radsimp puts *unnested* radicals in expr in *normal form*. ♣ Branches of multi-valued expression are chosen in a consistent way; but the rules used may not be consistent with rules used by other Maple commands. This can cause some spurious results.
**See also:** radnormal, simplify, rationalize
*LI = 170*

## rationalize(expr)

Puts expr, an expression containing radicals, in rational form.
**Output:** An expression is returned.
**Additional information:** This command needs to be defined by readlib(rationalize) before being invoked.
**See also:** radsimp, simplify, radnormal

## rhs(eqn)

Returns the righthand side of equation eqn.
**Output:** An expression is returned.
**Argument options:** (ineq) to return the righthand side of an inequality. ♣ (range) to return the righthand side of a range.
**Additional information:** In all the above cases, the result is equivalent to the second operand of the input.
**See also:** lhs, op
**FL = 202  *LI = 127***

## simplify(expr)

Simplifies the expression expr.
**Output:** An expression is returned.
**Argument options:** (expr, option$_1$, option$_2$, ..., option$_n$) to simplify expr using only simplification procedures for functions of type option$_1$ through option$_n$. Valid options are *atsign*, *Ei*, *exp*, *GAMMA*, *hypergeom*, *ln*, *polar*, *power*, *radical*, *RootOf*, *sqrt*, and *trig*. See the individual command entries for each option for more details. ♣ (expr, {eqn$_1$, ..., eqn$_n$}) or (expr, [eqn$_1$, ..., eqn$_n$]) to simplify expr with respect to side relations eqn$_1$ through eqn$_n$. See the on-line help page for simplify[siderels] for more information. ♣ (expr, *assume=real*) or (expr, *assume=positive*) to assume that all the variable in expr are of the given type. ♣ ([expr$_1$, ..., expr$_n$]) or ({expr$_1$, ..., expr$_n$}) to create a list or set of simplified expressions, respectively.
**Additional information:** simplify also works on each component of a range, series, equation, or function. ♣ If simplify is called with only one parameter, a search of expr is automatically made to determine what type of components are present, and the appropriate simplification routines are called. ♣ Keep in mind that what is "simplified" can be highly subjective. One person's idea of what is simple may be another person's idea of what is elaborate.
**See also:** factor, expand, collect, combine, normal, convert, radsimp, *simplify(atsign)*, *simplify(Ei)*, *simplify(exp)*, *simplify(GAMMA)*, *simplify(hypergeom)*, *simplify(ln)*, simplify(*polar*), simplify(*power*), simplify(*radical*), simplify(*RootOf*), *simplify(sqrt)*, *simplify(trig)*
**FL = 39–40  *LI = 190, 194***

## simplify(expr, '@')

This command is identical to simplify(*atsign*).
**See also:** simplify(*atsign*)

### simplify(expr, *atsign*)

Simplifies expr, an expression containing the composition operator @.

**Output:** If a simplification is possible, a Maple function (i.e., in -> notation) is returned. Otherwise, expr is returned.

**Additional information:** This option to simplify is particularly useful for composition of inverses. If you wish to set a certain function $fnc_1$ as the inverse of another function $fnc_2$, create an entry in the invfunc table with invfunc[$fnc_2$] := $fnc_1$.

**See also:** simplify, simplify(@)

*LI = 191*


### simplify(expr, *polar*)

Simplifies expr, an expression containing complex valued expressions (i.e., occurrences of the polar command).

**Output:** An expression is returned.

**Additional information:** The simplifications performed are polar(a, b)^r → polar(a^r, r*b), if r is real and polar(a, b)*polar(c, d) → polar(a*c, b+d). ♣ As well, sums of polar commands with floating-point values are simplified.

**See also:** simplify, *polar*, *evalc*


### simplify(expr, *power*)

Simplifies expr, an expression containing powers, exponentials, and logarithms.

**Output:** An expression is returned.

**Additional information:** The basic simplifications for powers, exponentials, and logarithms are performed.

**See also:** simplify, simplify(*ln*), combine, *exp*, *ln*

**FL = 39**  *LI = 192*


### simplify(expr, *radical*)

Simplifies expr, an expression containing radicals.

**Argument options:** (expr, *radical*, *symbolic*) to force simplifications based on assumptions about the domain of expr. See the help pages for simplify[radical] and sqrt for more information.

**See also:** sqrt, simplify, combine, radsimp

*LI = 193*


### simplify(expr, *RootOf*)

Simplifies expr, an expression containing occurrences of RootOf.

**Output:** An expression is returned.

**Additional information:** Polynomials and inverse polynomials in RootOfs and $n^{th}$ roots of RootOfs are simplified.
**See also:** simplify, simplify(*sqrt*), *RootOf*
*LI = 193*

### sort(list)

Sort the elements of list in lexicographical order.
**Output:** A list is returned.
**Argument options:** (list, *lexorder*) or (list, *string*) to produce the same result. ♣ ([num$_1$, ..., num$_n$]), (list, '<'), or (list, *numeric*) to sort numeric values in numeric order. ♣ (list, *address*) to sort elements by their internal addresses. ♣ (list, *fnc*) to sort elements by the boolean valued function fnc, which takes two parameters and returns true if the first parameter precedes the second parameter in the ordering.
**Additional information:** For information on how sort works on polynomials, see the entry for sort(poly).
**See also:** *sort(poly)*
*LI = 204*

### subs(expr$_{old}$ = expr$_{new}$, expr)

Substitutes expr$_{new}$ for expr$_{old}$ in expr.
**Output:** An expression is returned.
**Argument options:** (eqn$_1$, eqn$_2$, ..., eqn$_n$, expr), where the eqn$_i$ are of the form exprold$_i$ = exprnew$_i$, to substitute for n different subexpressions of expr *sequentially*. ♣ ({eqn$_1$, ..., eqn$_n$}, expr) or ([eqn$_1$, ..., eqn$_n$], expr) to substitute for n different subexpressions of expr *simultaneously*.
**Additional information:** Because of the nature of Maple's data storage, subs is not always guaranteed to find expr$_{old}$ in expr. ♣ expr$_{old}$ must be a operand (on any level) of expr to be found by subs. This is called *syntactic substitution*. ♣ All occurrences of expr$_{old}$ found in expr are replaced. ♣ Substitution is *not* followed by evaluation. To ensure evaluation, use eval. ♣ To substitute for one first-level occurrence of expr$_{old}$, use subsop.
**See also:** op, subsop, asubs, trigsubs, eval
**FL = 19−21** *LA = 22, 80−82* *LI = 206*

### subsop(n = expr$_{new}$, expr)

Substitutes expr$_{new}$ for the $n^{th}$ first-level operand of expr. n must represent a valid operand of expr.
**Output:** An expression is returned.
**Argument options:** (eqn$_1$, eqn$_2$, ..., eqn$_m$, expr), where the eqn$_i$

are of the form $n_i = \text{expr}_i$, to substitute for m different operands of expr simultaneously.

**Additional information:** subsop allows you to substitute for a subexpression without affecting every other identical subexpression in expr. subs cannot do this.

**See also:** op, nops, subs

**FL = 81–83** *LA = 82–83* **LI = 207**

### trigsubs(expr$_t$)

Returns all trigonometric expression, equivalent to the trigonometric expression expr$_t$, from the internal table of identities.

**Output:** A list of trigonometric expression is returned.

**Argument options:** (eqn), where eqn is a trigonometric identity, to determine if eqn is currently in the internal table of identities. If it is, the string 'found' is returned. Otherwise, the string 'not found' is returned. ♣ (eqn, expr$_t$), where eqn is a trigonometric identity and expr$_t$ is a trigonometric expression, to apply the identity eqn to expr. If eqn is not in the internal table, an error message is returned. ♣ (0) to return the set of trigonometric functions currently known to trigsubs.

**Additional information:** This command needs to be defined by readlib(trigsubs) before being invoked.

**See also:** subs, asubs, subsop

*LI = 326*

# Plotting

## Introduction

One of the most interesting uses of computer algebra systems is scientific visualization. Displaying expressions pictorially extends our understanding of their nature and significance. Maple provides a large range of plotting commands that accommodate many of your visualization needs.

In plotting, the basic concept is to provide an expression in one or two unknowns and ranges over which *each* unknown is evaluated. Maple then samples a meaningful set (or grid) of points and displays the results.

Of course, the quality with which images appear on your screen are directly proportionate to the graphics capabilities of your monitor. When Maple is installed, the device type for your monitor should be set for you, making the transition to plotting as seamless as possible. If you have any problems viewing the plots in this chapter, make sure that your terminal is properly configured in your Maple session. Also, having many plot windows open at one time can seriously limit your system's available resources.

## Two-Dimensional Plots

Expressions in one unknown can be plotted using the plot command. There are many types of two-dimensional plots available. Here are a few of the more frequently-used ones.

Expressions can be plotted.

```
> plot(x^2-3, x=-4..4);
```

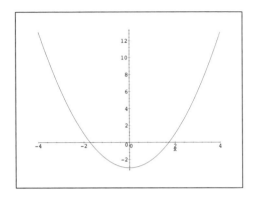

```
> plot(x*sin(x), x=-3*Pi..3*Pi);
```

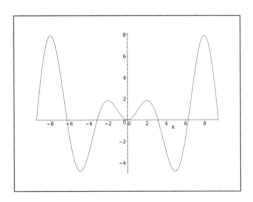

Parametric expressions can be plotted.

```
> plot([sin(t), cos(t), t=0..2*Pi]);
```

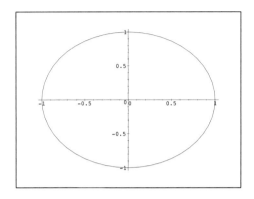

Lists of $x$ and $y$ values can be joined by lines.

```
> plot([3, 3, 6, 0, 3, -3, 0, 0, 3, 3], x=-2..10);
```

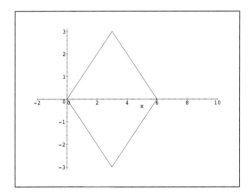

## 386 Plotting

Multiple functions can be plotted on one set of axes.

```
> plot({x^2, exp(x), x}, x=0..3);
```

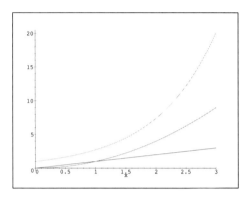

```
> plot({seq(cos(x*i), i=1..4)}, x=-Pi..Pi);
```

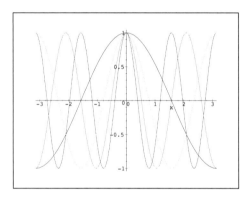

Besides these standard plots, several other types of two-dimensional plots are available in the plots package.

Polar plots can be specified.

```
> plots[polarplot](2*t);
```

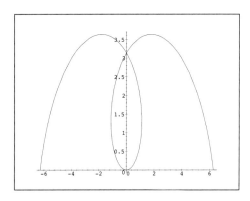

Two-dimensional vector fields can be plotted.

```
> plots[fieldplot]([sin(y),cos(x)], x=-10..10, y=-10..10,
> arrows=SLIM);
```

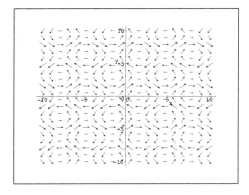

And implicit functions can be plotted in two dimensions.

```
> plots[implicitplot](x^2/25+y^2/9=1, x=-6..6, y=-6..6,
> scaling=CONSTRAINED);
```

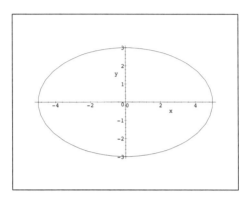

Several optional parameters are available for two-dimensional plots. The *numpoints* option allows you to specify that more sample points be taken, giving a smoother curve. The default is 49. You can specify what color the plot should be drawn in with the *color* option. The *axes* option allows you to specify what type of axes (*FRAME*, *BOXED*, *NORMAL*, or *NONE*) are used. With the *xtickmarks* and *ytickmarks* options, you can control the number of marks that Maple uses along each axis. The *style* option allows you to choose between different styles of interpolation between sampled points (e.g., *LINE*, *POINT*), while the *symbol* option allows you to choose the type of point symbol to be used. A title can be added to the plot with the *title* option.

The following is a plot showing some of these options in use.

```
> plot(x^3+2*x^2-3*x-1, x=-3..3, axes=FRAME,
> style=POINT, symbol=CIRCLE);
```

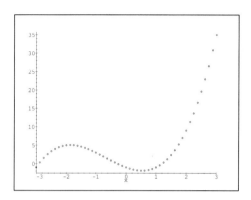

## Three-Dimensional Plots

Expressions in two unknowns can be plotted using the plot3d command. Several types of three-dimensional plots are available. Here are a few of the more popular ones.

Expressions can be plotted.

```
> plot3d((x^2-y^2)/(x^2+y^2), x=-2..2, y=-2..2);
```

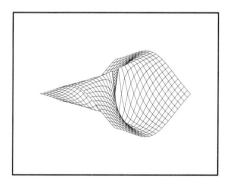

Parametric expressions can be plotted.

> plot3d([x*sin(x), x*cos(y), x*sin(y)], x=0..2*Pi, y=0..Pi)

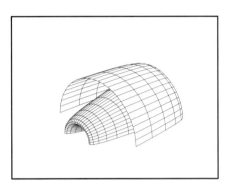

Multiple plots can be specified for one set of axes.

> plot3d({x+y^2, -x-y^2}, x=0..3, y=0..3);

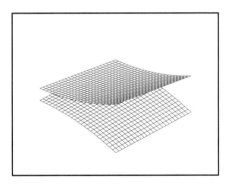

Besides these standard plots, several other types of three-dimensional plots are available in the plots package.

Plots can be specified in spherical coordinates.

```
> plots[sphereplot]((1.3)^z * sin(theta), z=-1..2*Pi,
> theta=0..Pi);
```

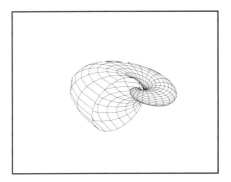

Curves can be drawn in three-dimensional space.

```
> plots[spacecurve]([t*cos(t), t*sin(t), t], t=0..7*Pi);
```

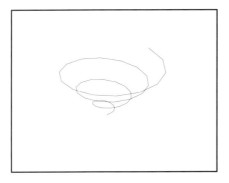

There are several additional options available for three-dimensional plots. The *grid* option allows you to specify the rectangular grid size for the sample points. The default is $25 \times 25$. You can specify the style with which the surface is rendered (e.g., *PATCH*, *WIREFRAME*, *POINT*) with the *style* option. Different coloring schemes can be specified with the *color* and *shading* options. Either ambient or directional lights can be applied to the surface with the *ambientlight* and *light* options, respectively. The *orientation* option lets you specify from which point in space you are to view the surface. Labelling of the plot can be handled with the *title*, *labels*, and *tickmarks* options.

The following is a plot that shows some of these options in use.

```
> plot3d((x^2-y^2)/(x^2+y^2), x=-2..2, y=-2..2,
> style=CONTOUR, title='saddle');
```

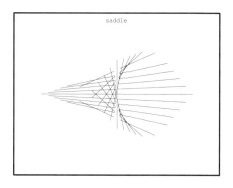

# Animation

Maple also performs basic animation. Two- and three-dimensional animations are produced with the commands animate and animate3d from the plots package. Be forewarned, though, that performing animations takes a lot of CPU time and memory space. If your system is in short supply of either of these, be careful when trying animation. (Be sure to save any important work before you try.)

To have animation, there must be an extra unknown in the expression being animated—the variable of animation. Give this variable a range of its own (specified as the last range) and specify a number of frames to be produced. Other than these two things, animate and animate3d are very similar to plot and plot3d, respectively.

Each animation appears in its own window, complete with motion controls much like those found on a VCR.

If you have the proper graphics terminal, please try the following four animation examples.

## Two-Dimensional Animation

```
> plots[animate](sin(x*t),x=-10..10, t=1..2, frames=25);
> plots[animate]([u*sin(t),u*cos(t),t=-Pi..Pi], u=1..8,
> view=[x=-8..8,y=-8..8]);
```

### Three-Dimensional Animation

```
> plots[animate3d](cos(t*x)*sin(t*y), x=-Pi..Pi,
> y=-Pi..Pi, t=1..2);
> plots[animate3d]((1.3)^x * sin(u*y), x=-1..2*Pi,
> y=0..Pi, u=1..8, coords=spherical);
```

## Parameter Types Specific to This Chapter

| | |
|---|---|
| 2DP, 2DPlot | a two-dimensional plot structure |
| 3DP, 3DPlot | a three-dimensional plot structure |
| device | an output device |

## Command Listing

### interface(*plotdevice* = device)

Sets the current plot device (i.e., display, printer, or file format) to device. All subsequent plots are rendered for that device.
**Output:** A NULL expression is returned.
**Argument options:** (*plotdevice*) to return the current plot device.
**Additional information:** The available devices are *mac* for Macintosh terminals, *x11* for x11 terminals, *tek* for Tektronix terminals (see preplot and postplot for special considerations about Tektronix terminals), *regis* for Regis terminals, *char* for character plots, *vt100* for VT100 line terminals, *i300* for imagen 300 laser printers, *ln03* for DEC LN03 laser printers, *hplj* for HP Laserjet printers, *hpgl* for HP GL printers, *ps* for PostScript printers or files, *cps* for Color PostScript printers or files, *pic* for troff pic file format, and *unix* for UNIX *plot* command file format. ♣ Each Maple platform is set up, either automatically or during installation, to have the proper device as default. ♣ To change the device for every session, place the appropriate interface command in your initialization file. ♣ If you are only changing the plot device temporarily, be sure to change it back to its original value when you are ready. ♣ Multiple options (e.g., *plotdevice* and *plotoutput*) can be set within one interface command. plotsetup provides some common pre-set combinations of interface options.
**See also:** plot, plot3d, plotsetup, interface(*plotoutput*), interface(*preplot*), interface(*postplot*)
*LA = 154* **LI = 116**

### interface(*plotoutput* = name)

Sets the current output file name for channelling plots. All subsequent plots are written to name.
**Output:** A NULL expression is returned.
**Argument options:** (*plotoutput*) to return the current output file name.
**Additional information:** The special file name *terminal* directs plots to the current monitor. ✦ If you supply a file name corresponding to a location in your system, the output is written there, overwriting any file that currently exists. Therefore, if you wish to write more than one plot, be sure to change name between each plot. ✦ Each platform of Maple is specifically set up to have *terminal* as the default. ✦ To change the file name for every session, place the appropriate interface command in your initialization file. ✦ If you are only changing the file name temporarily, be sure to change it back to its original value when you are ready. ✦ Multiple options (e.g., *plotdevice* and *plotoutput*) can be set within one interface command. plotsetup provides some common pre-set combinations.
**See also:** plot, plot3d, plotsetup, interface(*plotdevice*), interface(*preplot*), interface(*postplot*)
*LA = 155* **LI = 116**

### interface(*postplot* = [int$_1$, ..., int$_n$])

Sends variables int$_1$ through int$_n$ to exit graphics mode on displays running Tektronix emulation.
**Output:** A NULL expression is returned.
**Argument options:** (*postplot*) to return the current variables.
**Additional information:** Before the plot is performed, you need to send *preplot* variables to put the display in graphics mode. ✦ plotsetup provides some common pre-set combinations of interface options. ✦ For more information on *preplot* and *postplot* variables, see the on-line help page for plot[setup] or the documentation for your terminal emulation.
**See also:** plot, plot3d, plotsetup, interface(*plotdevice*), interface(*plotoutput*), interface(*preplot*)

### interface(*preplot* = [int$_1$, ..., int$_n$])

Sends variables int$_1$ through int$_n$ to enter graphics mode on displays running Tektronix emulation.
**Output:** A NULL expression is returned.
**Argument options:** (*preplot*) to return the current variables.
**Additional information:** After the plot is performed, you need to

send *postplot* variables to take the display out of graphics mode.
♣ plotsetup provides some common pre-set combinations of interface options. ♣ For more information on *preplot* and *postplot* variables, see the on-line help page for plot[setup] or the documentation for your terminal emulation.
**See also:** plot, plot3d, plotsetup, interface(*plotdevice*), interface(*plotoutput*), interface(*postplot*)
*LA = 155*

## op(2DPlot)

Displays the internal structure of two-dimensional PLOT data structure 2DP and allows you access to its operands.
**Output:** An expression sequence is returned. The operands are function calls representing the specifics of the plot.
**Argument options:** (*1*, 2DP) to extract a call to AXIS(*HORIZONTAL*). ♣ (*2*, 2DP) to extract a call to AXIS(*VERTICAL*). ♣ (*3*, 2DP) to extract a call to RANGE(*HORIZONTAL*). ♣ (*4*, 2DP) to extract either a call to RANGE(*VERTICAL*), if a horizontal range has been specified, or a call to CURVE detailing the points sampled and the sampling procedure. ♣ If a vertical range was supplied, then the CURVE call is the fifth operand. ♣ The remaining operands are calls to functions dealing with other options to the plot.
**Additional information:** By manipulating operands of a PLOT data structure, you can easily change option values. ♣ To substitute for an operand, use the subsop command. ♣ More ambitious manipulation can also be done, up to and including creating PLOT structures from scratch. ♣ For more information on the, see the on-line help page for plot[structure].
**See also:** plot, op(3DP), *subsop*

## op(3DPlot)

Displays the internal structure of a three-dimensional PLOT3D data structure 3DP and allows you access to its operands.
**Output:** An expression sequence is returned. The operands are function calls representing the specifics of the plot.
**Argument options:** (*1*, 3DP) to extract a call to FUNCTION detailing information about the function being plotted and the $x$ and $y$ ranges. ♣ (*2*, 3DP) to extract a call to CURVE detailing the points sampled. ♣ (*3*, 3DP) to extract a call to LABELS. ♣ The remaining operands are calls to functions dealing with other options to the plot.
**Additional information:** By manipulating operands of a PLOT3D

data structure, you can easily change option values. ♣ To substitute for an operand, use the subsop command. ♣ More ambitious manipulation can also be done, up to and including creating PLOT3D structures from scratch. ♣ For more information, see the on-line help page for plot3d[structure].
**See also:** plot3d, op(2DP), *subsop*

### plot(expr, var=a..b)

Displays a two-dimensional plot of expr, an expression in variable var only, for values of var ranging from a to b.
**Output:** A two-dimensional plot is displayed.
**Argument options:** (expr, a..b) to produce the same result. ♣ (expr, var=a..infinity), (expr, var=-infinity..b), or (expr, -infinity..infinity) to create infinity plots. A logarithmic scale is used to display the range. ♣ (expr) to plot expr across a range chosen by the system. ♣ (expr, var=a..b, $a_v..b_v$) to only plot values of expr that fall within vertical range $a_v$ to $b_v$. ♣ (fnc, a..b) to plot fnc, where fnc contains only Maple procedures or operators. ♣ ([$expr_1$, $expr_2$, var=a..b]) to display the parametric plot defined by $expr_1$ and $expr_2$ as var ranges from a to b. ♣ ([$x_1$, $y_1$, ..., $x_n$, $y_n$], var=a..b) or ([[$x_1$, $y_1$], ..., [$x_n$, $y_n$]], var=a..b) to plot n points, [$x_1$, $y_1$] through [$x_n$, $y_n$], connected in that order. ♣ ({$expr_1$, ..., $expr_n$}, var=a..b) to plot the expressions $expr_1$ through $expr_n$ on the same set of axes. ♣ (expr, var=a..b, *discont=true*) to specify that there are known discontinuities over the range a..b. The point-joining algorithm reacts appropriately. ♣ (expr, var=a..b, option) to manipulate an optional parameter of the plot. option can be one or more of the following: *coords = polar* to specify that expr is given in polar coordinates, *title* = str to print the title str on the plot, *font* = [$option_1$, $option_2$, n], *titlefont* = [$option_1$, $option_2$, n], *axesfont* = [$option_1$, $option_2$, n], or *labelfont* = [$option_1$, $option_2$, n], to specify the font specifics for text objects within the plot, the plot title, the axes tickmarks, and the axes labels, respectively. $option_1$ specifies the font type; i.e., *TIMES, COURIER,* or *HELVETICA*. $option_2$ specifies the font style; i.e., *ROMAN, BOLD, ITALIC,* or *BOLDITALIC*. n specifies the font size. *axes = $option_2$* to specify *FRAME, BOXED, NORMAL,* or *NONE* axes, *scaling = $option_2$* to specify whether the display of the plot should be *UNCONSTRAINED* or *CONSTRAINED*, *style = $option_2$* to specify *LINE* or *POINT* style, *numpoints = n* to specify that n points should be sampled on expr (default is 49), *resolution = n* to specify a resolution of n for your display (default is 200), *symbol = $option_2$* to specify the symbol used in point plots is a *BOX, CROSS, CIRCLE, DIAMOMD,* or *POINT*, *linestyle = $option_2$* to specify

the style of line used in line plots, *thickness* = n to specify the thickness of lines used in line plots, *color* = option$_2$ to specify the curve should be *blue*, *black*, *green*, etc., and *xtickmarks* = n$_x$ and/or *ytickmarks* = n$_y$ to indicate the minimum number of tickmarks appearing along the $x$ and/or $y$ axes.

**Additional information:** Many specialty two-dimensional plots are available in the plots package. ♣ All expressions must have no more than *one* unknown in them. ♣ All ranges must evaluate to numeric values. ♣ Some default values and choices available for plotting options change from platform to platform; try a simple plot to determine those for your system. Default settings can be altered with plots[setoptions]. ♣ To plot different types of two-dimensional plots or plots with different ranges on the same set of axes, use the plots[display] command. ♣ The $x$ and $y$ axes are automatically labelled with names supplied in the lefthand side of the range equations. ♣ A list of predefined colors can be found in the on-line help page for plot[color].

**See also:** plot3d, plots[animate], plots[conformal], plots[densityplot], plots[display], plots[fieldplot], plots[gradplot], plots[implicitplot], plots[logplot], plots[loglogplot], plots[polarplot], plots[polygonplot], plots[setoptions], plots[textplot]

**FL = 26−28, 155−169** *LA = 155−162* **LI = 148−164**

## plot3d(expr, var$_1$=a$_1$..b$_1$, var$_2$=a$_2$..b$_2$)

Displays a three-dimensional plot of expr (an expression in var$_1$ and var$_2$) for values of var$_1$ ranging from a$_1$ to b$_1$ and values of var$_2$ ranging from a$_2$ to b$_2$.

**Output:** A three-dimensional plot is displayed.

**Argument options:** (expr, var$_1$=a$_1$..b$_1$, var$_2$=a$_2$..b$_2$, a$_z$..b$_z$) to only plot values of expr that fall within range a$_z$ to b$_z$. ♣ (fnc, a$_1$..b$_1$, a$_2$..b$_2$) to plot fnc, where fnc contains only Maple procedures or operators. ♣ ([expr$_1$, expr$_2$, expr$_3$], var$_1$ = a$_1$..b$_1$, var$_2$=a$_2$..b$_2$) to display the parametric plot defined by expr$_1$, expr$_2$, and expr$_3$, as var$_1$ ranges from a$_1$ to b$_1$ and var$_2$ ranges from a$_2$ to b$_2$. ♣ ({expr$_1$, ..., expr$_n$}, var$_1$ = a$_1$..b$_1$, var$_2$=a$_2$..b$_2$) to plot expressions expr$_1$ through expr$_n$ on the same set of axes. ♣ (expr, var$_1$=a$_1$..b$_1$, var$_2$=a$_2$..b$_2$, option) to manipulate an optional parameter of the plot. option can be one or more of the following: *coords* = option$_2$ to specify that *CARTESIAN*, *SPHERICAL*, or *CYLINDRICAL* coordinates are being used (default is *CARTESIAN*), *title* = str to print the title str on the plot. *font* = [option$_1$, option$_2$, n], *titlefont* = [option$_1$, option$_2$, n], *axesfont* = [option$_1$, option$_2$, n], or *labelfont* = [option$_1$, option$_2$, n], to specify the font specifics for text objects within the plot, the plot title, the axes tickmarks, and the axes labels, respectively. option$_1$ specifies the font type; i.e., *TIMES*, *COURIER*, or *HELVETICA*. option$_2$ specifies the font style; i.e., *ROMAN*, *BOLD*, *ITALIC*, or *BOLDITALIC*. n specifies the font size. *axes* = option$_2$ to specify *FRAME*, *BOXED*, *NORMAL*, or *NONE* axes, *scaling* = option$_2$ to specify whether the display of the plot should be *UNCONSTRAINED* or *CONSTRAINED*, *orientation* = [angle$_\theta$, angle$_\phi$] to specify the angle in degrees from which the plot should be viewed, *view* = a$_z$..b$_z$ or *view* = [a$_x$..b$_x$, a$_y$..b$_y$, a$_z$..b$_z$] to crop the surface along the $z$-axis or the $x$, $y$, and $z$ axes, respectively, *projection* = n, where n is a value between 0 and 1, to specify the perspective that should be used, *style* = option$_2$ to specify the rendering style *POINT*, *LINE*, *HIDDEN*, *PATCH*, *WIREFRAME*, *CONTOUR*, *PATCHNOGRID*, or *PATCHCONTOUR*, *shading* = option$_2$ to specify the shading style *Z*, *XY*, *XYZ*, *Z_GREYSCALE*, *Z_HUE*, or *NONE*, *grid* = [int$_1$, int$_2$] to specify that an int$_1$ × int$_2$ grid of equally-spaced points is sampled (default is [25, 25]), *numpoints* = n to specify that n equally spaced points in a grid of $\sqrt{n} \times \sqrt{n}$ be sampled, *symbol* = option$_2$ to specify the symbol used in point plots is a *BOX*, *CROSS*, *CIRCLE*, *DIAMOMD*, or *POINT*, *linestyle* = option$_2$ to specify the style of line used in line plots, *thickness* = n to specify the thickness of lines used in line plots, *color* = option$_2$ to specify the surface should be *blue*, *black*, *green*, etc. only, *light* = [angle$_\phi$, angle$_\theta$, num$_r$, num$_g$, num$_b$] to apply directional lighting with light source at angles angle$_\phi$ and angle$_\theta$ and red, green, and blue intensities num$_r$, num$_g$, and num$_b$ between 0 and 1, *ambientlight* = [num$_r$, num$_g$, num$_b$] to apply ambient lighting (i.e., coming from all directions at once), *tickmarks* = [int$_1$, int$_2$, int$_3$] to indicate the minimum number of tickmarks appearing along the $x$, $y$, and $z$ axes, and *labels* = [str$_1$, str$_2$, str$_3$] to indicate the labels used along the $x$, $y$, and $z$ axes, respectively.

**Additional information:** Many specialty three-dimensional plots are available in the plots package. ♣ All expressions must have no more than *two* unknowns in them. ♣ All ranges must evaluate to numeric values. ♣ Default values and choices available for plotting options may change from platform to platform; try a simple plot to determine those for your system. ♣ Default settings can be altered with plots[setoptions3d]. ♣ To plot different types of three-dimensional plots or plots with different ranges on the same set of axes, use the plots [display3d] command. ♣ Unless the *labels* option is used, the $x$, $y$, and $z$ axes are automatically labelled with names supplied in the lefthand side of the range equations. ♣ For the *prospective* option, a value of 0 represents wide-angle

perspective and a value of 1 represents orthogonal perspective (i.e., no projection). The three special values *FISHEYE*, *NORMAL*, and *ORTHOGONAL* represent the values 0, 0.5, and 1, respectively.
♣ A list of predefined colors can be found in the on-line help page for plot[color].
**See also:** plot, plots[animate3d], plots[contourplot], plots[cylinderplot], plots[display3d], plots[fieldplot3d], plots[gradplot3d], plots[implicitplot3d], plots[matrixplot], plots[pointplot], plots[polygonplot3d], plots[polyhedraplot], plots[setoptions3d], plots[spacecurve], plots[sparsematrixplot], plots[sphereplot], plots[surfdata], plots[textplot3d] plots[tubeplot]
*FL = 169−184* **LA = 9, 162−169**

## plots package

Provides commands for specialty plots in two and three dimensions.
**Additional information:** To access the command fnc from the plots package, use the long form of the name plots[fnc], or with (plots, fnc) to load in a pointer for fnc only, or with(plots) to load in pointers to all the commands. ♣ In all plots package commands, expressions plotted must have exactly the appropriate number of unknowns and ranges must evaluate to numeric values. ♣ Many of these plots can also be produced with calls to plot and plot3d.

### plots[animate](expr, var$_h$=a$_h$..b$_h$, var$_f$=a$_f$..b$_f$)

Displays a two-dimensional animation of expression expr as horizontal variable var$_h$ varies from a$_h$ to b$_h$. Equally spaced frames are calculated as frame variable var$_f$ varies from a$_f$ to b$_f$.
**Output:** A two-dimensional animation is displayed.
**Argument options:** (fnc, var$_h$=a$_h$..b$_h$, var$_f$=a$_f$..b$_f$) to animate fnc, where fnc contains only procedures stated in terms of operators. ♣ (expr, var$_h$=a$_h$..b$_h$, var$_f$=a$_f$..b$_f$, *frames* = n) to specify the number of frames. Default is 16. ♣ (expr, var$_h$=a$_h$..b$_h$, var$_f$=a$_f$..b$_f$, option) to manipulate options to the plots. For more information on available options, see the plot command.
**Additional information:** With the exception of plots of Maple procedures, all types of single and multiple two-dimensional plots can be animated. See plot for available alternatives. ♣ After all the frames are calculated, an animation control window appears.

- ♣ Because the calculations required are equal in total to those for many individual plots, it may take some time for your animation to compute. ♣ If your system does not have sufficient memory or disk space, the animation may fail—try again with fewer frames or a smaller number of points. ♣ Animations may be built up frame by frame with the *insequence* option of plots[display].

**See also:** plot, plots[display], plots[animate3d]

### plots[animate3d](expr, var$_1$=a$_1$..b$_1$, var$_2$=a$_2$..b$_2$, var$_f$=a$_f$..b$_f$)

Displays a three-dimensional animation of expression expr as variables var$_1$ varies from a$_1$ to b$_1$ and var$_2$ varies from a$_2$ to b$_2$. Equally spaced frames are calculated as frame variable var$_f$ varies from a$_f$ to b$_f$.

**Output:** A three-dimensional animation is displayed.

**Argument options:** (fnc, var$_1$=a$_1$..b$_1$, var$_2$=a$_2$..b$_2$, var$_f$=a$_f$..b$_f$) to plot fnc, where fnc contains only procedures stated in terms of operators. ♣ (expr, var$_1$=a$_1$..b$_1$, var$_2$=a$_2$..b$_2$, var$_f$=a$_f$..b$_f$, *frames* = n) to specify the number of frames. Default is 8. ♣ (expr, var$_1$=a$_1$..b$_1$, var$_2$=a$_2$..b$_2$, var$_f$=a$_f$..b$_f$, option) to manipulate options to the plots. For more information on available options, see the plot3d command.

**Additional information:** With the exception of plots of Maple procedures, all types of single and multiple three-dimensional plots can be animated. See plot3d for available alternatives. ♣ After all the frames are calculated, an animation control window appears. ♣ Because the calculations required are equal in total to those for many individual plots, it may take some time for your animation to compute. ♣ If your system does not have sufficient memory or disk space, the animation may fail—try again with fewer frames or a smaller grid size. ♣ Animations may be built up frame by frame with the *insequence* option of plots[display3d].

**See also:** plot3d, plots[display3d], plots[animate]

### plots[conformal](expr, var=cmplx$_{a1}$..cmplx$_{b1}$, cmplx$_{a2}$..cmplx$_{b2}$)

Displays a two-dimensional mapping of expression expr in var (typically z) and for values in the given ranges from the plane into a curved grid determined by the images of standard grid lines of the plane under expr.

**Output:** A two-dimensional plot is displayed.

**Argument options:** (expr) to plot the mapping of expr with default values for the ranges. The first range is $0..1+(-1)^{1/2}$ and the second range is calculated to completely enclose the resulting conformal lines. ♣ (fnc, var=cmplx$_{a1}$..cmplx$_{b1}$, cmplx$_{a2}$..cmplx$_{b2}$) to plot the mapping of fnc, where fnc is a Maple procedure. ♣ (eqn, var$_1$=a$_1$..b$_1$, var$_2$=a$_2$..b$_2$, option) to specify options to the plot. The option *grid* = [int$_x$, int$_y$] allows you to set the number of grid lines mapped in the $x$ and $y$ directions. Default is [11, 11]. ♣ The option *numxy* = [int$_x$, int$_y$] allows you to set the number of points sampled for each grid line in the $x$ and $y$ directions. Default is [15, 15]. ♣ Other options applicable to conformal plots are *style*, *xtickmarks*, and *ytickmarks*. For more information on these options, see the plot command.

**Additional information:** The result is a set of curves intersecting at right angles at the points where expr is analytic.

**See also:** plot

**FL = 164−165, 168**

### plots[contourplot](expr, var$_1$=a$_1$..b$_1$, var$_2$=a$_2$..b$_2$)

Displays a three-dimensional contour plot of expr, an expression in var$_1$ and var$_2$, for values of var$_1$ ranging from a$_1$ to b$_1$ and var$_2$ ranging from a$_2$ to b$_2$.

**Output:** A three-dimensional plot is displayed.

**Additional information:** Input to plots[contourplot] produces results identical to the plot3d command with the option *style = CONTOUR*. ♣ For more information, see the entry for plot3d.

**See also:** plot3d

### plots[cylinderplot](expr, var$_1$=a$_1$..b$_1$, var$_2$=a$_2$..b$_2$)

Displays a three-dimensional cylindrical plot of radial expression expr as var$_1$ ranges from a$_1$ to b$_1$ and var$_2$ ranges from a$_2$ to b$_2$.

**Output:** A three-dimensional plot is displayed.

**Argument options:** ([expr$_r$, expr$_\theta$, expr$_z$], var$_1$ = a$_1$..b$_1$, var$_2$ = a$_2$..b$_2$) to plot the parametric representation, where expr$_r$, expr$_\theta$, and expr$_z$ are functions in var$_1$ and var$_2$ representing the radius, $\theta$, and $z$ coordinate, respectively. ♣ (expr, var$_1$=a$_1$..b$_1$, var$_2$=a$_2$..b$_2$, option) to specify options to the plot. For more information on available options, see the plot3d command.

**Additional information:** To plot different types of plots or plots with different ranges on the same set of axes, use plots[display3d].

**See also:** plot3d, plots[tubeplot], plots[sphereplot], plots[polarplot]

### plots[densityplot](expr, var$_1$=a$_1$..b$_1$, var$_2$=a$_2$..b$_2$)

Displays a two-dimensional density plot for the surface represented by expression expr in var$_1$ and var$_2$, for values in the given ranges.

**Output:** A two-dimensional plot is displayed.

**Argument options:** (fnc, $var_1=a_1..b_1$, $var_2=a_2..b_2$) to plot the density plot of fnc, where fnc is a Maple procedure or operator. ♣ ($\{expr_1, ..., expr_n\}$, $var_1=a_1..b_1$, $var_2=a_2..b_2$) to plot the density plots of $expr_1$ through $expr_n$ on the same set of axes. ♣ (eqn, $var_1=a_1..b_1$, $var_2=a_2..b_2$, option) specifies options to the plot. ♣ The option *grid* = [$int_1$, $int_2$] specifies the grid size used. ♣ Default is [25, 25]. The option *colorstyle* = $option_2$ specifies the coloring scheme *HUE* or *RGB* is used. ♣ For more information on available options, see the plot command.

**Additional information:** To plot different types of plots or plots with different ranges on the same set of axes, use plots[display].

**See also:** plot

### plots[display]($\{2DP_1, ..., 2DP_n\}$)

Displays n two-dimensional plot structures $2DP_1$ through $2DP_n$ on the same set of axes.

**Output:** A two-dimensional plot is displayed.

**Argument options:** ([$2DP_1, ..., 2DP_n$]) to achieve the same results. ♣ ([$2DP_1, ..., 2DP_n$], *insequence=true*) to display the n plots in sequence as an animation. ♣ ($\{2DP_1, ..., 2DP_n\}$, option) to specify options to be used for the plots. For more information on available options, see the plot command.

**Additional information:** If the plot structures specified with the *insequence* option are animations, then the n animations are run sequentially. ♣ If there is a conflict in options supplied in the plots[display] command and in the individual plot structures, the former always prevail. ♣ If there are multiple labels or titles supplied within the plot structures, plots[display] chooses from among them at random. ♣ You cannot mix infinity and non-infinity plots. ♣ If you mix log or log-log plots with non-log plots, the non-log plots control the combined plot structure.

**See also:** plot, plots[animate], plots[logplot], plots[loglogplot], plots[replot]

**FL = 159 – 162**

### plots[display3d]($\{3DP_1, ..., 3DP_n\}$)

Displays n three-dimensional plot structures $3DP_1$ through $3DP_n$ on the same set of axes.

**Output:** A three-dimensional plot is displayed.

**Argument options:** ([$3DP_1, ..., 3DP_n$]) to achieve the same results. ♣ ([$3DP_1, ..., 3DP_n$], *insequence=true*) to display the n plots in sequence as an animation. ♣ ($\{3DP_1, ..., 3DP_n\}$, option) to specify options to be used for the plots. For more information on

available options, see the plot3d command.
**Additional information:** If the plot structures specified with the *insequence* option are animations, then the n animations are run sequentially. ♣ If there is a conflict in options supplied in the plots[display3d] command and in the individual plot structures, the former always prevail. ♣ If there are multiple labels or titles supplied within the plot structures, plots[display3d] chooses from among them at random.
**See also:** plot3d, plots[animate3d], plots[replot]
**FL = 159–162**

**plots[fieldplot]([expr$_1$, expr$_2$], var$_1$=a$_1$..b$_1$, var$_2$=a$_2$..b$_2$)**
Displays a two-dimensional vector field for expressions expr$_1$ and expr$_2$, both in var$_1$ and var$_2$, for values in the given ranges.
**Output:** A two-dimensional plot is displayed.
**Argument options:** ([fnc$_1$, fnc$_2$], var$_1$=a$_1$..b$_1$, var$_2$=a$_2$..b$_2$) to display the vector field of fnc$_1$ and fnc$_2$, which contain only Maple procedures or operators. ♣ ([expr$_1$, expr$_2$], var$_1$=a$_1$..b$_1$, var$_2$ = a$_2$..b$_2$, option) to specify options to the vector field. ♣ The option *arrows* = option$_2$ allows you to set the arrow style to *LINE*, *THIN*, *SLIM*, or *THICK*. Default is *THIN*. ♣ The option *grid* = [int$_1$, int$_2$] allows you to set the grid size of the field. Default is [20, 20]. ♣ For more information on other available options, see the plot command.
**Additional information:** Gradient vector fields can be plotted with plots[gradplot]. ♣ To plot different types of two-dimensional plots or plots with different ranges on the same set of axes, use plots [display].
**See also:** plot, plots[gradplot], plots[fieldplot3d]

**plots[fieldplot3d]([expr$_1$, expr$_2$, expr$_3$], var$_1$=a$_1$..b$_1$, var$_2$=a$_2$..b$_2$, var$_3$=a$_3$..b$_3$)**
Displays a three-dimensional vector field for expr$_1$, expr$_2$, and expr$_3$, all expressions in var$_1$, var$_2$, and var$_3$, for values in the given ranges.
**Output:** A three-dimensional plot is displayed.
**Argument options:** ([fnc$_1$, fnc$_2$, fnc$_3$], var$_1$=a$_1$..b$_1$, var$_2$=a$_2$..b$_2$, var$_3$=a$_3$..b$_3$) to display the vector field of fnc$_1$, fnc$_2$, and fnc$_3$, which contain only Maple procedures or operators. ♣ ([expr$_1$, expr$_2$, expr$_3$], var$_1$=a$_1$..b$_1$, var$_2$=a$_2$..b$_2$, var$_3$=a$_3$..b$_3$, option) to specify options to the vector field. ♣ The option *arrows* = option$_2$ allows you to set the arrow style to *LINE*, *THIN*, *SLIM*, or *THICK*. Default is *THIN*. ♣ The option *grid* = [int$_1$, int$_2$, int$_3$] allows you to set the grid size of the field. Default is [8, 8, 8]. ♣ For

more information on available options, see the plot3d command.
**Additional information:** Gradient vector fields can be plotted with plots[gradplot3d]. ♣ To plot different types of three-dimensional plots or plots with different ranges on the same set of axes, use the plots[display3d] command.
**See also:** plot3d, plots[gradplot3d], plots[fieldplot]

### plots[gradplot](expr, var$_1$=a$_1$..b$_1$, var$_2$=a$_2$..b$_2$)

Displays a two-dimensional gradient vector field for expression expr in var$_1$ and var$_2$ for values in the given ranges.
**Output:** A two-dimensional plot is displayed.
**Argument options:** (expr, var$_1$=a$_1$..b$_1$, var$_2$=a$_2$..b$_2$, option) to specify options of the gradient vector field. ♣ The option *arrows* = option$_2$ allows you to set the arrow style to *LINE*, *THIN*, *SLIM*, or *THICK*. Default is *THIN*. ♣ The option *grid* = [int$_1$, int$_2$] allows you to set the grid size of the gradient field. Default is [20, 20]. ♣ For more information on available options, see the plot command.
**Additional information:** Straight vector fields can be plotted with plots[fieldplot]. ♣ To plot different types of two-dimensional plots or plots with different ranges on the same set of axes, use plots [display].
**See also:** plot, plots[fieldplot], plots[gradplot3d]

### plots[gradplot3d](expr, var$_1$=a$_1$..b$_1$, var$_2$=a$_2$..b$_2$, var$_3$=a$_3$..b$_3$)

Displays a two-dimensional gradient vector field for expression expr in var$_1$, var$_2$, and var$_3$ for values in the given ranges.
**Output:** A three-dimensional plot is displayed.
**Argument options:** (expr, var$_1$=a$_1$..b$_1$, var$_2$=a$_2$..b$_2$, var$_3$=a$_3$..b$_3$, option) to specify options to the gradient vector field. The option *arrows* = option$_2$ allows you to set the arrow style to *LINE*, *THIN*, *SLIM*, or *THICK*. Default is *THIN*. ♣ The option *grid* = [int$_1$, int$_2$] allows you to set the grid size of the gradient field. Default is [8, 8, 8]. ♣ For more information on available options, see the plot3d command.
**Additional information:** Straight vector fields can be plotted with plots[fieldplot3d]. ♣ To plot different types of three-dimensional plots or plots with different ranges on the same set of axes, use plots[display3d].
**See also:** plot3d, plots[fieldplot3d], plots[gradplot]

## plots[implicitplot](eqn, var$_1$=a$_1$..b$_1$, var$_2$=a$_2$..b$_2$)

Displays a two-dimensional implicit plot of equation eqn in var$_1$ and var$_2$, for values in the given ranges.

**Output:** A two-dimensional plot is displayed.

**Argument options:** (expr, var$_1$=a$_1$..b$_1$, var$_2$=a$_2$..b$_2$) to display an implicit plot of equation expr = 0. ♣ (fnc, var$_1$=a$_1$..b$_1$, var$_2$ = a$_2$..b$_2$) to implicitly plot fnc, where fnc contains only Maple procedures or operators. ♣ (eqn, var$_1$=a$_1$..b$_1$, var$_2$=a$_2$..b$_2$, option) to specify options to the plot. ♣ The option *grid* = [int$_1$, int$_2$] allows you to set the grid size of the plot. Default is [25, 25]. ♣ For more information on available options, see the plot command.

**Additional information:** To plot different types of two-dimensional plots or plots with different ranges on the same set of axes, use plots[display].

**See also:** plot, plots[implicitplot3d]

## plots[implicitplot3d](eqn, var$_1$= a$_1$..b$_1$, var$_2$=a$_2$..b$_2$, var$_3$=a$_3$..b$_3$)

Displays a three-dimensional implicit plot of equation eqn in var$_1$, var$_2$, and var$_3$ for values in the given ranges.

**Output:** A three-dimensional plot is displayed.

**Argument options:** (expr, var$_1$=a$_1$..b$_1$, var$_2$=a$_2$..b$_2$, var$_3$=a$_3$..b$_3$) to display an implicit plot of equation expr = 0. ♣ (fnc, var$_1$ = a$_1$..b$_1$, var$_2$=a$_2$..b$_2$, var$_3$=a$_3$..b$_3$) to implicitly plot fnc, where fnc contains only Maple procedures or operators. ♣ (eqn, var$_1$=a$_1$..b$_1$, var$_2$=a$_2$..b$_2$, var$_3$=a$_3$..b$_3$, option) to specify options to the plot. ♣ The option *grid* = [int$_1$, int$_2$] allows you to set the grid size of the plot. Default is [10, 10, 10]. ♣ For more information on available options, see the plot3d command.

**Additional information:** To plot different types of three-dimensional plots or plots with different ranges on the same set of axes, use plots[display3d].

**See also:** plot3d, plots[implicitplot]

## plots[loglogplot](expr, var=a..b)

Displays a two-dimensional logarithmic plot of expr, an expression in var, for var ranging from a to b. Both the vertical and horizontal axes are given in logarithmic scale.

**Output:** A two-dimensional plot is displayed.

**Argument options:** (expr, var=a..b, a$_v$..b$_v$) to plot only values of expr that fall within the vertical range a$_v$ to b$_v$. ♣ (expr, var=a..b, option) to specify options to the plot. For more information on available options, see the plot command.

**Additional information:** plots[loglogplot] supports the same plot types as plot. ♣ The $x$ and $y$ axes are automatically labelled with names supplied in the lefthand side of the range equations.
**See also:** plot, plots[logplot]

### plots[logplot](expr, var=a..b)

Displays a two-dimensional logarithmic plot of expr, an expression in var, for var ranging from a to b. The vertical axis is given in logarithmic scale.
**Output:** A two-dimensional plot is displayed.
**Argument options:** (expr, var=a..b, $a_v$..$b_v$) to plot only values of expr that fall within the vertical range $a_v$ to $b_v$. ♣ (expr, var=a..b, option) to specify options to the plot. For more information on available options, see the plot command.
**Additional information:** plots[logplot] supports the same plot types as plot. ♣ The $x$ and $y$ axes are automatically labelled with names supplied in the lefthand side of the range equations.
**See also:** plot, plots[loglogplot]

### plots[matrixplot](M)

Displays a three-dimensional surface plot representing matrix M, where the $x$ and $y$ grid is represented by the matrix indices and the $z$ values by the corresponding matrix entries.
**Output:** A three-dimensional plot is displayed.
**Argument options:** (M, option) to specify options to the matrix plot. ♣ The option *heights = histogram* specifies that the plot be rendered as a three-dimensional histogram with ''tower'' heights equalling matrix entries. ♣ The option *gap* = expr, where expr is between 0 and 0.5, specifies the relative size of the gap around each tower. Default is 0. ♣ The option *color* = fnc to specify the two-argument function defining how the plot is to be colored. ♣ For more information on available options, see the plot3d command.
**Additional information:** All elements of M must evaluate to numeric values.
**See also:** plot3d, plots[sparsematrixplot], plots[surfdata]

### plots[odeplot](proc, [var₁, var₂], a..b)

Displays a two-dimensional plot of proc, the result of a call to dsolve(*numeric*), as the input to proc goes from a to b. The values of $var_1$ are plotted as the $x$-component and the values of $var_2$ are plotted as the $y$-component.
**Output:** A two-dimensional plot is displayed.
**Argument options:** (proc, [$var_1$, $var_2$, $var_3$], a..b) to display a

three dimensional plot, where $var_1$, $var_2$, and $var_3$ are plotted as the $x$, $y$, and $z$ components, respectively. ♣ (proc, [$var_1$, $var_2$], a..b, option) or (proc, [$var_1$, $var_2$, $var_3$], a..b, option) to specify options to the plot. For more information on available options, see the plot and plot3d commands, respectively.

**Additional information:** Both $var_1$ and $var_2$ must represent either functions solved for in dsolve or the variable of those functions. ♣ If only functions are represented in the variable list, a *phase plot* is produced. ♣ If no range is supplied, a default value $-10..10$ is used. ♣ For more information and examples, see the on-line help page.

**See also:** plot, plot3d, dsolve(*numeric*), DEtools[DEplot], DEtools[DEplot1], DEtools[DEplot2]

### plots[pointplot]([$pt_1$, ...,$pt_n$], var=a..b)

Displays the three-dimensional plot of points $pt_1$ through $pt_n$.
**Output:** A three-dimensional plot is displayed.
**Argument options:** ({$pt_1$, ...,$pt_n$}) to display the same plot. ♣ ([$pt_1$, ...,$pt_n$], option) to specify options to the plot. For more information on available options, see the plot3d command.
**Additional information:** All points must evaluate to lists containing three numeric values. ♣ To plot different types of three-dimensional plots or plots with different ranges on the same set of axes, use plots[display3d].
**See also:** plot3d, plots[polyhedraplot]

### plots[polarplot]([$expr_r$, $expr_a$, var=a..b])

Displays a two-dimensional parametric polar plot defined by the radius $expr_r$ and angle $expr_a$, both expressions in var, as var ranges from a to b.
**Output:** A two-dimensional plot is displayed.
**Argument options:** ($expr_r$, var=a..b) to display a polar plot where $expr_r$ represents the radius. If no range is supplied, $-\pi$ to $\pi$ is used. ♣ ([$expr_r$, $expr_a$, var=a..b], option) to specify options to the plot. For more information on available options, see the plot command.
**Additional information:** To plot different types of two-dimensional plots or plots with different ranges on the same set of axes, use plots[display].
**See also:** plot, plots[cylinderplot], plots[tubeplot], plots[sphereplot]
**FL = 163−164**

## plots[polygonplot]([pt$_1$, ..., pt$_n$])

Creates a two-dimensional polygon with n vertices, consisting of straight lines connecting in sequence points pt$_1$ through pt$_n$. pt$_n$ is connected to pt$_1$ to complete the circuit.

**Output:** A two-dimensional plot is displayed.

**Argument options:** ([[pt$_{1,1}$, ..., pt$_{1,n1}$], ..., [pt$_{m,1}$, ..., pt$_{m,nm}$]]) or ({[pt$_{1,1}$, ..., pt$_{1,n1}$], ..., [pt$_{m,1}$, ..., pt$_{m,nm}$]}) to create a plot of m polygons with n1 through nm vertices, respectively. ♣ ([pt$_1$, ..., pt$_n$], option) to specify options to the plot. For more information on available options, see the plot command.

**Additional information:** All points must evaluate to lists containing two numeric values. ♣ To plot different types of two-dimensional plots or plots with different ranges on the same set of axes, use plots[display].

**See also:** plot, plots[polygonplot3d], plots[polyhedraplot]


## plots[polygonplot3d]([pt$_1$, ..., pt$_n$])

Creates a three-dimensional polygon with n vertices, consisting of straight lines connecting in sequence points pt$_1$ through pt$_n$. pt$_n$ is connected to pt$_1$ to complete the circuit.

**Output:** A three-dimensional plot is displayed.

**Argument options:** ([[pt$_{1,1}$, ..., pt$_{1,n1}$], ..., [pt$_{m,1}$, ..., pt$_{m,nm}$]]) or ({[pt$_{1,1}$, ..., pt$_{1,n1}$], ..., [pt$_{m,1}$, ..., pt$_{m,nm}$]}) to create a plot of m polygons with n1 through nm vertices, respectively. ♣ ([pt$_1$, ..., pt$_n$], option) to specify options to the plot. For more information on available options, see the plot3d command.

**Additional information:** All points must evaluate to lists of three numeric values. To plot different types of three-dimensional plots or plots with different ranges on the same set of axes, use plots[display3d].

**See also:** plot3d, plots[polyhedraplot], plots[polygonplot]


## plots[polyhedraplot](pt, *polyscale* = n, *polytype* = **type**)

Creates a three-dimensional polyhedron of type type and scale n centered at point pt.

**Output:** A three-dimensional plot is displayed.

**Argument options:** ([pt$_1$, ..., pt$_n$], *polyscale* = n, *polytype* = type) to create a point plot, where points are represented by polyhedra of type type.

**Additional information:** All points must evaluate to lists of three numeric values. ♣ The types of polyhedra available are *tetrahedron*, *octahedron*, *hexahedron*, *dodecahedron*, and *icosahedron*. ♣ The default polyhedron type is *tetrahedron* and the default scale is 1. ♣ To plot different types of three-dimensional

plots or plots with different ranges on the same set of axes, use plots[display3d].
**See also:** plot3d, plots[polygonplot], plots[pointplot]

### plots[replot](2DP, option)

Redisplays a two-dimensional plot structure 2DP.
**Output:** A two-dimensional plot is displayed.
**Argument options:** (3DP, option) to redisplay a three-dimensional plot structure.
**Additional information:** option represents the new options to the plot. The new value overwrites the old if there is any conflict.
✦ plots[replot] is often used in conjunction with the *view* option to zoom in on particular areas of a plot. Keep in mind that the values for a plot are *not* recomputed by plots[replot], so zooming in does not increase the amount of detail. ✦ plots[display] and plots[display3d] can be used to produce similar results.
**See also:** plot, plot3d, plots[display], plots[display3d]

### plots[setoptions](option$_1$ = expr$_1$, ..., option$_n$ = expr$_n$)

Controls default options for two-dimensional plots by setting options option$_1$ through option$_n$ equal to expr$_1$ through expr$_n$, respectively.
**Output:** A NULL output is returned.
**Additional information:** All subsequent two-dimensional plots in the current session use the new option values (until, of course, plots[setoptions] is used again). ✦ If you want to customize defaults for *all* your sessions, place a plot[setoptions] command in your initialization file. ✦ For more information on available options, see the plot command.
**See also:** plot, plots[setoptions3d], plots[textplot]

### plots[setoptions3d](option$_1$ = expr$_1$, ..., option$_n$ = expr$_n$)

Controls default options for three-dimensional plots by setting options option$_1$ through option$_n$ equal to expr$_1$ through expr$_n$, respectively.
**Output:** A NULL output is returned.
**Additional information:** All subsequent three-dimensional plots in the current session use the new option values (until, of course, plots[setoptions3d] is used again). ✦ If you want to customize defaults for *all* your sessions, place a plot[setoptions3d] command in your initialization file. ✦ For more information on available options, see the plot3d command.
**See also:** plot3d, plots[setoptions], plots[textplot3d]

## plots[spacecurve]([expr$_1$, expr$_2$, expr$_3$], var=a..b)

Displays the three-dimensional curve defined by parametric representation expr$_1$, expr$_2$, and expr$_3$ (all expressions in var), as var ranges from a to b.

**Output:** A three-dimensional plot is displayed.

**Argument options:** ([expr$_1$, expr$_2$, expr$_3$, var=a..b]) to produce the same results. ♣ ([pt$_1$, ..., pt$_n$], var=a..b) or ([pt$_1$, ..., pt$_n$, var=a..b]) to display the spacecurve connecting the points in three-element lists pt$_1$ through pt$_n$. ♣ ([expr$_1$, expr$_2$, expr$_3$], var=a..b, *numpoints* = n) to specify the number of points sampled along the curve. Default is 50. ♣ ([expr$_1$, expr$_2$, expr$_3$], var=a..b, option) to specify options to the plot. For more information on available options, see the plot3d command.

**Additional information:** The plots[tubeplot] command can be used to plot a tube wrapped about a spacecurve. ♣ To plot different types of three-dimensional plots or plots with different ranges on the same set of axes, use plots[display3d].

**See also:** plot3d, plots[pointplot], plots[tubeplot]

**FL = 182**

## plots[sparsematrixplot](M)

Displays a two-dimensional plot of sparse matrix M, where every nonzero element of M is represented by a point at the $x, y$-coordinate corresponding to its row and column placement.

**Output:** A two-dimensional plot is displayed.

**Argument options:** (M, option) to specify options to the matrix plot. For more information on available options, see the plot command.

**Additional information:** All elements of M must evaluate to numeric values.

**See also:** plot3d, plots[sparsematrixplot], plots[surfdata]

## plots[sphereplot](expr, var$_1$=a$_1$..b$_1$, var$_2$=a$_2$..b$_2$)

Displays a three-dimensional spherical plot of radial expression expr as var$_1$ ranges from a$_1$ to b$_1$ and var$_2$ ranges from a$_2$ to b$_2$.

**Output:** A three-dimensional plot is displayed.

**Argument options:** ([expr$_r$, expr$_\theta$, expr$_\phi$], var$_1$=a$_1$..b$_1$, var$_2$ = a$_2$..b$_2$) to plot the parametric representation, where expr$_r$, expr$_\theta$, and expr$_\phi$ are functions in var$_1$ and var$_2$ representing the radius, $\theta$, and $\phi$ coordinates, respectively. ♣ (expr, var$_1$=a$_1$..b$_1$, var$_2$=a$_2$..b$_2$, option) to specify options to the plot. For more information on available options, see the plot3d command.

Plotting 411

**Additional information:** To plot different types of three-dimensional plots or plots with different ranges on the same set of axes, use plots[display3d].
**See also:** plot3d, plots[tubeplot], plots[cylinderplot], plots[polarplot]
**FL = 180−181**

## plots[surfdata]([[pt$_{1,1}$, ..., pt$_{1,n}$], ..., [pt$_{m,1}$, ..., pt$_{m,n}$]])

Displays the three-dimensional surface defined by the list of lists of points provided. Each point is a list of three numeric values.
**Output:** A three-dimensional plot is displayed.
**Argument options:** (convert(M, *listlist*)) to convert a matrix of points M to a list of lists and then plot its surface. ♣ ([[pt$_{1,1}$, ..., pt$_{1,n}$], ..., [pt$_{m,1}$, ..., pt$_{m,n}$]], option) to specify options to the plot. For more information on available options, see the plot3d command.
**Additional information:** To plot different types of three-dimensional plots or plots with different ranges on the same set of axes, use plots[display3d].
**See also:** plot3d, plots[matrixplot], plots[sparsematrixplot]

## plots[textplot]([expr$_x$, expr$_y$, str])

Creates a set of $x, y$ axes with text string str centered both horizontally and vertically around point [expr$_x$, expr$_y$].
**Output:** A two-dimensional plot is displayed.
**Argument options:** ([[expr$_{x1}$, expr$_{y1}$, str$_1$], ..., [expr$_{xn}$, expr$_{yn}$, str$_n$]]) or ({[expr$_{x1}$, expr$_{y1}$, str$_1$], ..., [expr$_{xn}$, expr$_{yn}$, str$_n$]}) to create a plot with strings str$_1$ through str$_n$ at their respective $x, y$-coordinates. ([expr$_x$, expr$_y$, str], *align* = option), where option is one of or a set of *BELOW*, *ABOVE*, *LEFT*, or *RIGHT* to align the text accordingly about the point. If both *ABOVE* and *BELOW* or *LEFT* and *RIGHT* are specified, then *ABOVE* and *RIGHT* always take precedence. ♣ For more information on the available options, see the plot command.
**Additional information:** The results are usually used in combination with other two-dimensional plots to label key values on a curve. ♣ plots[display] is used to facilitate the merging of plots.
**See also:** plot, plots[display], plots[textplot3d]

## plots[textplot3d]([expr$_x$, expr$_y$, expr$_z$, str])

Creates a set of $x, y, z$ axes with text string str centered both horizontally and vertically around point [expr$_x$, expr$_y$, expr$_z$].
**Output:** A three-dimensional plot is displayed.

**Argument options:** ([[expr$_{x1}$, expr$_{y1}$, expr$_{z1}$, str$_1$], ..., [expr$_{xn}$, expr$_{yn}$, expr$_{zn}$, str$_n$]]) or ({[expr$_{x1}$, expr$_{y1}$, expr$_{z1}$, str$_1$], ..., [expr$_{xn}$, expr$_{yn}$, expr$_{zn}$, str$_n$]}) to create a plot with strings str$_1$ through str$_n$ at their respective $x, y, z$-coordinates. ♣ ([expr$_x$, expr$_y$, expr$_z$, str], *align* = option), where option is one of or a set of *BELOW*, *ABOVE*, *LEFT*, or *RIGHT* to align the text accordingly about the point.

**Additional information:** Note that perspective has no effect upon text—all text is printed at an equivalent font size. ♣ The results are usually used in combination with other three-dimensional plots to label key values on a curve. ♣ plots[display3d] is used to facilitate the merging of plots.

**See also:** plot3d, plots[display3d], plots[textplot]

### plots[tubeplot]([expr$_1$, expr$_2$, expr$_3$], var=a..b, *radius* = r)

Displays the three-dimensional tubeplot defined by the spacecurve defined by expr$_1$, expr$_2$, and expr$_3$ (all expressions in var) as var ranges from a to b and by the radius r of the tube.

**Output:** A three-dimensional plot is displayed.

**Argument options:** ([expr$_1$, expr$_2$, expr$_3$], var=a..b]) to display the same plot. ♣ ([pt$_1$, pt$_2$, ...,pt$_n$], var=a..b) or ([pt$_1$, pt$_2$, ...,pt$_n$], var=a..b]) to display the spacecurve connecting the points in three element lists pt$_1$ through pt$_n$. ♣ ([expr$_1$, expr$_2$, expr$_3$], var=a..b, *radius* = r, *numpoints* = n) to specify the number of points sampled along the spacecurve. Default is 50. ♣ ([expr$_1$, expr$_2$, expr$_3$], var=a..b, *radius* = r, *tubepoints* = n) to specify the number of points to be sampled around the tube. Default is 10. ♣ ([expr$_1$, expr$_2$, expr$_3$], var=a..b, option) to specify options to the plot. For more information on available options, see the plot3d command.

**Additional information:** The radius r can be either numerically-valued or an expression in var. ♣ To plot different types of three-dimensional plots or plots with different ranges on the same set of axes, use plots[display3d].

**See also:** plot3d, plots[spacecurve]

*LA = 169*

### plotsetup(device)

Sets all the necessary interface plot options for standard use of supported output device of type device.

**Output:** A NULL expression is returned.

**Argument options:** (device, terminal) to further specify the supported terminal type being used.

**Additional information:** The available devices are *mac* for Macintosh terminals, *x11* for x11 terminals, *tek* for Tektronix terminals, *regis* for Regis terminals, *dumb* for character plots, *vt100* for VT100 line terminals, *i300* for imagen 300 laser printers, *ln03* for DEC LN03 laser printers, *hplj* for HP Laserjet printers, *hpgl* for HP GL printers, *ps* for PostScript printers or files, *cps* for Color PostScript printers or files, *pic* for troff pic file format, and *unix* for UNIX *plot* command file format. ♣ The available terminals that can be used in combination with the *tek* device are *vt###* (where ### is the number of the terminal type), *kd500g*, *kd404g*, and *wy99gt*. ♣ If a file is being written to, Maple informs you of the default file name being used.

**See also:** plot, plot3d, interface(*plotdevice*), interface(*plotoutput*), interface(*preplot*), interface(*postplot*)

**FL = 28−31** *LI = 163*

### type(expr, *PLOT*)

Determines whether expr is a PLOT data structure.
**Output:** A boolean value of either true or false is returned.
**See also:** op(PLOT), type(PLOT3D)
*LI = 236*

### type(expr, *PLOT3D*)

Determines whether expr is a PLOT3D data structure.
**Output:** A boolean value of either true or false is returned.
**See also:** op(PLOT3D), type(PLOT)
*LI = 236*

# Programming and System Commands

## Introduction

Over ninety percent of Maple's built-in commands are programmed in Maple's own Pascal-like programming language. Procedural programming is the very heart of Maple. While many useful and exciting things can be done in Maple without doing any programming at all, you will surely find yourself delving deeper and deeper into the programming possibilities as you become adept at using Maple.

There are two major reasons to use Maple's programming language. The most straightforward reason is to automate repetitive calculations that you are doing in "command-line" mode. The more important reason is to create your own Maple commands, augmenting the existing library. The first of these reasons can be explained quite thoroughly here, the latter is well beyond this tutorial.

This chapter takes a look at the programming constructs used in Maple, as well as the existing commands that are most often used within the bodies of other Maple procedures.

In addition, this chapter examines various system commands for doing such things as reading from and writing to external files, performing debugging and tracing of procedures, determining the current status of a session, and setting values that control how Maple interacts with both you and your computer.

## Basic Programming Constructs

Maple offers the basic programming constructs found in most other programming languages. Following is a short tour of each of them.

## if/then/else/fi

Maple handles conditional statements with the if/then/else/fi construct. Basically, you provide Maple with conditions (that evaluate to either true or false) and Maple branches accordingly.

```
> if 9/23 < 13/33 then 13/31 else 9/23 fi;
```

$$\frac{13}{31}$$

The above example compares two fractions and prints out the larger. Note the fi (if backwards) at the end—this is the terminator for the structure. Any amount of code can be inserted after each of then and else; you need not restrict yourself to a single expression as above. Note also that the entire structure ends in a semicolon. This is not a coincidence; the if/then/else/fi structure is a Maple expression as much as any command or polynomial, and it must end with a valid terminator.

An extra element, elif, can be added to make it if/then/elif/then ... /else/fi. Any number of these secondary conditions can be provided.

```
> if isprime(221) then p elif numtheory[issqrfree](221)
> then s else neither fi;
```

$$s$$

## for/from/by/to/do/od

Maple handles repetition with three different constructs. The first, for/from/by/to/do/od, takes an upper and lower bound on a repetition variable and a step value, then performs the statements between do and od the appropriate number of times. The repetition variable can be included in the calculations, but does not strictly have to be. Again, as with if/then/else/fi, any number of statements can be included between do and od.

```
> for i from 1 to 11 by 2 do print(i!) od;
```

$$1$$
$$6$$
$$120$$
$$5040$$

362880

39916800

## for/from/by/while/do/od

Repetition can also be done with the for/from/by/while/do/od structure. Maple loops through the statement(s) between do and od as long as the condition stated after while is true.

```
> for x from 1 by 4 while ithprime(x) < 100 do
> print(x, ithprime(x)); od;
```

$$1, 2$$
$$5, 11$$
$$9, 23$$
$$13, 41$$
$$17, 59$$
$$21, 73$$
$$25, 97$$

```
> x := 'x':
```

## for/in/do/od

Repetition can also be achieved with the for/in/do/od structure. Maple loops through the statement(s) between do and od once for each operand of the expression following in.

```
> i := 'i':
> for i in 4*x-3*y-6 do i/2 od;
```

$$2x$$

$$-\frac{3}{2}y$$

$$-3$$

## Procedures

You have now learned how to use the basic constructs of Maple's programming language. But the above examples only work when they are typed in fully. The major advantage in programming is that you can set up procedures that *automate* these calculations for you. To create a procedure that you can use over and over again with different input, you must use the proc/local/global/options/end structure.

When you define a procedure, you extend Maple's functionality. And if you save the procedure appropriately, this extension can be carried over to other Maple sessions. In the following example, a very simple procedure, largestfactor, takes integer n and finds its largest prime factor (and its multiplicity).

```
> largestfactor := proc(n:integer)
>     op(nops(ifactor(n)), ifactor(n));
> end;
largestfactor := proc(n:integer)
op(nops(ifactor(n)),ifactor(n)) end
```

Basically, we have pulled off the last operand (the placement of which is equal to the number of operands) of the integer factorization of n. Let's try using it a few times.

```
> largestfactor(2387);
```

$$(31)$$

```
> largestfactor(118277523);
```

$$(23)^2$$

Of course, this is a extremely simplified example of a Maple procedure. Many procedures in the Maple library are hundreds of lines long and contain extensive type testing, error checking, and alternative algorithms.

For a more detailed explanation of Maple programming, refer to your other Maple manuals.

# Extending Existing Commands

While most programming is designed to add *new* commands to Maple, there is another very important area of programming that should not be overlooked. Many of Maple's more standard commands (and some that are are not so standard) can be extended to operate over new, user-defined data types. The trick is to program the functionality as the command 'fnc/newtype', where fnc is the existing command and newtype the new data type.

For example, let's say that you create a new type that is represented by the unevaluated command call mytype(a, b, var), where a and b are expressions in var. Furthermore, let's assume that the integral of mytype(a, b, var) equals int(a + b, var). By programming the appropriate functionality in the procedure 'int/mytype' and saving it in the library directory under that name, you update the workings of int. From then on, whenever an expression passed to int contains a call to mytype, your new procedure is invoked.

While there are many such existing Maple commands that can be extended in this manner, there are many more that cannot. The best clue to whether a command is extensible is if it already is able to handle a wide variety of unevaluated commands as input. For a better feel, take a look at the source code for the command, if possible.

# Determining Data Types

Maple is capable of creating a wide variety of data structures. Because any of these structures could be returned to you as output to a command, it is essential that you have a good grasp of the tools that allow you to work with them. In many cases you need to identify a structure before passing it to another command.

Each object in Maple has a type associated with it. These types range from integer, to list, to '+' (a sum of objects). Understanding the type of an object is often imperative for using that object in a calculation—many Maple commands only take specific object types as input. As well, when investigating the elements that make up an object, it is also important to know its type. whattype allows us to determine an object's type.

```
> whattype(34/57);
```

$$fraction$$

420    Programming

```
> whattype([1,2,3,4,5]);
```

$$list$$

```
> whattype((x+3)*(y-4));
```

$$*$$

Keep in mind that while you can always query the type of an object, there is no way to stipulate that a certain variable is always of a certain type (i.e., there is no strong typing in Maple). For example, there is no way to specify that variable j must always be of type integer.

In many cases, especially in programming with Maple, there are times when you want to do different calculations depending on the type of a variable. type allows you to query whether a variable is of a specified type.

```
> greetings := 'hello there':
> type(greetings, integer);
```

$$false$$

```
> type(greetings, string);
```

$$true$$

One detail about data types that may confuse you is that some types are *compound*. That is, they are constructed out of other, more basic, data types. This in itself is not difficult to understand, but the effect on the whattype and type commands can be misleading. What is important to remember is that type allows you to search for certain compound types, but whattype only displays basic data types.

In the following example, g is assigned to a polynomial with integer coefficients. type shows that g is indeed recognized as a polynomial, but whattype declares g to be a sum.

```
> g := 3*x^2 + 2*x - 1:
> type(g, polynom);
```

$$true$$

```
> whattype(g);
```

$$+$$

Even finer details can be determined with type (e.g., that g has integer coefficients).

```
> type(g, polynom(integer));
```

$$true$$

Two other useful commands are hastype, that tells you whether an object contains a subobject of the given type, and has, that tells you if a certain subobject is contained within an object.

```
> hastype((x+1/2)*exp(3), fraction);
```

$$true$$

```
> hastype(x^2+3*x+5, '*');
```

$$true$$

```
> has(x^2+3*x+5, 3);
```

$$true$$

```
> has(x^2+3*x+5, 2*x);
```

$$false$$

While these examples are fairly obvious, hastype and has are invaluable when dealing with very large objects.

## Parameter Types Specific to This Chapter

| | |
|---|---|
| str$_{fmt}$ | a formatting string |
| pckg | a package name |

# Command Listing

### addressof(expr)

Calculates a pointer to the place in memory that holds expression expr.
**Output:** An integer address is returned.
**Additional information:** To determine what Maple object lies at a certain memory location, use pointto. ♣ For more information on this and other ''hardware'' commands, see the on-line help page.
**See also:** pointto, assemble, disassemble
*LI = 11*

### anames()

Returns an expression sequence containing *all* the names currently assigned to values other than their own names.
**Output:** An expression sequence is returned.
**Additional information:** The result of anames includes names of constants and procedure used within other commands. ♣ There will be names that you have not seen before, and they will most likely outnumber the names you have seen before.
**See also:** assigned, unames

### appendto(filename)

Appends all subsequent output to file filename.
**Output:** A NULL expression is returned.
**Additional information:** When appendto is activated, no output appears on your terminal. ♣ Like the writeto command, a call to appendto(*terminal*) resets the output stream to the terminal.
**See also:** save, open, write, writeln, writeto

### args[n]

Returns, within a procedure, the $n^{th}$ argument (parameter) passed to that procedure.
**Output:** An expression is returned.
**Argument options:** args[i..j] to return an expression sequence consisting of the $i^{th}$ through the $j^{th}$ parameters.
**Additional information:** Typically, more parameters than are stipulated in the proc command of a procedure can be passed to that procedure. The args construct is useful in determining the values of those extra parameters. ♣ Use nargs to determine the number of parameters passed to a procedure.
**See also:** nargs, proc/local/options/end
**FL = 124** *LA = 115−116*

## assemble(int$_1$, ..., int$_n$)

Assembles integer addresses int$_1$ through int$_n$ into a Maple object.
**Output:** An integer value representing a pointer to the new object is returned.
**Additional information:** Before using this command, you *must* be very familiar with the internal representation of Maple data; otherwise you could very easily cause serious errors. ♣ To break an object down into its component parts, use disassemble. ♣ For more information on this and other "hardware" commands, see the on-line help page.
**See also:** disassemble, pointto, addressof
*LI = 11*

## assigned(name)

Determines whether name is already assigned to a value other than its own name.
**Output:** A boolean value of true or false is returned.
**Argument options:** (T[expr]) or (A[int$_1$, ..., int$_n$]) to determine whether the subscripted element of table T or n-dimensional array A is assigned a value other than its own name. ♣ (fnc(expr$_1$, ..., expr$_n$)) to determine whether the function fnc, evaluated at the parameters expr$_1$ through expr$_n$, has a value other than its own name. These may appear in the remember table for fnc.
**Additional information:** assigned is most often used to check whether an object is already assigned to some other value before using it as a programming variable.
**See also:** evaln, anames, unames
*LI = 13*

## break

A programming construct for exiting prematurely from a loop.
**Output:** Not applicable.
**Additional information:** This construct forces an exit from the innermost for/do/od loop in which it is contained. The statement directly after the for/do/od structure is then evaluated. ♣ If break is found within any other type of construct, an error message is returned. ♣ To break out of the current iteration of a for/do/od loop only, use the next construct.
**See also:** for/from/by/to/do/od, for/from/by/while/do/od, for/in/do/od, next, proc/local/options/end
**FL = 120–121**

## close()

Closes the currently open file.
**Output:** A NULL expression is returned.

**Additional information:** Using close is basically the same as using writeto(*terminal*). It is recommended that you stick to the writeto form. ♣ This command needs to be defined by readlib(write) before being invoked.
**See also:** open, write, writeln, writeto, appendto

### convert(filename, *hostfile*)

Converts filename, a Maple file name, into the file name as used on the host system.
**Output:** A string is returned.
**Additional information:** This command provides an interface with Maple's UNIX style file names for users on other platforms.
**See also:** read, save
**FL = 143** *LA = 36, 127−128* **LI = 41**

### cost(expr)

Calculates the operation count, or cost, of numerical evaluation of algebraic expression expr.
**Output:** A polynomial with non-negative integer coefficients and variable names additions, multiplications, divisions, functions, subscripts, and assignments is returned.
**Argument options:** (A) to return the total cost of all algebraic expressions in array A.
**Additional information:** This command needs to be defined by readlib(cost) before being invoked. ♣ If you wish to determine a *weighted cost*, assign numeric values to the variables additions, multiplications, etc. ♣ Use optimize to attempt to save operation cost on algebraic expressions.
**See also:** *optimize*
*LA = 26−27* **LI = 273**

### define(option(name))

Defines an operator name (typically beginning with the character &) over abstract algebraic object option. option can be either *group* or *linear*.
**Output:** A NULL expression is returned.
**Argument options:** (option($name_1$, ..., $name_n$)) to define n operators over the same abstract algebraic object. (oper, $str_1$, $str_2$, ..., $str_n$), where oper is an already defined operator, to assign properties $str_1$ through $str_n$ to an operator.
**Additional information:** For more information about defining operators and specific properties that can be assigned, see the on-line help pages for define, define[operator], define[linear], define[group], and define[forall].
*LA = 134* **LI = 58−61**

### disassemble(int)

Disassembles the Maple object at integer addresses int into its component parts.
**Output:** An expression sequence of integer addresses is returned.
**Additional information:** Before using this command, you *must* be very familiar with the internal representation of Maple data; otherwise you could very easily cause serious errors. ♣ To assemble an object from its component parts, use assemble. ♣ For more information on this and other "hardware" commands, see the on-line help page.
**See also:** assemble, pointto, addressof
*LI = 11*


### done

This command is equivalent to the quit command and terminates your Maple session.
**See also:** quit


### ERROR(expr$_1$, ..., expr$_n$)

Forces an explicit return from within a procedure because of an error. The result of the procedure is set to NULL.
**Output:** Not applicable.
**Additional information:** An error message is displayed, which starts with the string Error, (in fnc), where fnc is the procedure name, and then has the values of expressions expr$_1$ through expr$_n$ printed in order. ♣ Most often the expr$_i$ are strings containing additional information about the error or value of variables that caused the error to be triggered. ♣ The remaining statements in the procedure's statement block, the procedure that called it, the procedure that called it, and so on, are ignored and the result NULL is immediately returned. ♣ To trap the error, preventing it from propagating to the top, use the traperror command.
**See also:** RETURN, NULL, traperror, lasterror, proc/local/options/end
**FL = 128−131** *LA = 85, 123−124* **LI = 73**


### extract(str)

Extracts the maximal element from priority queue str.
**Output:** A two-element list is returned representing the priority of the maximal element and the element itself, respectively.
**Additional information:** This command needs to be defined by readlib(priqueue) before being invoked. ♣ Before a priority queue

can be accessed, it must first be defined with initialize. ♣ Elements are sorted in the queue in numerical order depending on their priorities. ♣ To insert an element into a priority queue, use insert. ♣ The number of entries in a priority queue can be accessed with str[0].
**See also:** initialize, insert, heap
*LI = 310*

### FAIL

A value that can be passed back as the result of a procedure to indicate that the computation was abandoned or failed altogether.
**Output:** Not applicable.
**Additional information:** FAIL is *not* meant to indicate that a procedure has no solution. For that, use NULL, an empty set { }, or an empty list [].
**See also:** RETURN, ERROR, NULL, traperror, lasterror, proc/local/options/end

### for/from/by/to/do/od

A programming construct for looping through several iterations.
**Output:** Not applicable.
**Additional information:** This construct allows you to execute a block of statements for several iterations of the variable following for. ♣ After the first iteration of the block of statements following do, the iteration variable, which has starting value determined by the value after from, is incremented by the value after by. At that time, it is ascertained whether the iteration variable is within the bound represented by the value following to. If so, the block of statements is executed again and the iteration variable is updated and rechecked. If not, the construct is exited. ♣ The entire structure must be terminated with od. ♣ If the keyword next is executed within this construct, the remaining statements in the block are skipped and the iteration variable is updated. ♣ If the keyword break is executed within this construct, the entire construct is exited. ♣ Keep in mind that this structure is valid without a from and/or by. Both are set to a default value of 1. ♣ Other keywords

can be used in conjunction with for, do, and od. See the entries for for/in/do/od and for/from/by/while/do/od for more information.
**See also:** for/in/do/od, for/from/by/while/do/od, proc/local/options/end, seq
**FL = 116−117**


### for/from/by/while/do/od

A programming construct for looping through several iterations.
**Output:** Not applicable.
**Additional information:** This construct allows you execute a block of statements for several iterations of the variable following for. ✦ After the first iteration of the block of statements following do, the iteration variable, which has starting value determined by the value after from, is incremented by the value after by. At that time, the iteration variable is checked to see if it meets the requirements of the boolean statement that follows while. If so, the block of statements is executed again. If not, the construct is exited. ✦ This structure must be terminated with od. ✦ If the keyword next is executed within this construct, the remaining statements in the block are skipped. ✦ If the keyword break is executed within this construct, the construct is exited. ✦ Keep in mind that this structure is valid without a from and/or by. Both are set at a default value of 1. ✦ Other keywords can be used in conjunction with for, do, and od. See the entries for for/from/by/to/do/od and for/in/do/od for more information.
**See also:** for/from/by/to/do/od, for/in/do/od, proc/local/options/end, seq
**FL = 116−117**


### for/in/do/od

A programming construct for looping through several iterations.
**Output:** Not applicable.
**Additional information:** This construct allows you execute a block of statements for several iterations of the variable following for. ✦ The block of statements following do, is executed with the iteration variable replaced by the first operand of the expression following in. The same block of statements is then executed with the iteration variable replaced by each subsequent operand of that expression. ✦ This structure must be terminated with od. ✦ If the keyword next is executed within this construct, the remaining statements in the block are skipped and the next operand is substituted for the iteration variable. ✦ If the keyword break is executed

within this construct, the entire construct is exited. ♣ Other keywords can be used is conjunction with for, do, and od. See the entries for for/from/by/to/do/od and for/from/by/while/do/od for more information.

**See also:** for/from/by/to/do/od, for/from/by/while/do/od, proc/local/options/end, seq

**FL = 118** *LA = 18*

### forget(fnc, expr)

Removes the entry for fnc(expr) from the remember table of fnc.

**Output:** A NULL statement is returned.

**Argument options:** (fnc) to remove *all* entries from the remember table of fnc.

**Additional information:** Remember that expr, the index of the remember table entry, can be any type of expression including an expression sequence. ♣ To assign a value to the remember table of fnc, simply enter the command fnc(expr) := value.

**See also:** options, remember, proc/local/options/end

*LI = 288*

### freeze(expr)

Replaces expression expr with a variable name of the form _R0, _R1, etc.

**Output:** A variable name is returned.

**Additional information:** To *release* a frozen expression, use thaw. ♣ These commands are primarily used when you want a certain subexpression to be unaffected when its larger expression is acted upon.

**See also:** thaw, frontend

*LI = 289*

### frontend(proc, [expr$_1$, ..., expr$_n$])

Freezes each argument expr$_1$ through expr$_n$ before passing it to command proc.

**Output:** An expression representing the result of proc is returned.

**Argument options:** (proc, [expr$_1$, ..., expr$_n$], [set$_1$, set$_2$]) to specify that set$_1$ and set$_2$, which contain types and expression, respectively, are *not* to be frozen. The default value is [{'+','*'},{}].

**Additional information:** You cannot use frontend on any command that actively assigns a value to one or more of its parameters. ♣ frontend only freezes expressions for *one* invocation of proc. ♣ For more information, see the on-line help page.

**See also:** freeze, thaw

*LI = 96*

## Gauss

Initiate a specialized Maple programming environment that revolves around parameterized domains.

**Output:** Not applicable.

**Additional information:** The Gauss programming environment is loaded by calling with(Gauss). ✤ To learn more about parameterized domains and *domain of computation* programming, see the on-line help pages for Gauss, Gauss[coding], and Gauss[domain]. For examples, see the on-line help page for Gauss[example].

## gc()

Performs garbage collection, attempting to reclaim all data to which no references are made.

**Output:** A NULL expression is returned. As well, a *bytes used* message is automatically printed.

**Argument options:** (n) to specify that garbage collection is to happen automatically after every n words used. ✤ (0) to specify that bytes used messages are not to be printed after subsequent calls to gc. Keep in mind that this has no effect on the frequency of garbage collection.

**Additional information:** The current frequency of garbage collection can be accessed through status[5].

**See also:** status, words

**FL = 187−188**

## heap[new](fnc, expr$_1$, expr$_2$, ..., expr$_n$)

Creates a new heap structure which includes elements expr$_1$ through expr$_n$ and uses the boolean command fnc to specify the total ordering of the heap.

**Output:** A table structure representing the heap is returned.

**Argument options:** heap[insert](expr, H) to insert element expr into heap H. ✤ heap[extract](H) to extract an element from heap H. ✤ heap[empty](H) to determine if heap H is empty. ✤ heap[size](H) to determine the number of elements in heap H. ✤ heap[max](H) to return the maximum element, that is, the next element to be extracted, from heap H without actually extracting it from the heap.

**Additional information:** This command needs to be defined by readlib(heap) before being invoked.

**See also:** initialize, insert, extract

*LI = 291*

### history()

Activates the Maple *history* substructure.
**Output:** A NULL expression is returned.
**Additional information:** This command needs to be defined by readlib(history) before being invoked. ♣ After history() has been entered, each subsequent input line is prompted by a string in the sequence O1 :=, O2 :=, etc. The outputs for each input are then assigned to names O1, O2, etc. ♣ While the history facility is operating, enclosing any input in a timing() command also displays the amount of CPU time used to run that command. ♣ The history facility provides more access to previous outputs than the ", "", and """ operators can; and while it is in operation, these operators do not work. ♣ To terminate the history facility, enter off. ♣ To clear variable names O1, O2, etc., enter clear.
**See also:** time, showtime
**FL = 186−187** *LI = 292*


### if/then/elif/then/else/fi

A programming construct for selection of different paths.
**Output:** Not applicable.
**Additional information:** This construct allows you to branch off the results of boolean statements placed after if and elif. ♣ If the boolean statement following if evaluates to true, the block of statements following the subsequent then is executed and the construct is exited. If the boolean statement following if evaluates to false, the boolean statement following elif is tested. If that is true, the block of statements following the subsequent then is executed and the construct is exited. ♣ This procedure continues until an elif boolean statement has been found to be true or until the last elif boolean statement is found to be false and the block of statements following else is executed. ♣ This structure must be terminated with fi. ♣ Keep in mind that this structure is valid without an elif and/or else.
**See also:** for/from/by/to/do/od, for/from/by/while/do/od, for/in/do/od, proc/local/options/end
**FL = 118−120**


### indets(expr)

Determines all indeterminates present in expr.
**Output:** A set of variable names is returned.
**Argument options:** (expr, name) to return all subexpressions of expr of type name. These may include some subexpressions that were not found by the single-parameter case of indets.

**Additional information:** Unevaluated functions or commands with indeterminate parameters (e.g., cos(x), g(x,y,z)) are considered to be indeterminates as well.
**See also:** *coeffs*
*LI = 109*

### infolevel[fnc] := n

Sets the information level for command fnc to a level of n for interaction with userinfo commands.
**Output:** The input assignment is echoed and n is returned.
**Argument options:** infolevel[all] := n to set the information level for *all* Maple commands to n.
**Additional information:** In the existing Maple library, userinfo commands have been liberally distributed with level 1 reserved for information the user must be told, 2 and 3 reserved for more general information, and 4 and 5 reserved for even more minute detail.
**See also:** userinfo, printlevel, trace
*LA = 180*

### initialize(name)

Creates an empty priority queue and assigns it to name.
**Output:** A value of 0 is returned.
**Additional information:** This command needs to be defined by readlib(priqueue) before being invoked. ♣ To insert an element into a priority queue, use the insert command. ♣ To extract an element from a priority queue, use the extract command.
**See also:** insert, extract, heap
*LI = 310*

### insert([num, expr], str)

Adds a record for element expr, with priority set by numeric value num, to priority queue str.
**Output:** A NULL expression is returned.
**Additional information:** This command needs to be defined by readlib(priqueue) before being invoked. ♣ Before a priority queue can be accessed, it must first be defined with the initialize command. ♣ The element is sorted in the queue in numerical order depending on the value of num. ♣ To extract an element from a priority queue, use the extract command. ♣ The number of entries in a priority queue can be accessed with str[0].
**See also:** initialize, extract, heap
*LI = 310*

### interface(echo = int)

Sets the level for echoing to the terminal to int. int must be one of 0, 1, 2, 3, or 4.
**Output:** A NULL expression is returned.
**Argument options:** (*echo*) to return the current value of the option. ♣ (*echo = 0*) to specify that absolutely no echoing is to be done. ♣ (*echo = 1*) to echo when the input or output is not from/to the terminal. This is the default value. ♣ (*echo = 2*) to echo whenever reading from a file or writing to a file. ♣ (*echo = 3*) to echo if a read statement is in effect. ♣ (*echo = 4*) to echo everything possible.
**Additional information:** To echo means to have the result of some action taken displayed to your terminal. ♣ Having interface (*echo*) always set to 4 causes far too much information to be displayed. ♣ If the interface(*quiet*) option is set to *true*, then no echoing is done regardless of the value of interface(*echo*).
**See also:** read, interface(*quiet*), interface(*verboseproc*)
*LI = 116*


### interface(indentamount = int)

Specifies the number of spaces that items are indented if they must be broken across more than one line.
**Output:** A NULL expression is returned.
**Argument options:** (*indentamount*) to return the current value of the option.
**Additional information:** The first line is not indented, but each subsequent line is indented int spaces. ♣ The default value is typically preset for each specific terminal type.


### interface(iris)

Returns user interface information for the version of Maple you are using.
**Output:** A string is returned.
**See also:** interface(*version*)


### interface(labeling = false)

Disables labeling of common subexpressions in output.
**Output:** A NULL expression is returned.
**Argument options:** (*labeling*) to return the current value of the option. ♣ (*labeling = true*) to specify that labeling is to be used. This is the default.
**Additional information:** Labeling of common subexpressions

makes very large expressions much easier to read and does *not* affect the internal structure of the expression. ♣ It is suggested that you leave labeling enabled in most cases.
**See also:** interface(*labeling*)

### interface(*labelwidth* = int)

Specifies that labeling of a common subexpression in output occurs only if that subexpression, as stored in memory, is at least int characters long.
**Output:** A NULL expression is returned.
**Argument options:** (*labelwidth*) to return the current value of the option.
**Additional information:** Setting value of interface(*labelwidth*) higher decreases the number of labels and increase the display size of large expressions.
**See also:** interface(*labeling*)

### interface(*prettyprint* = int)

Specifies how expressions are displayed.
**Output:** A NULL expression is returned.
**Argument options:** (*prettyprint*) to return the current value of the option. ♣ (*prettyprint = 0*) to specify that line printing through the lprint command is to be used. ♣ (*prettyprint = 1*) to specify that pretty-printing is to be used. ♣ (*prettyprint = 2*) to specify that typographical output or real mathematical representation is to be used, if possible for your terminal.
**Additional information:** While typographical output of expressions is easier to read, in most cases such output cannot be cut and pasted back as input to another command.
**See also:** print, lprint
*LA = 44*

### interface(*prompt* = str)

Specifies that the characters in str be used for the prompt which appears before every input line on the terminal.
**Output:** A NULL expression is returned.
**Argument options:** (*prompt*) to return the current value of the option.
**See also:** interface(*quiet*)
*LI = 116*

### interface(*quiet* = *true*)

Suppresses all auxiliary printing to the terminal.
**Output:** A NULL expression is returned.

**Argument options:** (*quiet*) to return the current value of the option. ♣ (*quiet = false*) to specify no suppression should be done. This is the default.

**Additional information:** If interface(*quiet*) is set to true, all values of interface(*echo*) are overridden and no logo, garbage collection messages, bytes used messages, or prompts are displayed.

**See also:** interface(*echo*), interface(*prompt*), gc, status

*LI = 116*

### interface(*screenheight* = int)

Specifies that the screen height of the terminal being used is int lines of characters.

**Output:** A NULL expression is returned.

**Argument options:** (*screenheight*) to return the current value of the option.

**Additional information:** Setting interface(*screenheight*) affects display of expressions and character plots. ♣ The default value for this option is typically preset for each specific terminal type.

**See also:** interface(*screenwidth*)

### interface(*screenwidth* = int)

Specifies that the screen width of the terminal being used is int characters.

**Output:** A NULL expression is returned.

**Argument options:** (*screenwidth*) to return the current value of the option.

**Additional information:** Setting interface(*screenwidth*) affects display of expressions and character plots. ♣ The default value for this option is typically preset for each specific terminal type.

**See also:** interface(*screenheight*)

### interface(*verboseproc* = int)

Sets the level for how procedures are to be printed to screen when read. int must be one of 0, 1, or 2.

**Output:** A NULL expression is returned.

**Argument options:** (*verboseproc*) to return the current value of the option. ♣ (*verboseproc = 0*) to specify that only a one-line skeleton of the procedure is displayed. ♣ (*verboseproc = 1*) to print all of user-defined procedures but only the skeleton of procedure from the library. This is the default value. ♣ (*verboseproc = 2*) to print all of *every* procedure read.

**Additional information:** The most common reason for increasing interface(*verboseproc*) to 2 is that you can then use print to view

the code for a Maple library function. Keep in mind, though, that you cannot see the code for commands in the kernel.
**See also:** read, print, interface(*echo*)
**FL = 152–153** *LA = 187–188*

### interface(*version*)

Returns version information for the version of Maple you are using.
**Output:** A string is returned.
**Additional information:** The same information, in a more understandable format, can often be found in the logo or splash screen that is displayed when Maple is started.
**See also:** interface(*iris*)

### interface(*wordsize*)

Returns the number of bits in a machine word for your system.
**Output:** A positive integer is returned.

### lasterror

A variable assigned to the last error message encountered in the session.
**Output:** Not applicable.
**Additional information:** When you trap an error with traperror, make sure to check that the string returned is an error and not just the normal result of an evaluation by testing whether the result of traperror is identical to lasterror. ♣ If you do not trap an error, it propagates itself all the way up to the top level of computation.
**See also:** traperror, ERROR
**FL = 129–131** *LI = 216*

### libname := str

Sets the variable libname, specifying the location in your system's file space of the directory containing the standard Maple library, to a value of str.
**Output:** The assignment is echoed.
**Argument options:** libname to display the current value of the variable. ♣ libname := $str_1$, ..., $str_n$ to specify n library directories.
**Additional information:** Be very careful to specify the correct directory. ♣ Multiple libraries are useful when you have large amounts of your own Maple code. ♣ When multiple libraries are specified and Maple is looking for a file, it searches the libraries

sequentially from str$_1$ to str$_n$. This allows for you to set up overrides for the standard Maple library routines, if desired.
**See also:** args, proc/local/options/end
*LA = 127−128*

### lprint(expr$_1$, ..., expr$_n$)

Displays expressions expr$_1$ through expr$_n$ in one-dimensional format (i.e., more or less as you would enter them as input).
**Output:** Each expression is printed on its own line (or lines if it is long enough), and subsequent expressions are separated by three blank lines.
**Additional information:** The expressions are displayed as a "side-effect" of lprint. The actual output is a NULL statement. This means that the " operators cannot recall the displayed expressions. ♣ The interface(*prettyprint*) command allows you to set the default display mode to lprint.
**See also:** print, printf, interface(*prettyprint*)
**FL = 35−36** *LI = 131*

### nargs

Represents, within a procedure, the number of arguments (parameters) passed to that procedure.
**Output:** A non-negative integer is returned.
**Additional information:** More parameters than are stipulated in the proc command of a procedure can be passed to that procedure. nargs is useful in determining how many arguments were passed. ♣ Use args[n] to determine the individual parameters passed.
**See also:** args, proc/local/options/end
**FL = 124** *LA = 116*

### next

A programming construct for proceeding directly to the next iteration of a looping structure.
**Output:** Not applicable.
**Additional information:** This construct forces an exit from the current iteration of the innermost for/do/od loop in which it is contained. The iteration variable is incremented and/or the termination test is performed and then the next iteration of the loop is started, if appropriate. ♣ If next is found within any other type of construct, an error message is returned. ♣ To break out of a for/do/od loop entirely, use break.
**See also:** for/from/by/to/do/od, for/from/by/while/do/od, for/in/do/od, break, proc/local/options/end

### NULL

Represents the null expression sequence.
**Output:** Not applicable.
**Additional information:** NULL is typically meant to indicate that a procedure has no solution. ♣ When NULL is returned as output, it is not displayed and is not retrievable by the ", " ", or " " " operators. ♣ To indicate that a procedure failed in its computations, use FAIL.
**See also:** RETURN, ERROR, FAIL, proc/local/options/end
*LA = 48*

### open(filename)

Opens a file filename for output.
**Output:** A NULL expression is returned.
**Additional information:** This command needs to be defined by readlib(write) before being invoked. ♣ Using open is basically the same as using writeto(filename). It is recommended that you stick to the writeto form. ♣ After performing an open command, all subsequent output is written to filename instead of to the terminal. ♣ When you are finished writing to the file, issue the close() command. ♣ Only one file can be open at the same time.
**See also:** close, write, writeln, writeto, appendto

### operator

Specifies that the operator option is to be used within a procedure.
**Output:** Not applicable.
**Additional information:** If the operator option is specified in the options sequence of a procedure, the procedure is considered a Maple operator (i.e., a function in -> or < | > notation).
**See also:** proc/local/options/end

### parse(str)

Parses string str as a Maple expression, adds a semicolon to the end (if no colon or semicolon is present), and returns it unevaluated.
**Output:** An unevaluated expression is returned.
**Argument options:** (str, *statement*) to parse str as a Maple expression and evaluate it. The evaluated result is returned.
**Additional information:** If the string ends in a colon, the output from parse is suppressed. ♣ If the parse command itself ends in a colon, the result is still displayed, unless the string ended in a colon. ♣ In either form of the parse command, if the parsed string contains more than one complete Maple expression, an error message is displayed and the first complete expression is returned. ♣ The

Maple expression cannot contain an assignment. ♣ Strings that represent partial Maple expressions cannot be parsed. A proc command is considered as *one* expression by parse. ♣ When combined with readline, parse can execute data pulled from within a text file.
**See also:** readline, readstat, sscanf, *convert(string)*

### pointto(int)

Determines what Maple object, if any, is located at integer address int.
**Output:** If an object is stored at int, the object is returned. If no object is stored at int, NULL is returned. If int is an invalid address, an error message is returned.
**Additional information:** To determine the address of a Maple object, use addressof. ♣ For more information on this and other "hardware" commands, see the on-line help page.
**See also:** addressof, assemble, disassemble
*LI = 11*

### print(expr$_1$, ..., expr$_n$)

Displays expressions expr$_1$ through expr$_n$ in two-dimensional format.
**Output:** Subsequent expressions are separated by a comma and a blank space.
**Additional information:** The expressions are displayed as a "side-effect" of print. The actual output is a NULL statement. This means that the " operators cannot recall the printed expressions. ♣ The interface(*prettyprint*) command allows you to set the default display mode to print. ♣ For more information on print, see the on-line help page.
**See also:** lprint, printf, interface(*prettyprint*),
interface(*screenwidth*), interface(*screenheight*),
interface(*verboseproc*)
**FL = 35–36, 119**  *LI = 168*

### printf(str$_{fmt}$, expr$_1$, ..., expr$_n$)

Displays expressions expr$_1$ through expr$_n$ in accordance with format string str$_{fmt}$.
**Output:** A NULL expression is returned.
**Additional information:** This command is modelled on the C language command of the same name. ♣ The format string is made up of special flags, justification information, width specifications, precisions, and type details for the expressions to be printed. ♣

Using printf may allow you to put your Maple data in a form more acceptable as input to other applications. ♣ If you do not follow the printf command with an lprint() command when printing to the terminal, your next input prompt is appended to the end of the formatted output. ♣ For more information on printf and for examples of its uses, see the on-line help page.
**See also:** lprint, print, sscanf

### printlevel := n

Sets the information level for debugging purposes.
**Output:** The input assignment is echoed and n is returned.
**Argument options:** printlevel := 0 to have the lowest level of echoing. ♣ printlevel := 1 for normal operation. ♣ printlevel := 5 to display information about procedures nested to a depth of 1. ♣ printlevel := 5*posint to display information about procedures nested to a depth of posint.
**Additional information:** Initially printlevel is set to 1. ♣ If you want to see a lot of information, set printlevel to 1000. ♣ This facility is being replaced by the userinfo command and infolevel. It is recommended that you use the newer facility for most cases.
**See also:** userinfo, infolevel, trace
**FL = 148–150** *LA = 175–178*

### proc(name$_1$, ..., name$_n$)/local/global/options/end

A programming construct for creating Maple commands.
**Output:** The procedure entered is returned.
**Additional information:** This construct allows you create your own Maple procedures (commands) and extend the power of the system. ♣ The parameters that you pass into the proc command are names that the local parameters assume within the procedure body. When you subsequently invoke the procedure, substitute specific expressions for each variable. ♣ Following local, specify (separated with commas and terminated with a semicolon) any variables considered local to the procedure. ♣ Following global, specify (separated with commas and terminated with a semicolon) any variables considered global. ♣ Following options, specify (separated with commas and terminated with a semicolon) any combination of *remember*, *system*, *operator*, and *trace*. See the individual entries for these options for more details. ♣ The statements found between options and end are executed each time the procedure is invoked. ♣ This structure must be terminated with end. ♣ If a RETURN command is encountered within this construct, the construct is exited and the value passed to RETURN is the result of

the procedure. If an ERROR command is executed within this construct, the construct is exited, the error message is displayed, and the result of the procedure is NULL. Otherwise, the result of the procedure is the result of the last statement executed within the procedure. ♣ Typically, only the final result of a procedure is displayed when a procedure is invoked. If you wish to always have some other value displayed from within the procedure, enclose it in a call to print or lprint. ♣ Keep in mind that this structure is valid without local and/or options. Both are set at a default value of NULL.
**See also:** for/in/do/od, for/from/by/while/do/od, remember, system, operator, trace, RETURN, ERROR, print, lprint, userinfo
**FL = 121–131, 134–136** *LA = 27–28, 113–128*

### procbody(proc)

Creates a neutralized form of procedure proc.
**Output:** An neutralized procedure is returned.
**Additional information:** This command needs to be defined by readlib(procbody) before being invoked. ♣ The major use of procbody is in creating procedures that create or manipulate other procedures. ♣ For examples of use, see the on-line help page.
**See also:** procmake, proc/local/options/end
*LI = 310*

### procmake(proc$_{nf}$)

Creates an executable form of proc$_{nf}$, a procedure in neutral form.
**Output:** A procedure is returned.
**Additional information:** This command needs to be defined by readlib(procmake) before being invoked. ♣ The major use of procmake is in creating procedures that create or manipulate other procedures. ♣ For examples of its uses, see the on-line help page.
**See also:** procbody, proc/local/options/end
*LI = 311*

### quit

Terminates the current Maple session.
**Output:** Not applicable.
**Additional information:** This keyword forces termination of the current Maple session. ♣ If you have not saved the work you are doing, it is lost. No ''Are you sure you want to quit?'' message is displayed, so be sure you really want to exit before using quit.

* You do not need to terminate quit with a semicolon or colon.
**See also:** restart
**FL = 9**

### read(filename)

Reads the file filename as input.
**Output:** The information in filename is returned.
**Argument options:** read filename to produce the same result.
* ('filename.m') to read in a file assumed to be in Maple's internal .m format.

**Additional information:** To read in files from the Maple library, use readlib.
**See also:** save, readlib, readdata, readline, readstat, interface(*echo*), convert(*hostfile*)
**FL = 106−109** *LA = 39−40, 127−128*

### readdata(filename, option, posint)

Creates a list of lists by reading raw data (i.e., data not in Maple expression format) from file filename. The data is of type option, where option can be either *integer* or *float*, and is arranged in posint columns.
**Output:** A list of lists, where each internal list represents one row of the data, is returned.
**Additional information:** This command needs to be defined by readlib(readdata) before being invoked. * If there are not an equal number of data elements in each line of the file, an irregular list of lists (i.e., where the internal lists have different lengths) may be produced. * If there are more than posint data elements in a row, only the first posint elements are read. * Once the data is read in, you may be able to transform it to a matrix with the convert(*matrix*) command. * To read in files that are already in Maple expression format, use the read command.
**See also:** read, readline, *convert(matrix)*, *convert(vector)*

### readlib(fnc)

Reads the Maple library command fnc into memory.
**Output:** An abbreviated form of procedure fnc is returned.
**Additional information:** To read in files you have created yourself, use the read command. * Most of Maple's existing commands are automatically read in when they are first called. Others are in specialty packages that are loaded with with. The few that remain must be read in with readlib. * To display the full procedure fnc, alter the value of the echo option of the interface command.
**See also:** read, with, interface(*echo*)
**FL = 110−111** *LA = 127−128* **LI = 174**

### readline(filename)

Reads the next line from file filename into Maple as a string.
**Output:** A string is returned.
**Argument options:** () to read a line from the current input stream (i.e., the terminal) as a string. ♣ (terminal) to read a line from the top level input stream. If Maple is being used interactively, this is interpreted as the terminal. If Maple is being called from batch mode, this is interpreted as the batch input file.
**Additional information:** If filename represents a file that has not previously been read from, readline opens that file for reading and reads the first line. ♣ If there are no more lines to be read from filename, then it is closed and a value of 0 is returned. ♣ Input lines longer than 499 characters are read in pieces. ♣ To read in complete files, use read.
**See also:** read, readstat, parse, sscanf

### readstat(str)

Prompts the user by displaying string str and then reads one statement from the input stream.
**Output:** The statement entered after the prompt is returned.
**Additional information:** If Maple is being used interactively, the input stream is interpreted as from the terminal. If Maple is being called from batch mode, it is interpreted as from the batch input file. ♣ The response must be a complete Maple statement including termination character, or readstat continues to prompt for completion. ♣ If the response is an assignment, that assignment is echoed but the value officially returned is the righthand side of the assignment. ♣ This command offers a convenient way to design procedures that prompt the user for input at any time.
**See also:** read, readline
*LI = 175*

### remember

Specifies that the remember option is to be used within a procedure.
**Output:** Not applicable.
**Additional information:** For *every* Maple command a remember table is available that remembers results of that command for certain combinations of parameters. ♣ If the remember option is specified in the options sequence of a procedure, then every time that procedure is subsequently called with a new set of parameters, an entry is placed in the remember table. ♣ This option is particularly useful in recursive commands, as remembering past values

can greatly increase the speed at which future calculations are done.
♣ To forget an entry in a remember table, use forget.
**See also:** forget, system, proc/local/options/end
**FL = 133–136, 191–192**

### restart

Restarts the Maple kernel, clearing the internal memory.
**Output:** Not applicable.
**Additional information:** restart removes the meaning of all variables and procedures loaded into your session to that point. ♣ Any initiallization file(s) you might have are re-entered. ♣ No "Are you sure you want to restart?" message is displayed, so be sure you really want to clear the session before using restart. ♣ No parameter braces, ( ), are necessary, but a terminator must be supplied.
**See also:** quit

### RETURN(expr)

Forces an explicit return from within a procedure. The result of the procedure is set to expr.
**Output:** Not applicable.
**Argument options:** ($expr_1$, ..., $expr_n$) to return the expression sequence $expr_1$ through $expr_n$.
**Additional information:** The remaining statements in the procedure's statement block are ignored and the result expr is immediately returned. ♣ expr can be *any* type of Maple expression.
♣ If RETURN is found within any type of construct other than proc/local/options/end, an error message is returned.
**See also:** ERROR, FAIL, NULL, proc/local/options/end
**FL = 125–126** *LA = 123–124* *LI = 179*

### save(filename)

Saves the current state of the Maple session in file filename.
**Output:** A NULL expression is returned.
**Argument options:** save filename to produce the same result.
♣ ('filename.m') to save into a file in Maple's internal .m format. ♣ ($expr_1$, ..., $expr_n$, filename) to save just the values of expressions $expr_1$ through $expr_n$ in file filename.
**Additional information:** To read in a saved file, use read.
**See also:** read, writeto, appendto, convert(*hostfile*)
**FL = 106–109** *LA = 127*

### seq(expr, var=a..b)

Creates a sequence of expressions that represents expr with the values a, a+1, ..., b substituted into it for variable var.
**Output:** An expression sequence is returned.
**Argument options:** (expr, var=list) to create a sequence where the elements of list are substituted for var.
**Additional information:** This command is a much faster alternative to simple for-loops. ♣ Enclosing the entire seq command in brackets ([]) or braces ({}) results in a list or set, respectively.
**See also:** for/from/by/to/do/od, for/from/by/while/do/od, for/in/do/od
**FL = 79−80** *LA = 16, 48−49* **LI = 185**


### showtime()

Activates the Maple timing substructure.
**Output:** A NULL expression is returned.
**Additional information:** This command needs to be defined by readlib(showtime) before being invoked. ♣ After showtime() has been entered, each subsequent input line is prompted by a string from the sequence O1 :=, O2 :=, etc. ♣ The outputs for each input are then assigned to names O1, O2, etc.; and after each output a message is displayed that lists the amount of time that command took to run and the number of words of memory it used. ♣ To terminate the showtime facility, enter off. ♣ To clear the variable names O1, O2, etc., enter clear.
**See also:** time, history
**FL = 186−187**


### sscanf(str, $str_{fmt}$)

Scans string str for Maple data occurring in it in accordance with formatting specification string $str_{fmt}$.
**Output:** A NULL expression is returned.
**Additional information:** This command is modelled on the C language command of the same name. ♣ The format string is made up of special flags, width specifications, and type details for the expression being scanned. ♣ Typically, a call to sscanf is preceded by a readline command, which reads a string from an input file. ♣ Using sscanf allows you to read data into Maple as it is output by other applications. ♣ For more information on sscanf and for examples of its use, see the on-line help page.
**See also:** readline, parse, printf

### ssystem(str)

Performs the system command represented by str at the system level without exiting Maple.
**Output:** A two-element list is returned. The first element contains the return code for the system command. The second element contains a string representing the result of the system command.
**Argument options:** (str, int) to terminate the system command str if not completed after int seconds. If time runs out, the second element of the result is the string *Timeout*.
**Additional information:** This command is not available on all Maple platforms.
**See also:** system

### status

Displays the values of eight system status variables.
**Output:** An expression sequence is returned.
**Argument options:** status[1] to display the total number of words of memory used during the current session. ♣ status[2] to display the number of actual words allocated during the current session. Some words get reused. ♣ status[3] to display the number of CPU seconds used during the current session. ♣ status[4] to display how many words are requested before a new *bytes used* message is displayed. ♣ status[5] to display the current word increment for automatic garbage collection. ♣ status[6] to display the number of words returned in the last garbage collection. ♣ status[7] to display the number of words available after the last garbage collection. ♣ status[8] to display the number of times garbage collection has occurred in the current session.
**Additional information:** The entries for status[1], status[2], and status[3] are displayed occasionally throughout your Maple session. The *words* values are translated to *bytes*, however. ♣ To stop these messages from occurring set the quiet option of the interface command to true.
**See also:** time, trace, gc, words, interface(*quiet*)
*LA = 184*

### stop

This command is equivalent to quit and terminates your Maple session.
**See also:** quit

### system

Specifies that the system option is to be used within a procedure.
**Output:** Not applicable.

**Additional information:** If the system option is specified in the options sequence of a procedure, the remember table of the procedure can be deleted whenever garbage collection is performed. Otherwise, the remember table remains intact throughout the session.
**See also:** ssystem, gc, remember, proc/local/options/end

### system(str)

Performs the system command represented by str at the system level without exiting Maple.
**Output:** The result of the system command is displayed and the return status of the system command is returned.
**Additional information:** This command is not available on all Maple platforms.
**See also:** ssystem
*LA = 189−190 LI = 211*

### thaw(var)

Replaces variable name var, which represents a previously frozen expression, with the expression it represents.
**Output:** An expression is returned.
**Additional information:** Before thaw can be used, freeze should be used. ♣ These commands are primarily used when you want a certain subexpression to be unaffected when its larger expression is being acted upon. ♣ If var is not of the form _R0, _R1, etc., var is returned as is.
**See also:** freeze, frontend
*LI = 289*

### time()

Displays total CPU time for the current session.
**Output:** An expression in seconds is returned.
**Additional information:** This command is most often used when timing specific commands. ♣ The best way to time a command fnc is to enter on one line st := time(): fnc: time()-st; . ♣ This value can also be accessed through status[3].
**See also:** showtime, status, words
**FL = 185** *LI = 215*

### trace

Specifies that the trace option is to be used within a procedure.
**Output:** Not applicable.

**Additional information:** If the trace option is specified in the options sequence of a procedure, all entry and exit calls created within the procedure are displayed, regardless of their depth, when the procedure is invoked.
**See also:** printlevel, proc/local/options/end

### trace(fnc$_1$, ..., fnc$_n$)

Commences tracing of the Maple commands fnc$_1$ through fnc$_n$.
**Output:** An expression sequence containing the names of the commands to be traced is returned.
**Additional information:** If a command is being traced, reaching both the entry point and exit point causes a tracing message to be printed. At the entry point, the incoming parameters are listed. At the exit point, the outgoing result is listed. ♣ To turn tracing off for certain commands, use untrace. ♣ Certain commands with special evaluation rules (such as, eval, evalhf, evalf, etc.) cannot be traced. ♣ Each subsequent call to trace adds to the commands being traced.
**See also:** printlevel, untrace
**FL = 150–152** *LA = 179*

### traperror(expr)

Evaluates expr, but traps any error that might be returned from that evaluation.
**Output:** If an error is trapped, the string corresponding to the error message is returned. Otherwise, the result of the evaluation is returned.
**Argument options:** (expr$_1$, ..., expr$_n$) to evaluate expr$_1$ through expr$_n$ and trap the first error that occurs. If an error occurs, the remaining expressions are *not* evaluated.
**Additional information:** The most recent error that occurred is always stored in the variable lasterror. ♣ When you trap an error with traperror make sure to check that the string returned is an error and not just the normal result of an evaluation by testing whether the result of traperror is identical to lasterror. ♣ If you do not trap an error it propagates itself all the way up to the top level of computation.
**See also:** lasterror, ERROR
**FL = 129–131** *LI = 216*

### type(expr, '*')

Determines whether expr is a product data type.
**Output:** A boolean value of either true or false is returned.

**Additional information:** The first-level operands of expr must be multiplied together in order for it to be a product. For example, a sum of products is not of type *. ♣ Divisions are also considered products.
**See also:** type('+'), type('^'), type('!'), *op*

### type(expr, '+')

Determines whether expr is a sum data type.
**Output:** A boolean value of either true or false is returned.
**Additional information:** The first-level operands of expr must be added together in order for it to be a product. For example, a product of sums is not of type +.
**See also:** type('*'), type('^'), type('!'), *op*

*LA = 88*

### type(expr, '^')

Determines whether expr is an exponential data type.
**Output:** A boolean value of either true or false is returned.
**Additional information:** The first operand of expr must be taken to the power of the second operand of expr. For example, a product of exponentials is not of type ^.
**See also:** type('*'), type('+'), type('!'), *op*

### type(expr, '**')

This command is identical to type('^').
**See also:** type('^')

### type(expr, '.')

Determines whether expr is an unevaluated concatenation.
**Output:** A boolean value of either true or false is returned.
**Additional information:** A concatenation remains unevaluated only if its righthand side is not a string, integer, or range.
**See also:** *cat*, type('*string*')

### type(expr, '..')

Determines whether expr is a range data type.
**Output:** A boolean value of either true or false is returned.

*LI = 227*

### type(expr, '!')

Determines whether expr is an unevaluated factorial.
**Output:** A boolean value of either true or false is returned.
**See also:** type('*'), type('+'), type('^')


### type(expr, *algebraic*)

Determines whether expr is an algebraic expression.
**Output:** A boolean value of either true or false is returned.
**Additional information:** expr is an algebraic expression if it tests positive as an integer, fraction, float, string, indexed, '+', '*', '^', '**', '!', '.', series, function, or uneval type.
**See also:** type(*integer*), type(*fraction*), type(*float*), type(*string*), type(*indexed*), type('+'), type('*'), type('^'), type('**'), type('!'), type('.'), type(*series*), type(*function*), type(*uneval*)
*LI = 219*


### type(expr, *algext*)

Determines whether expr is an algebraic extension (i.e., a RootOf expression).
**Output:** A boolean value of either true or false is returned.
**Argument options:** (expr, *algext*(typename)) to determine if expr is an algebraic expression over the coefficient domain typename.
**See also:** *RootOf*, type(*algnum*), type(*algnumext*), type(*algfun*)
*LI = 220*


### type(expr, *algfun*)

Determines whether expr is an algebraic function.
**Output:** A boolean value of either true or false is returned.
**Argument options:** (expr, *algfun*(typename)) to determine if expr is an algebraic function over the coefficient domain typename. ♣ (expr, *algfun*(typename, [$var_1$, ..., $var_n$])) or (expr, *algfun* (typename, {$var_1$, ..., $var_n$})) to check for an algebraic function over the variables $var_1$ through $var_n$.
**Additional information:** An algebraic function is an expression in a set of variables over a domain extended by RootOf expressions.
♣ For examples of algebraic functions, see the on-line help page for type[algfun].
**See also:** *RootOf*, type(*algext*), type(*algnumext*), type(*algnum*)
*LI = 221*

### type(expr, *algnum*)

Determines whether expr is an algebraic number.
**Output:** A boolean value of either true or false is returned.
**Additional information:** An algebraic number is either a rational number or a root of a univariate polynomial with algebraic number coefficients.
**See also:** *RootOf*, type(*algext*), type(*algnumext*), type(*algfun*)
*LI = 221*

### type(expr, *algnumext*)

Determines whether expr is an algebraic number extension.
**Output:** A boolean value of either true or false is returned.
**Additional information:** An algebraic number extension is a root of a univariate polynomial with algebraic number coefficients.
**See also:** *RootOf*, type(*algext*), type(*algnum*), type(*algfun*)
*LI = 222*

### type(expr, *anything*)

Returns true for any expression expr.
**Output:** A boolean value of true is returned.
**Additional information:** The only type of expression that does not return true is an expression sequence, which returns an error message. ♣ *anything* is used in structured types to represent a component that always passes.
**See also:** type(*nothing*)
*LI = 223*

### type(expr, *boolean*)

Determines whether expr is a boolean expression.
**Output:** A boolean value of either true or false is returned.
**Additional information:** expr is boolean if it tests positive as a relation or a logical, or if it is equivalent to true or false.
**See also:** type(*relation*), *type(logical)*, *true*, *false*
*LA = 74 LI = 225*

### type(expr, *complex*)

Tests whether expr is a complex constant.
**Output:** A boolean value of either true or false is returned.
**Argument options:** (expr, *complex*(typename)) to determine if expr is a complex constant whose real and imaginary parts are both of type typename.

**Additional information:** Complex constants cannot involve non-numeric constants such as true, false, FAIL, and infinity.
**See also:** type(*rational*), type(*radical*), type(*numeric*), Re, Im, evalc

### type(expr, *constant*)

Determines whether expr is a (possibly complex) constant expression.
**Output:** A boolean value of either true or false is returned.
**Additional information:** Unevaluated commands and irrationals can both be constants, providing they have numeric components.
**See also:** type(*realcons*), type(*numeric*)
**FL = 132**

### type(expr, *even*)

Determines whether expr is an even integer.
**Output:** A boolean value of either true or false is returned.
**See also:** type(*integer*), type(*odd*)
*LI = 227*

### type(expr, *float*)

Determines whether expr is a floating-point value.
**Output:** A boolean value of either true or false is returned.
**Additional information:** expr is a floating-point value if it consists of a sequence of digits with a decimal point or if it is an unevaluated call to Float.
**See also:** type(*integer*), type(*numeric*), Float
*LA = 70–71* **LI = 229**

### type(expr, *fraction*)

Determines whether expr is a fractional value.
**Output:** A boolean value of either true or false is returned.
**Additional information:** expr is a fractional value if it consists of a signed integer divided by a positive integer. ♣ Note that Maple automatically simplifies rational values.
**See also:** type(*integer*), type(*posint*), type(*rational*), type(*numeric*)
*LI = 229*

### type(expr, *heap*)

Determines whether expr is heap structure.
**Output:** A boolean value of either true or false is returned.

**Additional information:** expr must have been assigned to the results of a heap[new] command.
**See also:** heap

### type(expr, *integer*)

Determines whether expr is an integer.
**Output:** A boolean value of either true or false is returned.
**See also:** type(*posint*), type(*negint*), type(*nonnegint*), type(*odd*), type(*even*), type(*constant*), type(*numeric*)
*LA = 69*

### type(expr, *list*)

Determines whether expr is a list.
**Output:** A boolean value of either true or false is returned.
**Argument options:** (expr, *list*(typename)) to determine if expr is a list with all its elements in domain typename.
**See also:** type(*set*), *op(list)*
*LI = 249*

### type(expr, *listlist*)

Determines whether expr is a list of lists.
**Output:** A boolean value of either true or false is returned.
**Additional information:** Each inner list must have the same number of elements.
**See also:** type(*list*), *convert(listlist)*, *convert(matrix)*
*LI = 232*

### type('expr', *name*)

Determines whether expr is a name.
**Output:** A boolean value of either true or false is returned.
**Additional information:** To avoid evaluation of expr before type checking can be done, enclose expr in forward quotes. Even if expr is not a name, doing this does not affect adversely affect the outcome.
**See also:** type(*string*), *type(indexed)*, type(*uneval*)

### type(expr, *negative*)

Determines whether expr is a negative numeric value.
**Output:** A boolean value of either true or false is returned.
**See also:** type(*positive*), type(*numeric*), type(*nonneg*)
*LI = 238*

## type(expr, *negint*)

Determines whether expr is an integer less than 0.
**Output:** A boolean value of either true or false is returned.
**See also:** type(*posint*), type(*integer*), type(*nonnegint*)
*LI = 239*

## type(expr, *nonneg*)

Determines whether expr is a non-negative numeric value.
**Output:** A boolean value of either true or false is returned.
**See also:** type(*negative*), type(*positive*), type(*numeric*)
*LI = 238*

## type(expr, *nonnegint*)

Determines whether expr is an integer greater than or equal to 0.
**Output:** A boolean value of either true or false is returned.
**See also:** type(*posint*), type(*negint*), type(*integer*)
*LI = 239*

## type(expr, *nothing*)

Returns false for any expression expr.
**Output:** A boolean value of false is returned.
**Additional information:** The only type of expression that returns false is an expression sequence, which returns an error message.
✦ *nothing* is used in structured types to represent a component that never passes.
**See also:** type(*anything*)

## type(expr, *numeric*)

Determines whether expr is a numeric value.
**Output:** A boolean value of either true or false is returned.
**Additional information:** expr is numeric if it tests positive as an integer, fraction, or float.
**See also:** type(*integer*), type(*fraction*), type(*float*)
*LA = 86 LI = 234*

## type(expr, *odd*)

Determines whether expr is an odd integer.
**Output:** A boolean value of either true or false is returned.
**See also:** type(*integer*), type(*even*)
*LA = 87 LI = 227*

## type(expr, *operator*)

Tests whether expr is a functional operator.
**Output:** A boolean value of either true or false is returned.

**Additional information:** Functional operators are defined with the -> or < | > syntax.
*LI = 235*

### type(expr, *point*)

Determines whether expr is a point (i.e., an equation with a name on the lefthand side and an algebraic expression on the righthand side).
**Output:** A boolean value of either true or false is returned.
**Argument options:** ({expr$_1$, ..., expr$_n$}, *point*) to determine if expressions expr$_1$ through expr$_n$ are all valid points.
**See also:** *RootOf*, type(*algebraic*), type(*equation*)
*LI = 237*

### type(expr, *posint*)

Determines whether expr is an integer greater than 0.
**Output:** A boolean value of either true or false is returned.
**See also:** type(*negint*), type(*integer*), type(*nonnegint*)
*LI = 239*

### type(expr, *positive*)

Determines whether expr is a positive numeric value.
**Output:** A boolean value of either true or false is returned.
**See also:** type(*negative*), type(*numeric*), type(*nonneg*)
*LI = 238*

### type(expr, *procedure*)

Tests whether expr is a Maple procedure name.
**Output:** A boolean value of either true or false is returned.
**Additional information:** Both existing Maple procedures and those you program yourself are valid procedures.
**See also:** proc/local/options/end, *type(function)*
*LA = 76−77*

### type(expr, *radext*)

Determines whether expr is a radical extension.
**Output:** A boolean value of either true or false is returned.
**Argument options:** (expr, *radext*(typename)) to determine if expr is a radical extension with the base of the exponential in the domain typename.
**See also:** type(*radfunext*), type(*algext*), type(*radnum*), type(*radfun*), type(*radnumext*)
*LI = 240*

## type(expr, *radfun*)

Determines whether expr is a radical function.
**Output:** A boolean value of either true or false is returned.
**Argument options:** (expr, *radfun* (typename)) to determine if expr is a radical function over coefficient domain typename. ♣ (expr, *radfun*(typename, [$var_1$, ..., $var_n$])) or (expr, *radfun*(typename, {$var_1$, ..., $var_n$})) to check for a radical function over the variables $var_1$ through $var_n$.
**Additional information:** A radical function is an expression in a set of variables over a domain extended by RootOf expressions. ♣ For examples of radical functions, see the on-line help page for type[radfun].
**See also:** type(*radfunext*), type(*radext*), type(*radnumext*), type(*radnum*), type(*algfun*)
*LI = 241*

## type(expr, *radfunext*)

Determines whether expr is a radical function extension.
**Output:** A boolean value of either true or false is returned.
**Additional information:** A radical function extension is an algebraic function specified with radicals. ♣ For examples of radical function extensions, see the on-line help page for type[radfunext].
**See also:** type(*radext*), type(*radfun*), type(*radnumext*), type(*algfun*)
*LI = 241*

## type(expr, *radical*)

Determines whether expr is a radical expression.
**Output:** A boolean value of either true or false is returned.
**Additional information:** expr is an radical expression if it tests positive as type '^' and has a fractional power.
**See also:** type(*radnum*), type(*radext*), type(*radfun*), type(*radfunext*), type('^'), type(*fraction*)
*LI = 242*

## type(expr, *radnum*)

Determines whether expr is a radical number.
**Output:** A boolean value of either true or false is returned.
**Additional information:** A radical number is either a rational number or a combination of roots of rational numbers specified in terms of radicals.
**See also:** type(*radext*), type(*radnumext*), type(*radfun*), type(*algnum*)
*LI = 243*

**type(expr,** *radnumext***)**

This command is identical to type(*radnum*).
**See also:** *RootOf*, type(*radnum*), type(*algnumext*)


**type(expr,** *range***)**

This command is identical to type('..').
**See also:** type('..')
*LI = 244*


**type(expr,** *rational***)**

Determines whether expr is a rational value.
**Output:** A boolean value of either true or false is returned.
**Additional information:** expr is a rational value if it is either of type integer or fraction.
**See also:** type(*integer*), type(*fraction*), type(*numeric*)
*LI = 244*


**type(expr,** *realcons***)**

Determines whether expr is a real constant expression.
**Output:** A boolean value of either true or false is returned.
**Additional information:** A real constant is infinity, -infinity, or a value which, when acted upon by evalf, returns a floating-point value.
**See also:** type(*constant*), type(*float*), type(*numeric*), *evalf*
*LI = 246*


**type(expr,** *relation***)**

Determines whether expr is a relational expression.
**Output:** A boolean value of either true or false is returned.
**Additional information:** expr is relational if it tests positive as type '=', '<', '<=', or '<>',
**See also:** type('='), type('<'), type('<='), type('<>')
*LI = 225*


**type(expr,** *RootOf***)**

This command is the same as the type(*algext*) command.
**See also:** type(*algext*)
*LI = 246*

### type(expr, *set*)

Determines whether expr is a set.
**Output:** A boolean value of either true or false is returned.
**Argument options:** (expr, *set*(typename)) to determine if expr is a set with all its elements in domain typename.
**See also:** type(*list*), *op(set)*

### type(expr, *sqrt*)

Determines whether expr is a square root radical.
**Output:** A boolean value of either true or false is returned.
**Argument options:** (expr, *sqrt*(typename)) to determine if expr is a square root and a radical extension in domain typename.
**Additional information:** A square root is a radical whose exponent has a denominator of 2.
**See also:** type(*square*), type(*radext*), type(*radical*)
*LI = 250*

### type(expr, *square*)

Determines whether expr is a perfect square expression.
**Output:** A boolean value of either true or false is returned.
**Additional information:** A perfect square expression is one whose square root is free of radicals. ♣ Keep in mind that this command is only as effective as the sqrt command.
**See also:** type(*sqrt*), *sqrt*
*LI = 250*

### type(expr, *string*)

Determines whether expr is a string.
**Output:** A boolean value of either true or false is returned.
**See also:** type(*name*)

### type(expr, *TEXT*)

Determines whether expr is an object of type TEXT.
**Output:** A boolean value of either true or false is returned.
**Additional information:** For more information on TEXT data types, see the command entry for TEXT.
**See also:** *TEXT*, type(*string*)

### type(expr, *type*)

Determines whether expr is a valid type name.
**Output:** A boolean value of either true or false is returned.

**Argument options:** (expr$_1$(expr$_2$), *type*), (expr$_1$..expr$_2$, *type*), (expr$_1$ = expr$_2$, *type*), etc. to check for valid structured types.
*LI = 256*

### type(expr, *uneval*)

Determines whether expr is enclosed in forward quotes.
**Output:** A boolean value of either true or false is returned.
**See also:** type(*name*)

### unames()

Returns an expression sequence containing *all* the names (or strings) currently assigned to their own names and nothing else.
**Output:** An expression sequence is returned.
**Additional information:** Since strings are automatically thought of as unassigned names, all strings, file names, and error messages used in the session are displayed by unames.
**See also:** anames

### unassign('var$_1$', ..., 'var$_n$')

Unassigns all variables var$_1$ through var$_n$.
**Output:** A NULL expression is returned.
**Additional information:** the forward quotes around var$_1$ through var$_n$ are necessary to ensure that the variables are not evaluated before unassignment is attempted. ♣ After unassign, each variable listed is an unassigned name.
**See also:** *assign*
*LI = 327*

### untrace(fnc$_1$, ..., fnc$_n$)

Stops tracing of commands fnc$_1$ through fnc$_n$.
**Output:** An expression sequence containing the names of the commands no longer being traced is returned.
**Additional information:** To turn tracing on for certain commands, use trace.
**See also:** trace
*LA = 179*

### userinfo(n, fnc, expr$_1$, expr$_2$, ..., expr$_n$)

Stores information about the procedure fnc, that is displayed only if the information display level has been set to a value at least as great as n. When information is displayed, expressions expr$_1$ through

expr$_n$ are displayed following the command name fnc.
**Output:** A NULL expression is returned.
**Additional information:** To alter the information display level for one, several, or all Maple functions, use infolevel. ✳ In the existing Maple library, userinfo commands have been distributed liberally with level 1 reserved for information the user must be told, 2 and 3 reserved for more general information, and 4 and 5 reserved for even more minute detail.
**See also:** infolevel, printlevel, trace
*LA = 180* **LI = 258**

### with(pckg)

Sets up the commands in specialty package pckg for use.
**Output:** A list containing all command names defined by the with command is returned. As well, for any new command names that overwrite existing commands, a warning message is displayed.
**Argument options:** (pckg, fnc$_1$, fnc$_2$, ..., fnc$_n$), where fnc$_1$ through fnc$_n$ are names of commands in the pckg package, to define just those particular commands.
**Additional information:** For more information on exactly how the with command and Maple packages work, see the on-line help pages for with and package. ✳ You may be able to access the *Share Library*, an unsupported collection of Maple users' contributed code, by entering with(share). Try it and see.
**See also:** read, readlib
**FL = 105** *LA = 19* **LI = 260**

### words()

Displays total memory usage for the current session.
**Output:** An integer representing the total number of words used is returned.
**Argument options:** (n) to specify that *bytes used* messages are to be printed for every n words used. If n is less than or equal to 0, no such messages are printed.
**Additional information:** The number of bytes used and the frequency of *bytes used* messages can also be accessed through status[1] and status[4], respectively.
**See also:** status, time, gc
**FL = 186−187**

### write(expr$_1$, ..., expr$_n$)

Writes expressions expr$_1$ through expr$_n$ to a buffer.
**Output:** A NULL expression is returned.

**Additional information:** This command needs to be defined by readlib(write) before being invoked. ♣ The information is written to a buffer whose name is _buffer. ♣ Using write is not recommended. It is recommended that you stick to using writeto.
**See also:** close, open, writeln, writeto, appendto

### writeln(expr$_1$, ..., expr$_n$)

Writes expressions expr$_1$ through expr$_n$ to a buffer.
**Output:** A NULL expression is returned.
**Additional information:** This command needs to be defined by readlib(write) before being invoked. ♣ The information is written to a buffer whose name is _buffer. ♣ Using writeln is not recommended. It is recommended that you stick to using writeto.
**See also:** close, open, write, writeto, appendto

### writeto(filename)

Writes all subsequent output to file filename.
**Output:** A NULL expression is returned.
**Argument options:** (*terminal*) to return the output stream to the terminal.
**Additional information:** Whatever was in filename before the writeto command is lost. ♣ If you want to append to a file, use appendto.
**See also:** save, open, write, writeln, appendto

# Miscellaneous

## Introduction

This final chapter of *The Maple Handbook* is devoted to all the commands that did not find a home in the previous eleven chapters. This is not to say that all the functions contained herein are unimportant. Some are esoteric, relating to areas for which few of you will ever have a practical use. Many others are used by everyone, and are here simply because they do not fit comfortably into any larger group.

Since there is no theme that unites all of the commands in this chapter, we begin with a brief tour of three of the more commonly used packages included.

## Group Theory

Over two dozen commands for manipulating permutation groups are contained in the group package. The most basic command, permgroup, allows you to define a permutation group by stating its degree and the set of permutations defining it.

```
> with(group):
> mygroup := permgroup(7, {[[1,2], [3,4,6,7]],
> [[5,6,7]]});
```

$mygroup := permgroup(7, \{[[1,2],[3,4,6,7]],[[5,6,7]]\})$

Each list of lists represents a permutation, and their internal lists represent cycles in that permutation. After a permutation group has been defined, operations can be performed upon it including finding its center, calculating the orbit about one of its points, and determining whether it is abelian.

```
> center(mygroup);
```

$$permgroup(\,7, \{\ \})$$

```
> orbit(mygroup, 4);
```

$$\{\,3, 4, 5, 6, 7\,\}$$

```
> isabelian(mygroup);
```

$$false$$

# Boolean Logic

The logic package handles expressions containing inert boolean operators &and, &or, &not, &nand, &nor, &xor, &iff, and &implies. For example, bequal tests if two statements are logically equivalent and tautology tests if a statement is always true.

```
> with(logic):
> a := 'a': b := 'b': c := 'c': d := 'd':
> bequal(a &iff (a &and b), a &implies b);
```

$$true$$

```
> tautology((a &and b) &or (a &nor b));
```

$$false$$

# Rational Generating Functions

Most of the commands in the genfunc package begin with the characters rgf_. Three commands that create and manipulate rational generating functions are shown below.

```
> with(genfunc):
> rgf_encode(3^n, n, z);
```

$$\frac{1}{1-3\,z}$$

```
> rgf_term(", z, 50);
```

$$71789798769 1852588770249$$

```
> rgf_pfrac((1-z)/(-6-z+z^2), z);
```

$$-\frac{3}{5}\frac{1}{z+2} - \frac{2}{5}\frac{1}{z-3}$$

## Parameter Types Specific to This Chapter

| | |
|---|---|
| exprb | a boolean expression |
| rgf | a rational generating function |
| ext | an algebraic extension |
| pg | a permutation group |
| subpg | a permutation subgroup |
| grelg | a generators and relations group |
| subgrelg | a generators and relations subgroup |
| met | a metric |

## Command Listing

### var$_1$ &and var$_2$

Defines a boolean expression for the inert *and* operator &and.
**Argument options:** (expr$_1$) &and (expr$_2$), where expr$_1$ and expr$_2$ are boolean expressions.
**Additional information:** In order for var$_1$ &and var$_2$ to evaluate as true, both var$_1$ and var$_2$ must be true.
**See also:** &or, &not, &iff, &nor, &nand, &xor, &implies, *and*

### var$_1$ &iff var$_2$

Defines a boolean expression for the inert *if and only if* operator &iff.
**Argument options:** (expr$_1$) &iff (expr$_2$), where expr$_1$ and expr$_2$ are boolean expressions.
**Additional information:** var$_1$ &iff var$_2$ implies that if var$_1$ is true then so is var$_2$ and also implies that if var$_2$ is true then so is var$_1$.
**See also:** &and, &or, &not, &nor, &nand, &xor, &iff, *if*

### var₁ &implies var₂

Defines a boolean expression for the inert *if* operator &implies.
**Argument options:** (expr₁) &implies (expr₂), where expr₁ and expr₂ are boolean expressions.
**Additional information:** var₁ &implies var₂ implies that if var₁ is true then so is var₂.
**See also:** &and, &or, &not, &nor, &nand, &xor, &implies

### var₁ &nand var₂

Defines a boolean expression for the inert *not and* operator &nand.
**Argument options:** (expr₁) &nand (expr₂), where expr₁ and expr₂ are boolean expressions.
**Additional information:** In order for var₁ &nand var₂ to evaluate as true, either var₁ or var₂ must be false. ♣ This operator is the opposite of the &and operator.
**See also:** &and, &or, &not, &iff, &nor, &xor, &implies

### var₁ &nor var₂

Defines a boolean expression for the inert not or operator &nor.
**Argument options:** (expr₁) &nor (expr₂), where expr₁ and expr₂ are boolean expressions.
**Additional information:** In order for var₁ &nor var₂ to evaluate as true, neither var₁ nor var₂ (or both) can be true. ♣ This operator is the opposite of the &or operator.
**See also:** &and, &or, &not, &iff, &nand, &xor, &implies

### &not var

Defines a boolean expression for the inert *negation* operator &not.
**Argument options:** &not (expr), where expr is a boolean expressions.
**Additional information:** &not changes true values to false and vice versa.
**See also:** &and, &or, &iff, &nor, &nand, &xor, &implies

### var₁ &or var₂

Defines a boolean expression for the inert *inclusive or* operator &or.
**Argument options:** (expr₁) &or (expr₂), where expr₁ and expr₂ are boolean expressions.
**Additional information:** In order for var₁ &or var₂ to evaluate as true, either var₁ or var₂ (or both) must be true.
**See also:** &and, &not, &iff, &nor, &nand, &xor, &implies, *or*

## var₁ &xor var₂

Defines a boolean expression for the inert *exclusive or* operator &xor.

**Argument options:** (expr₁) &xor (expr₂), where expr₁ and expr₂ are boolean expressions.

**Additional information:** In order for var₁ &xor var₂ to evaluate as true, either var₁ or var₂ (but not both) must be true.

**See also:** &or, &and, &not, &iff, &nor, &nand, &implies

## about(var)

Determines what is currently assumed about the value of unknown variable var.

**Output:** A string explaining the current state of var is returned.

**Additional information:** If no properties have been assumed on var, about returns unevaluated. ♣ Other commands that allow you to query or alter information about a variable are is, isgiven, assume, and addproperty. ♣ For more information on about, see the on-line help page.

**See also:** assume, is, isgiven, addproperty

## alias(name = expr)

Sets up an alias of expression expr to name. From then on, all occurrences of name are recognized as expr.

**Output:** An expression sequence of all currently known aliases is returned.

**Argument options:** (eqn₁, ..., eqnₙ), where eqnᵢ is of the form nameᵢ = exprᵢ, to define n aliases. ♣ (name = name) to erase the alias for name.

**Additional information:** The only alias defined when the system starts is I, which is aliased to sqrt(-1). ♣ A common use of aliases is in representing long RootOf expressions. ♣ You cannot define one alias in terms of another. ♣ When writing a procedure, *local variables* are not affected by aliases.

**See also:** macro, *RootOf*

**FL = 106, 201−202** *LA = 14, 181−182* **LI = 6**

## amortization(num$_a$, num$_i$, num$_p$)

Computes an amortization schedule for a loan of amount num$_a$, interest rate num$_i$, and periodic payment num$_p$.

**Output:** An array representing the amortization schedule is returned. The array in indexed from 0 to the number of payments, and each element contains a list. Each list has five values representing the payment number, amount of the payment, interest portion of that payment, principal portion of that payment, and balance owing.

**Argument options:** (eqn$_1$, eqn$_2$, eqn$_3$), where the equations are any three of *amount*=num$_1$, *interest*=num$_2$, *periods*=num$_3$, and *payment*=num$_4$, to determine the fourth equation in the set.

**Additional information:** This command needs to be defined by readlib(finance) before being invoked.

**See also:** finance, blackscholes

*LI = 287*

### assume(ineq$_{\text{var}}$)

Assumes that the value of unknown variable var is restricted by inequation ineq$_{\text{var}}$.

**Output:** A NULL expression is returned.

**Argument options:** (var, a..b), to assume that the value of var lies in the closed range from a to b. ♣ (var, *Open*(a)..*Open*(b)), to assume that the value of var lies in the open range from a to b. ♣ (var, name), where name is a type or a property like *integer* or *rational*, respectively, to assume that var is of that kind. ♣ (var, *AndProp*(name$_1$, ..., name$_n$)), to assume that var conforms to *all* the properties represented by name$_1$ through name$_n$. ♣ (var, *OrProp*(name$_1$, ..., name$_n$)), to assume that var conforms to at least *one* of the properties represented by name$_1$ through name$_n$.

**Additional information:** Calling assume on var wipes out all previous restrictions assumed on var. ♣ assume does not infer any particular value on var, it merely narrows the possible range of values that var *should* represent. ♣ Assuming a value or property on var does not prevent you from assigning var to some value falling outside of the restriction. ♣ Other commands that allow you to query or alter information about a variable are is, isgiven, about, and addproperty. ♣ For more information on assume, see the on-line help page.

**See also:** is, isgiven, about, addproperty

### argument(cmplx)

Determines the principal value of the argument of complex value cmplx.

**Output:** A real value is returned.

**Additional information:** If the argument is represented by a, then cmplx = polar(abs(x), a) holds.

**See also:** evalc, polar, conjugate, Re, Im, *abs*, *signum*

### blackscholes(num$_x$, num$_t$, num$_p$, num$_s$, num$_r$)

Computes, using the formula attributed to Black and Scholes, the present value of a call option on a stock, where num$_x$ represents the exercise price, num$_t$ represents the time to the exercise date, num$_p$ represents the current price of the stock, num$_s$ represents the standard deviation per period of the stock's continuously compounded rate of return, and num$_r$ represents the continuously compounded risk-free rate of interest.
**Output:** A floating-point value is returned.
**Argument options:** (num$_x$, num$_t$, num$_p$, num$_s$, num$_r$, name), to assign the *hedge ratio* to name, if present.
**Additional information:** This command needs to be defined by readlib(finance) before being invoked.
See also: finance, amortization
*LI = 287*


### bspline(n, name, list)

Computes the segment polynomials in variable name corresponding to the B-spline of degree n on the knot sequence represented by list, a list of n+2 numbers or variables.
**Output:** A list of two-element lists is returned. The first element of an internal list is an inequation specifying the range of the segment polynomial; the second element is an expression defining it.
**Argument options:** (n, name) to use the default knot sequence [0, 1, ..., n+1].
**Additional information:** This command needs to be defined by readlib(bspline) before being invoked. ♣ For more information on this command, see the on-line help page.
See also: spline
*LI = 265*


### c(expr$_1$, expr$_2$)

Represents the non-commutative multiplication between expr$_1$ and expr$_2$.
**Output:** An unevaluated c command is returned.
**Additional information:** This command needs to be defined by readlib(commutat) before being invoked. ♣ c expression can be converted to expressions in terms of &*. ♣ Simplifications, conversions, and expansions can be done on both forms.
See also: commutat, simplify('&*'), simplify(c), convert('&*'), convert(c), *expand*

### C(expr)

Translates expression expr into an equivalent expression in the C programming language.
**Output:** A string is returned.
**Argument options:** (expr, *double*) to specify that double precision should be used. ♣ (expr, *optimized*) to perform common subexpression optimization on the result using subexpression names t0, t1, etc. ♣ (expr, *filename* = str) to write the results to a file named str instead of to the terminal. ♣ (array) or (list), where array or list consist of equations that are understood to represent assignment statements, to translate statements into C.
**Additional information:** This command needs to be defined by readlib(C) before being invoked. ♣ Currently, C does not work on Maple procedures.
**See also:** fortran, eqn, latex, optimize
**FL = 112** *LI = 266*

### cartan($A_1$, $A_2$, list, met)

Computes connection coefficients for arrays $A_1$ and $A_2$, list, and constant metric met using Cartan's structure equations.
**Output:** A three-dimensional array, named gamma, is returned.
**Additional information:** This command needs to be defined by readlib(cartan) before being invoked. ♣ The result of cartan is used in simp1, which simplifies the gamma array. riemann is then used to compute the Riemann tensor component array. Finally, simp2 simplifies the Riemann tensor component array. ♣ printcartan prints the non-zero entries of any array. ♣ For more information on all these commands, see the on-line help page for cartan.
**See also:** npspin, tensor, petrov
*LI = 267*

### ceil(expr)

Determines the smallest integer greater than or equal to real expression expr.
**Output:** If expr is a numeric value, then an integer is returned. Otherwise, ceil returns unevaluated.
**Argument options:** (cmplx) to determine the ceiling for a complex valued expression. See the on-line help page for details on the returned value. ♣ (*1*, expr) to represent the derivative of the ceil command at expr.
**Additional information:** The derivative of the ceil command is 0 wherever it is defined.
**See also:** floor, trunc, round, frac

### commutat(expr$_c$)

Displays expr$_c$, an expression in commutators, in *Lie bracket form*.
**Output:** An expression is returned.
**Additional information:** This command needs to be defined by readlib(commutat) before being invoked. ♣ The commutators represent non-commutative multiplications and can be represented either in c form or as expressions in &*. ♣ Simplifications, conversions, and expansions can be done on both forms.
**See also:** c, simplify('&*'), simplify(c), convert('&*'), convert(c), *expand*


### conjugate(cmplx)

Determines the complex conjugate of complex expression cmplx.
**Output:** A complex expression is returned.
**Additional information:** If expressions in terms of exp or polar are given, conjugate responds with the appropriately transformed command calls.
**See also:** evalc, polar, *exp*, argument, Re, Im


### convergs([a$_1$, ..., a$_m$], [b$_1$, ..., b$_n$])

Prints the convergents of the continued fraction represented by a$_1$ + b$_2$/(a$_2$ + b$_3$/(a$_3$ + ....
**Output:** A sequence of print statements are displayed. Each line contains two values, an integer representing which convergent is displayed next and that convergent itself, respectively. The *actual* output of this command is always NULL. That is, if you try to access the result with ", you do *not* get the print results.
**Argument options:** ([a$_1$, ..., a$_m$]) to use the default [1, 1, 1, ...] for the second list of values. ♣ ([a$_1$, ..., a$_m$], [b$_1$, ..., b$_n$], n) to print only to the n$^{th}$ convergent.
**Additional information:** This command needs to be defined by readlib(convergs) before being invoked.
**See also:** *convert(confrac)*


### convert(expr, '&*')

Converts expr, an expression in unevaluated c calls, to an expression in &* operators.
**Output:** An expression is returned.
**Additional information:** This command needs to be defined by readlib(commutat) before being invoked.
**See also:** commutat, c, simplify('&*'), simplify(c), convert(c)

## convert(exprb, 'and')

Converts all binary boolean operators (and, or, &and, &or, &nand, &nor, &xor, &iff, and &implies) in exprb to the and operator.
**Output:** A boolean expression is returned.
**Argument options:** ([exprb$_1$, ..., exprb$_n$], 'and') or ({exprb$_1$, ..., exprb$_n$}, 'and') to create a boolean expression for exprb$_1$ and exprb$_2$ and ... and exprb$_n$.
**Additional information:** The resulting boolean expression is, in most cases, *not* equivalent to exprb.
**See also:** convert('or'), type('and')


## convert(expr, c)

Converts expr, an expression in &* operators, to an expression in unevaluated c calls.
**Output:** An expression is returned.
**Additional information:** This command needs to be defined by readlib(commutat) before being invoked.
**See also:** commutat, c, simplify('&*'), simplify(c), convert('&*')


## convert(exprb, 'or')

Converts all binary boolean operators (and, or, &and, &or, &nand, &nor, &xor, &iff, and &implies) in exprb to the or operator.
**Output:** A boolean expression is returned.
**Argument options:** ([exprb$_1$, ..., exprb$_n$], 'or') or ({exprb$_1$, ..., exprb$_n$}, 'or') to create a boolean expression for exprb$_1$ or exprb$_2$ or ... or exprb$_n$.
**Additional information:** The resulting boolean expression is, in most cases, *not* equivalent to exprb.
**See also:** convert('and'), type('or')


## Digits := n

Sets the global variable Digits, specifying the number of digits used in floating-point calculations, to a value of n.
**Output:** The input is echoed.
**Argument options:** Digits to return the current value.
**Additional information:** The default value of Digits is 10. ✤ Digits sets the number of digits displayed in floating-point results.
**See also:** evalf
**FL = 21−23, 25, 141** *LA = 6*

### eqn(expr)

Translates expression expr into an equivalent expression in the *troff/eqn* document preparation language.
**Output:** A string is returned.
**Argument options:** (expr, *filename* = str) to write the results to a file named str instead of to the terminal.
**Additional information:** This command needs to be defined by readlib(eqn) before being invoked. ♣ Currently, eqn works on Maple integrals, limits, sums, products, and matrices. ♣ The *.EQ* and *.EN* lines typically seen in *eqn* files are not produced by eqn. ♣ The results of a call to eqn are not recallable by ".
**See also:** latex, C, fortran
*LI = 72*

### eval(expr, n)

Evaluates expression expr just n levels.
**Output:** An expression is returned.
**Argument options:** (expr) to evaluate expr fully.
**Additional information:** This command is rarely needed by most users. ♣ Full evaluation occurs by default on *most* Maple expressions, with the exception of local variables and parameters which have one-level evaluation. ♣ One particular use for eval is in evaluation of the results of a subs command. ♣ eval(expr, 1) determines to what the first value each variable in expr is assigned and substitutes in for those values. These values themselves may very well be assigned to further values, but that is ignored.
**See also:** *subs,* evalf, evalc, *evala*
**FL = 86** *LA = 82* **LI = 75**

### evalb($expr_b$)

Evaluates $expr_b$, a boolean equation or inequation containing relational operators =, <>, <, <=, >, or <=.
**Output:** If boolean evaluation can be performed, a value of true or false is returned. Otherwise, the expression is echoed in an equivalent form.
**Additional information:** Expressions containing relational operators > and >= are converted to expressions containing < and <=, respectively. ♣ $expr_b$ can also contain the values true and false strung together with operators such as and, or, and not. ♣ For more information on boolean operations, see the logic package.
**See also:** logic package
**FL = 119** *LA = 51* **LI = 77**

### evalc(expr$_c$)

Evaluates expr$_c$, an expression typically containing complex values, over the complex field.
**Output:** An expression is returned.
**Additional information:** Complex values can be represented either in the form a + b*I or polar(r, $\theta$), though the former representation is more common. ♣ expr$_c$ often contains calls to unevaluated commands like sin, exp, Zeta, etc. that have complex values with nondecimal valued components as parameters. evalc simplifies these commands as much as possible, but it is very likely that the result still contains unevaluated commands (not necessarily the same commands passed in). ♣ If complex values have symbolic components, evalc assumes that these components are real valued. ♣ Many standard commands have built-in capabilities for automatically dealing with complex values with floating-point components. In these cases, a complex value is returned without using evalc.
**See also:** polar, eval, evalf, Re, Im
*LI = 78*

### evalf(expr)

Evaluates symbolic expression expr numerically. Default accuracy is set by global variable Digits.
**Output:** A simplified expression with all possible non-numeric values converted to floats is returned.
**Argument options:** (expr, n) for evaluation to n digits accuracy. ♣ (Fnc) to evaluate inert function Fnc numerically, without first trying the symbolic route. ♣ (Fnc), from failed attempts at symbolic methods, to try numeric evaluation on unevaluated function Fnc.
**Additional information:** To speed up floating-point calculations, try the evalhf (*eval*uate in *h*ardware *f* loating-point) command.
**See also:** evalhf, *Digits*, 'evalf/int'
**FL = 21−23** *LA = 6, 20 LI = 79*

### evalhf(expr)

Evaluates symbolic expression expr numerically, using the hardware floating-point double-precision calculations available on your system. expr must evaluate to a single value or be an array data type.
**Output:** A Maple floating-point value is returned.
**Additional information:** Using evalhf speeds up your calculations in most cases; but you lose the definable accuracy you get

using evalf and Digits together. ♣ If Digits is set to a value of 15, evalf and evalhf produce very similar results. ♣ All dealings with hardware floating-point calculations are internal to evalhf; you never have to deal directly with them. ♣ For more information on evalhf, see the on-line help page and the files mentioned therein.
**See also:** evalf, *Digits*
*FL = 192−196  LI = 80−86*

### evaln(expr)

Creates an assignable name out of expr.
**Output:** A name is returned.
**Additional information:** expr can be a string, a subscript, a function call, or a concatenation. ♣ The effect is similar to enclosing expr in forward quotes. ♣ See the section *The Use of Quotes in Maple* in the chapter *Getting Started With Maple* for more information on forward quotes.
**See also:** *assigned*
*LI = 87*

### evalr($expr_r$)

Evaluates expr, typically a call to another command using ranges instead of variables, using range arithmetic.
**Output:** A range or sequence of ranges representing bounding values for $expr_r$ is returned.
**Argument options:** (expr), where expr is a normal call to the min, max, abs, or signum command, to return an exact value or range for that command.
**Additional information:** This command needs to be defined by readlib(evalr) before being invoked. ♣ $expr_r$ can contain a few different types of range declarations. Examples of these ranges are [a..b], [a..b, c..d, e..f], and [var, a..b], where for each range the lefthand side is less than or equal to the righthand side. ♣ If any standard variables are used in $expr_r$, they are replaced with the range [var, -infinity..infinity]. ♣ If there is more than one variable or range in the result, then a list of lists is returned.
**See also:** *solve, fsolve,* shake
*LI = 282*

### example(name)

Displays the *EXAMPLES* section of the on-line help page for Maple command or structure name.
**Output:** A NULL expression is returned.

**Argument options:** ('name') if name is a keyword such as quit. ♣ (name$_1$, name$_2$) to display information from a more specific help page, where name$_2$ is a subtopic of name$_1$ or name$_2$ is a command in the package name$_1$.

**Additional information:** The ???name syntax is much easier to use, and does not require a semicolon at the end of the statement as example(name) does.

**See also:** help, info, usage, related, tutorial

## finance(amount=num$_1$, interest=num$_2$, periods=num$_3$)

Determines the payment per period on a loan of num$_1$ with interest rate num$_2$ over num$_3$ periods. num$_1$, num$_2$, and num$_3$ must all be real or floating-point numeric values.

**Output:** An equation is returned.

**Argument options:** (eqn$_1$, eqn$_2$, eqn$_3$), where the equations are any three of *amount*=num$_1$, *interest*=num$_2$, *periods*=num$_3$, and *payment*=num$_4$, to determine the fourth equation in the set.

**Additional information:** This command needs to be defined by readlib(finance) before being invoked.

**See also:** amortization, blackscholes

*LI = 287*

## Float(int$_m$, int$_e$)

Represents the floating-point value whose mantissa and exponent are int$_m$ and int$_e$, respectively.

**Output:** Either a floating-point value or an unevaluated Float command is returned.

**Additional information:** The Float representation is used within Maple to store floating-point numbers. ♣ As well, Float is used to display numbers with very small or large magnitudes. ♣ The value of Digits determines how many digits are used in the mantissa.

**See also:** evalf, op(float)

*LA = 44*

## floor(expr)

Determines the largest integer less than or equal to real expression expr.

**Output:** If expr is a numeric value, then an integer is returned. Otherwise, floor returns unevaluated.

**Argument options:** (cmplx) to determine the floor for a complex valued expression. See the on-line help page for details on the returned value. ♣ (*1*, expr) to represent the derivative of the floor

command at expr.
**Additional information:** The derivative of the floor command is 0 wherever it is defined.
**See also:** ceil, trunc, round, frac

### fnormal(expr)

Performs floating-point normalization on algebraic expression expr.
**Output:** An expression in floating-point values is returned.
**Argument options:** (expr, n) to set the number of digits of floating-point evaluation to n. Default is the value of Digits. ♣ (expr, n, num) to set the error tolerance for normalization to num. Default is the value Float(1, -Digits+2).
**Additional information:** fnormal equates any value in expr less than a very small number to 0. ♣ fnormal also works on each element of a list, set, series, equation, or relation.
**See also:** *simplify*, evalf
*LI = 94*

### fortran(expr)

Translates expression expr into an equivalent expression in the *FORTRAN* programming language.
**Output:** A string is returned.
**Argument options:** (expr, *double*) to specify that double precision is used in the translation. ♣ (expr, *optimized*) to perform common subexpression optimization on the result, using subexpression names t0, t1, etc. ♣ (expr, *filename = str*) to write the results to a file named str instead of to the terminal. ♣ (expr, *mode = option*), where option is one of *single, double, complex, generic*, to specify how function names are to be translated. The default is *single*. ♣ (array) or (list), where array or list consist of equations understood to represent assignment statements, to translate the statements into *FORTRAN*.
**Additional information:** Currently, fortran does not work on Maple procedures, nor does it work very well with complex values. ♣ For more information on fortran, see the on-line help page.
**See also:** C, eqn, latex, optimize
**FL = 111-112** *LA = 26 LI = 95*

### frac(expr)

Determines the fractional part of real valued expression expr.
**Output:** If expr is an rational numeric value, an exact fraction is returned. If expr is an irrational numeric value, expr-trunc(expr) is returned. If expr is a floating-point value, a floating-point value is returned. Otherwise, frac returns unevaluated.

**Argument options:** (cmplx) to determine the fractional part of a complex valued expression. The result is equivalent to frac(Re(expr)) + I*frac(Im(expr)). ♣ (1, expr) to represent the derivative of the frac command at expr.

**Additional information:** The derivative of the frac command is 1 wherever it is defined.

**See also:** trunc, round, ceil, floor

*LI = 217*

### genfunc package

Provides commands for creating and manipulating rational generating functions.

**Additional information:** To access the command fnc from the genfunc package, use the long form of the name genfunc[fnc], or with(genfunc, fnc) to load in a pointer for fnc only, or with(genfunc) to load in pointers to all the commands.

### genfunc[rgf_charseq](rgf, $var_1$, fnc, $var_2$)

Determines the characteristic sequence of rgf, a rational generating function in terms of $var_1$. The results is written in terms of fnc, a function in $var_2$.

**Output:** An expression in fnc is returned.

**Additional information:** If rgf is a trivial generating function, then FAIL is returned. ♣ For more information on how the characteristic sequence is formed, see the on-line help page.

**See also:** genfunc[rgf_encode], genfunc[rgf_expand], genfunc[rgf_relate], genfunc[termscale]

### genfunc[rgf_encode](expr, $var_1$, $var_2$)

Computes the rational generating function, in terms of $var_2$, for closed-form expression expr with index variable $var_1$.

**Output:** A rational expression is returned.

**Argument options:** (expr, $var_1$, $var_2$, $var_1$ = n) to specify that the term at index n is the first nonzero term of the sequence defined by the generating function. ♣ (expr, $var_1$, $var_2$, [$int_1$ = $expr_1$, ..., $int_n$ = $expr_n$]) to specify that the term at index $int_i$ has the value of $expr_i$ in the sequence defined by the generating function.

**Additional information:** The result is assumed to define a sequence for all index values greater than or equal to 0. ♣ expr must be a closed-form expression. ♣ To ensure that expr is valid, use the type(*rgf_seq*) command.

**See also:** genfunc[rgf_expand], genfunc[rgf_seq], type(*rgf_seq*), *ztrans, laplace, fourier*

### genfunc[rgf_expand](rgf, var$_1$, var$_2$)

Computes the closed-form expansion of the var$_2$ term of rgf, a rational generating function in var$_1$.
**Output:** A rational expression in var$_2$ is returned.
**Additional information:** If the denominator of rgf cannot be factored over the rationals, an inert Sum command involving RootOfs is returned.
**See also:** genfunc[rgf_encode], genfunc[rgf_term], genfunc[rgf_hybrid], genfunc[rgf_charseq], genfunc[rgf_norm], genfunc[rgf_simp], *Sum*, *RootOf*

### genfunc[rgf_findrecur](posint, [expr$_1$, ..., expr$_n$], fnc, var)

Determines the recurrence relation of order posint for the sequence with consecutive values expr$_1$ through expr$_n$. The result is expressed in terms of function fnc in var.
**Output:** An equation with fnc(var) as its lefthand side is returned.
**See also:** genfunc[rgf_sequence]

### genfunc[rgf_hybrid](var, rgf$_1$, rgf$_2$, ..., rgf$_n$)

Computes the rational generating function resulting from the termwise multiplication of the sequences encoded by rgf$_1$ through rgf$_n$, rational generating functions in var.
**Output:** A rational generating function in var is returned.
**See also:** genfunc[rgf_encode], genfunc[rgf_expand]

### genfunc[rgf_norm](rgf, var)

Normalizes rgf, a rational generating function in terms of var.
**Output:** A rational expression in var is returned.
**Argument options:** (rgf, var, *factored*) to specify that factor is applied to the denominator of the result. ✦ (rgf, var, *factored*(expr$_1$, ..., expr$_n$)) or (rgf, var, *factored*({expr$_1$, ..., expr$_n$})) to specify algebraic extension(s) to use when factoring the denominator.
**Additional information:** The generating function is put in normal form by first expanding the numerator and denominator and then factoring the coefficient of the denominator's lowest order term.
**See also:** genfunc[rgf_encode], genfunc[rgf_factor], genfunc[rgf_simp], *factor*

### genfunc[rgf_pfrac](rgf, var)

Computes the partial fraction expansion of rgf, a rational generating function in terms of var.

**Output:** A rational expression in var is returned.
**Argument options:** (rgf, var, *no_RootOf*) to specify that the denominator of rgf is to be completely factored over the complex numbers, if possible. If not possible, an inert expression is returned.
**Additional information:** The denominator of rgf is always factored before the partial fraction is computed. ♣ If the denominator cannot be factored over the rationals, an inert Sum command containing RootOfs is returned.
**See also:** genfunc[rgf_encode], *convert(parfrac)*, *factor*

### genfunc[rgf_relate](rgf$_1$, var$_1$, fnc, var, rgf$_2$)

Relates rgf$_1$ and rgf$_2$, two rational generating functions in var$_1$ with common nonzero roots in their denominators. The result is written in terms of function fnc in var.
**Output:** An expression in var is returned.
**Argument options:** (rgf$_1$, var$_1$, fnc, var, rgf$_2$, var$_2$) to specify that rational generating function rgf$_2$ is in terms of var$_2$.
**Additional information:** If rgf$_1$ is a trivial generating function, then FAIL is returned.
**See also:** genfunc[rgf_encode], genfunc[rgf_charseq]

### genfunc[rgf_sequence](option, rgf, var)

Extracts information about the sequence encoded by rgf, a rational generating function in var.
**Output:** The results depend on the value chosen for option.
**Argument options:** (*'boundary'*, rgf, var) to return an expression sequence of equations representing boundary conditions for rgf. ♣ (*'coeffs'*, rgf, var) to return an ordered expression sequence representing the coefficients of the recurrence defining rgf. ♣ (*'delta'*, rgf, var) to return an expression sequence of equations representing the delta terms that, when added to the closed-form expression of the sequence encoded by rgf, completely define the sequence. ♣ (*'first'*, rgf, var) to return the index value of the first nonzero term in the sequence encoded by rgf. ♣ (*'firstcf'*, rgf, var) to return the minimum index value of the sequence encoded by rgf from which all terms in the sequence are encoded. ♣ (*'order'*, rgf, var) to return the order of rgf. ♣ (*'recur'*, rgf, var) to return the recurrence relation that defines the sequence encoded by rgf.
**See also:** genfunc[rgf_encode]

### genfunc[rgf_simp](expr, rgf, var$_1$, fnc, var$_2$)

Simplifies expression expr, an expression containing occurrences of rational generating function rgf in terms of var$_1$. fnc is the name

of the function encoded by rgf and has index variable $var_2$.
**Output:** An expression in fnc and $var_2$ is returned.
**Argument options:** (expr, rgf, $var_1$, fnc, $var_2$, posint * $var_2$ + int) to perform simplification relative to the term with pattern posint * $var_2$ + int, if possible. ♣ (expr, rgf, $var_1$, fnc, $var_2$, $fnc_c$) to perform the command $fnc_c$ on each coefficient of the result.
**Additional information:** Like terms are automatically collected in the result. ♣ For more information and examples of this command, see the on-line help page.
**See also:** genfunc[rgf_encode], genfunc[rgf_factor], genfunc[rgf_norm], *collect*

### genfunc[rgf_term](rgf, var, int)

Computes the value of the $int^{th}$ term of rgf, a rational generating function in var.
**Output:** A rational expression is returned.
**Additional information:** If rgf is a trivial generating function, then FAIL is returned.
**See also:** genfunc[rgf_encode], genfunc[rgf_expand]

### genfunc[termscale](poly, $var_1$, rgf, $var_2$)

Computes the result of multiplying rgf, a rational generating function in terms of $var_2$, by poly, a polynomial in $var_1$.
**Output:** An expression in $var_2$ is returned.
**See also:** genfunc[rgf_encode], genfunc[rgf_charseq], genfunc[rgf_expand], genfunc[rgf_relate]

### GF($int_p$, posint, poly)

Creates a table of commands and constants for performing arithmetic in the finite Galois Field with $int_p$^posint elements defined by the field extension GF($int_p$)[x]/(poly). poly is an irreducible polynomial of degree posint over the integers modulo $int_p$.
**Output:** A table of commands is returned.
**Additional information:** This command needs to be defined by readlib(GF) before being invoked. ♣ It is recommended that you complete the GF command with a colon instead of a semi-colon, because the table produced is several pages long. ♣ Once you create the table, say T, then you can use the many commands that are its entries. T[input] and T[output] convert between an integer in the range 1..$int_p$^posint and a corresponding polynomial. T[ConvertIn] and T[ConvertOut] convert between an element of the Galois Field and a Maple sum. T['+'], T['-'], T['*'], T['^'], T['inverse'], and T['/'] perform the stated operation on elements

of the Galois field. T[0] and T[1] represent the additive and multiplicative inverses. T[trace], T[norm], and T[order] compute values for elements of the Galois field. T[random] returns a random element. T[PrimitiveElement] returns a random primitive element. T[isPrimitiveElement] determines whether an element is a primitive element. T[extension] returns the polynomial extension being used for the Galois field. ♣ For more information and examples, see the on-line help page.
**See also:** *mod, modp1*
*LI = 290*

### group package

Provides standard commands in the realm of group theory.
**Additional information:** To access the command fnc from the group package, use the long form of the name group[fnc], or with (group, fnc) to load in a pointer for fnc only, or with(group) to load in pointers to all the commands. ♣ Since defining groups is not altogether intuitive, it is recommended that you read the on-line help page for group[permgroup] before proceeding. ♣ The results of many commands in the group package are unevaluated calls to permgroup, grelgroup, etc. This is simply the representation of the answer, not an indication that the calculation did not take place.

### group[areconjugate](pg, list$_1$, list$_2$)

Determines whether the cycles of permutation group pg, represented by elements list$_1$ and list$_2$, are conjugate.
**Output:** A boolean value of true or false is returned.
**Additional information:** list$_1$ and list$_2$ are each lists containing one internal list.
**See also:** group[permgroup], group[groupmember], group[RandElement]

### group[center](pg)

Calculates the center of permutation group pg.
**Output:** An unevaluated permgroup command is returned.
**Additional information:** The center of pg is the largest subgroup of pg such that every element present commutes with every element of pg. ♣ If the empty subgroup is determined to be the center, a permgroup command containing the empty set {} is returned.
**See also:** group[centralizer], group[normalizer]

## group[centralizer](pg, {list₁, ..., list_n})

Calculates the centralizer, which is in permutation group pg, of permutations list₁ through list₂.
**Output:** An unevaluated permgroup command is returned.
**Argument options:** (pg, list) to calculate the centralizer of the single permutation list.
**Additional information:** The centralizer of permutations list₁ through list_n is the largest subgroup of pg such that every element present commutes with every element of list₁ through list_n. ♣ The list_i are typically lists of two or more points in pg. ♣ If the empty subgroup is determined to be the centralizer, a permgroup command containing the empty set {} is returned.
**See also:** group[center], group[normalizer]

## group[core](pg₁, pg₂)

Calculates the largest normal subgroup of permutation group pg₂ contained in permutation group pg₁.
**Output:** An unevaluated permgroup command is returned.
**Additional information:** pg₁ must be a subgroup of pg₂ or an error message is generated.
**See also:** group[permgroup], group[issubgroup], group[NormalClosure]

## group[cosets](pg₁, pg₂)

Calculates the complete set of right coset representatives for subgroup pg₂ of permutation group pg₁.
**Output:** A set of cosets is returned. Each coset is a cycle (i.e., a list of a list or the empty list).
**Argument options:** (sbgrlg) to find the right coset representations of sbgrlg, a subgroup determined from a generators and relations group.
**See also:** group[cosrep], group[permgroup], group[grelgroup], group[subgrel]

## group[cosrep](list, pg)

Represents the permutation list, which is a list within a list, as a permutation in the group pg multiplied by a right coset representative for that group.
**Output:** A list with two elements is returned. The first element is a list within a list representing a permutation from pg. The second element is a list within a list representing a right coset representation.

**Argument options:** (list, sbgrlg) to represent the subgroup sbgrlg of a generators and relations group as an element in the subgroup multiplied by a right coset representative for that subgroup. The result has two lists as elements, the first being the subgroup element and the second being the right coset representative.
**See also:** group[cosets], group[permgroup], group[grelgroup], group[subgrel]

### group[derived](pg)

Calculates the derived subgroup of permutation group pg.
**Output:** An unevaluated permgroup command is returned.
**See also:** group[permgroup], group[DerivedS], group[LCS]

### group[DerivedS](pg)

Calculates the derived series of permutation group pg.
**Output:** A list of unevaluated permgroup commands is returned.
**Additional information:** This command can be used to find whether pg is solvable.
**See also:** group[permgroup], group[derived], group[LCS]

### group[grelgroup]({expr$_1$, ..., expr$_m$}, {list$_1$, ..., list$_n$})

Represents the group defined by generators expr$_1$ through expr$_m$ and relations list$_1$ through list$_n$ which typically contain two or more of either expr$_1$ through expr$_m$ or their inverses.
**Output:** An unevaluated grelgroup call is returned.
**Additional information:** The inverse of expr$_i$ is represented by $1/\text{expr}_i$ and the identity element by the empty list []. 
**See also:** group[permgroup], group[subgrel]

### group[groupmember](list, pg)

Determines whether the permutation represented by list is a member of permutation group pg.
**Output:** A boolean value of true or false is returned.
**Additional information:** list must be a list of lists.
**See also:** group[permgroup], group[RandElement]

### group[grouporder](pg)

Calculates the number of elements in permutation group pg.
**Output:** An integer value is returned.

**Argument options:** (grelg) to calculate the number of elements of generators and relations group grelg.
**See also:** group[permgroup], group[gelgroup], group[groupmember]

### group[inter](pg$_1$, pg$_2$)

Calculates the intersection of two permutation groups, pg$_1$ and pg$_2$.
**Output:** An unevaluated permgroup command is returned.
**Additional information:** If pg$_1$ and pg$_2$ are not of the same degree, an error message is generated.
**See also:** group[permgroup]

### group[invperm]([list$_1$, ..., list$_n$])

Calculates the inverse of the permutation represented by list$_1$ through list$_n$.
**Output:** A list of lists (in disjoint cycle notation) is returned.
**See also:** group[permgroup]

### group[isabelian](pg)

Determines whether permutation group pg is an abelian group.
**Output:** A boolean value of true or false is returned.
**See also:** group[permgroup], group[isnormal], group[issubgroup]

### group[isnormal](subpg)

Determines whether subpg is a normal subgroup.
**Output:** A boolean value of true or false is returned.
**Argument options:** (pg$_1$, pg$_2$) to determine whether pg$_2$ is a normal subgroup of the permutation group generated by the union of pg$_2$ with pg$_1$.
**See also:** group[permgroup], group[subgrel], group[isabelian], group[issubgroup]

### group[issubgroup](pg$_1$, pg$_2$)

Determines whether permutation group pg$_1$ is a subgroup of permutation group pg$_2$.
**Output:** A boolean value of true or false is returned.
**See also:** group[permgroup], group[isnormal], group[isabelian], group[NormalClosure], group[Sylow], group[core]

### group[LCS](pg)

Calculates the *Lower Central Series* of permutation group pg.
**Output:** A list of unevaluated permgroup commands is returned.
**Additional information:** This command can be used to determine if pg is *nilpotent*.
**See also:** group[permgroup], group[derived], group[DerivedS]

### group[mulperms]([list$_{1,1}$, ..., list$_{1,m}$], [list$_{2,1}$, ..., list$_{2,n}$])

Multiplies together the permutations in disjoint cycle notation represented by list$_{1,1}$ through list$_{1,m}$ and list$_{2,1}$ through list$_{2,n}$.
**Output:** A list of lists (in disjoint cycle notation) is returned.
**See also:** group[permgroup], group[invperm]

### group[NormalClosure](subpg, pg)

Calculates the normal closure of permutation group pg with respect to subgroup subpg.
**Output:** An unevaluated permgroup command is returned.
**Additional information:** The normal closure of pg with respect to subpg is the smallest normal subgroup of pg containing subpg.
**See also:** group[permgroup], group[isnormal], group[issubgroup], group[core]

### group[normalizer](pg$_1$, pg$_2$)

Calculates the normalizer of permutation group pg$_2$, which is a subgroup of permutation group pg$_1$.
**Output:** An unevaluated permgroup command is returned.
**Additional information:** The normalizer of pg$_2$ is defined as the largest subgroup of pg$_1$ in which pg$_2$ is a normal subgroup.
**See also:** group[permgroup], group[centralizer]

### group[orbit](pg, expr)

Calculates the orbit of point expr under the group action of permutation group pg.
**Output:** A set of points is returned.
**Argument options:** ({list$_1$, ..., list$_n$}, expr) to use a set of permutations that has not yet been made an official permutation group. The necessary syntax is identical to that used within group[permgroup].
**Additional information:** If expr is not a member of the group, it is returned in a single-element set.
**See also:** group[permgroup]

## group[permgroup](n, {list₁, ..., list_n})

Represents the degree n permutation group defined by generators list₁ through list_n, which each contain one or more internal lists representing cycles in that particular permutation.

**Output:** An unevaluated permgroup command is returned.

**Argument options:** (n, {name₁ = list₁, ..., name_n = list_n}) to name the individual generators.

**Additional information:** n should be an integer. ♣ Elements of the sublists are typically integers from 1 to n. ♣ The notation in which the permutations are listed is called *disjoint cycle notation*. ♣ Each list_i is a permutation made up of the product of the cycles (sublists) within it. ♣ The identity cycle is represented by the empty list [].

**See also:** group[grelgroup]

## group[permrep](subgrelg)

Represents subgrelg, a subgroup of a generators and relations group, as a permutation group.

**Output:** An unevaluated permgroup command is returned.

**Additional information:** The permutation group returned is a *homomorphic* image of the original generators and relations group, but it is not necessarily *isomorphic* to the original group.

**See also:** group[permgroup], group[grelgroup], group[subgrel], group[pres]

## group[pres](subgrelg)

Finds a presentation for subgrelg, a subgroup of a generators and relations group.

**Output:** An unevaluated grelgroup call is returned.

**Additional information:** A set of relations among subgrelg's generators, sufficient to define the subgroup, is calculated.

**See also:** group[grelgroup], group[subgrel], group[permrep]

## group[RandElement](pg)

Creates a random element of permutation group pg.

**Output:** A list of lists is returned.

**Argument options:** (grelg) to create a random element of generators and relations group grelg.

**See also:** group[permgroup], group[gelgroup], group[groupmember], group[areconjugate]

## group[subgrel]({name₁ = list₁, ..., name_n = list_n}, grelg)

Represents the subgroup of generators and relations group grelg defined by generator equations name₁=list₁ through name_n=list_n.

**Output:** An unevaluated subgrel call is returned.
**Additional information:** The list$_i$ are typically lists of two or more expressions that are either generators of grelg or their inverses.
♣ The inverse of a generator a is represented by 1/a.
**See also:** group[grel|group]

### group[Sylow](pg, p)

Calculates the Sylow-p-subgroup of permutation group pg. p must be a prime divisor of the order of pg.
**Output:** An unevaluated permgroup command is returned.
**Additional information:** The Sylow-p-subgroup of pg is the maximal p-group (i.e., with all cycles of length p) that is a subgroup of pg.
**See also:** group[permgroup], group[issubgroup]

### has(expr, expr$_s$)

Determines if expression expr contains subexpression expr$_s$.
**Output:** A boolean value of true or false is returned.
**Argument options:** (expr, [expr$_1$, ..., expr$_n$]) or (expr, {expr$_1$, ..., expr$_n$}) to determine if expr contains at least one of the subexpressions expr$_1$ through expr$_n$.
**Additional information:** expr can be *any* type of expression. ♣ expr is evaluated before the determination is done and has can only find a subexpression of expr if it is a proper operand of that evaluated expression. See the entry for subs for more details.
**See also:** hastype, *subs*, *member*, *op*
*LI = 104*

### hastype(expr, type)

Determines if expression expr contains a subexpression of type type.
**Output:** A boolean value of true or false is returned.
**Additional information:** hastype also searches for structured types.
**See also:** *whattype*, has
*LA = 86* *LI = 104*

### help(name)

Displays the on-line help page for Maple command or structure name.
**Output:** A NULL expression is returned.
**Argument options:** ('name') if name is a keyword such as quit.

♣ (name$_1$,name$_2$) to display a more specific help page, where name$_2$ is a subtopic of name$_1$ or name$_2$ is a command in the package name$_1$.
**Additional information:** The ?name syntax is much easier to use, and does not require a semicolon at the end of the statement as help(name) does. ♣ Some of the special help pages available include ?intro, ?library, and ?index.
**See also:** info, usage, example, related, tutorial
**FL = 5** *LA = 4−5*

### Im(cmplx)

Extracts the imaginary component of complex value cmplx.
**Output:** A real value is returned.
**Additional information:** If it is not possible to determine the imaginary component of cmplx, then Im is left unevaluated.
**See also:** Re, argument, evalc

### info(name)

Displays the first line of the on-line help page for Maple command or structure name.
**Output:** A NULL expression is returned.
**Argument options:** ('name') if name is a keyword such as quit.
♣ (name$_1$,name$_2$) to display information from a more specific help page, where name$_2$ is a subtopic of name$_1$ or name$_2$ is a command in the package name$_1$.
**Additional information:** The first line of a help page contains the structure name and a very brief synopsis of its use.
**See also:** help, usage, example, related

### set$_1$ intersect set$_2$

Computes the intersection of set$_1$ and set$_2$.
**Output:** A set is returned.
**Argument options:** 'intersect'(set$_1$, set$_2$) to produce the same results. intersect must be in backquotes because it is a keyword.
♣ name$_1$ intersect name$_2$ to represent the intersection of two unassigned names that may later be assigned to sets. The input is left unevaluated.
**Additional information:** The intersection of set$_1$ and set$_2$ contains the elements found in *both* set$_1$ and set$_2$.
**See also:** minus, union, member
**FL = 81** *LA = 17, 51*

### is(ineq$_{var}$)

Determines if variable var has been previously assumed to satisfy inequation ineq$_{var}$.

**Output:** A boolean value of true, false or FAIL is returned.

**Argument options:** (var, name), where name is a type or a property such as *integer* or *rational*, respectively, to determine if var is of that kind. ♣ (var, *AndProp*(name$_1$, ..., name$_n$)), to determine if var conforms to *all* the properties represented by name$_1$ through name$_n$. ♣ (var, *OrProp*(name$_1$, ..., name$_n$)), to determine if var conforms to at least *one* of the properties represented by name$_1$ through name$_n$.

**Additional information:** The properties specified in is need not have been assumed *exactly* as stated. Maple does some interpolation, if possible. For example, if var is assumed greater than 0, then is(var > -4) is true. ♣ Other commands that allow you to query or alter information about a variable are isgiven, about, assume, and addproperty. ♣ For more information on is, see the on-line help page for assume.

**See also:** assume, isgiven, about, addproperty

### isgiven(ineq$_{var}$)

Determines if variable var has been previously assumed with the exact inequation ineq$_{var}$.

**Output:** A boolean value of true or false is returned.

**Argument options:** (var, name), where name is a type or a property, to determine if var has been assumed to be of that property. ♣ (var, *AndProp*(name$_1$, ..., name$_n$)), to determine if var has been assumed to be of *all* the properties represented by name$_1$ through name$_n$. ♣ (var, *OrProp*(name$_1$, ..., name$_n$)), to determine if var has been assumed to be of at least *one* of the properties represented by name$_1$ through name$_n$.

**Additional information:** Other commands that allow you to query or alter information about a variable are is, about, assume, and addproperty. ♣ For more information on isgiven, see the on-line help page for assume.

**See also:** assume, is, about, addproperty

### latex(expr)

Translates expression expr into an equivalent expression in the LaTeX document preparation language.

**Output:** A string is returned.

**Argument options:** (expr, *filename* = str) to write the results to a file named str instead of to the terminal.

Miscellaneous 489

**Additional information:** Currently, latex works on Maple integrals, limits, sums, products, and matrices. ♣ The commands to invoke the LaTeX math environment are not produced. ♣ The results of a call to latex are not recallable by ".
**See also:** eqn, C, fortran
**FL = 113** *LI = 123–125*

### length(str)

Returns the number of characters, including blanks and special characters, in string str.
**Output:** A positive integer is returned.
**Argument options:** (int) to return the number of decimal digits in the absolute value of integer int. The length of 0 is defined to be 0. ♣ (expr) to return the size, calculated by summing the length (in words) of each operand of expr.
**See also:** search, substring, cat
**FL = 215** *LI = 126*

### lexorder(str$_1$, str$_2$)

Determines whether string str$_1$ occurs before string str$_2$ in lexicographical order.
**Output:** A boolean value of true or false is returned.
**Additional information:** To a certain extent, this command relies on the underlying ordering of your computer's character set.
**See also:** *sort(list)*
*LI = 126*

### logic package

Provides commands for computations with logical expressions.
**Additional information:** To access the command fnc from the logic package, use the long form of the name logic[fnc], or with (logic, fnc) to load in a pointer for fnc only, or with(logic) to load in pointers to all the commands. ♣ The logical expressions used in logic are created with the boolean operators &and, &or, &not, &iff, &nor, &nand, &xor, and &implies.

### logic[bequal](exprb$_1$, exprb$_2$)

Tests whether two boolean expressions, exprb$_1$ and exprb$_2$, are equivalent.
**Output:** A boolean value of either true or false is returned.
**Argument options:** (exprb$_1$, exprb$_2$, name) to assign to name a valuation of the variables in exprb$_1$ and exprb$_2$ showing that the

two expressions are not equivalent. name is assigned a set containing an equation for each variable if the expressions are not equivalent, or NULL if they are equivalent.
**Additional information:** Without the third parameter, logic[bequal] runs significantly faster for complicated expressions.
**See also:** logic[bsimp], logic[satisfy], logic[tautology]

### logic[bsimp](exprb)

Simplifies boolean expression bsimp to an irreducible sum of prime implicants.
**Output:** A boolean expression is returned.
**Additional information:** While the boolean expression returned is an irreducible (minimal) sum, it is not necessarily a *minimum* sum.
♣ Note that simplify does *not* simplify boolean expressions.
**See also:** logic[distrib], logic[bequal]

### logic[canon](exprb, {$var_1$, ..., $var_n$})

Converts the boolean expression exprb into canonical form.
**Output:** A boolean expression is returned.
**Argument options:** (exprb, {$var_1$, ..., $var_n$}, MOD2) to convert to modulo 2 canonical form. ♣ (exprb, {$var_1$, ..., $var_n$}, CNF) to convert to canonical conjunctive normal form. ♣ (exprb, {$var_1$, ..., $var_n$}, DNF) to convert to canonical disjunctive normal form. This option is the default.
**Additional information:** When using the MOD2 option, the variables $var_1$ through $var_n$ have no effect on the result, but must be supplied nonetheless.
**See also:** logic[distrib], logic[convert]

### logic[convert](exprb, *frominert*)

Converts exprb, a boolean expression containing inert operators, into an expression containing system operators.
**Output:** A boolean expression with system operators is returned.
**Argument options:** (exprb, *toinert*) to convert from system operators to inert operators. ♣ (exprb, MOD2) to convert to modulo 2 format. ♣ (exprb, MOD2, *expanded*) to convert to fully expanded modulo 2 format.
**Additional information:** Inert operators &and, &or, and &not are converted to system operators and, or, and not, and vice versa.
**See also:** *convert[mod2]*, logic[environ], logic[canon]

## logic[distrib](exprb)

Expands boolean expression exprb into a sum of products form.
**Output:** A boolean expression is returned.
**Argument options:** (exprb, {$var_1$, ..., $var_n$}) to put exprb into *canonical disjunctive normal form* with respect to variables $var_1$ through $var_n$.
**Additional information:** Expansion into a sums of products form is achieved by application of the distributive laws and DeMorgan's law, and does not guarantee that the result is either minimal or canonical.
**See also:** logic[canon], logic[bsimp], logic[bequal]

## logic[dual](exprb)

Constructs the dual of boolean expression exprb.
**Output:** A boolean expression is returned.
**Additional information:** In the dual of a boolean expression, all occurrences of true are replaced with false and vice versa, and all occurrences of &and are replaced with &or and vice versa. All other boolean operators are left unchanged. ♣ If you have turned on automatic simplification via the logic[environ] command, the dual of the simplified form of exprb is returned.
**See also:** logic[environ]

## logic[environ](n)

Sets the automatic simplification level for boolean expressions to n. n must be 0, 1, 2, or 3.
**Output:** A NULL expression is returned.
**Argument options:** (0) for no automatic simplification. ♣ (1) to use the associative property to remove redundant parentheses. ♣ (2) to use the associative property, properties a &and a → a and a &or a → a, and a knowledge of true and false in simplifying expressions. ♣ (3) to convert expressions into unexpanded modulo 2 form. ♣
**See also:** logic[convert], *convert[mod2]*

## logic[randbool]({$var_1$, ..., $var_n$})

Creates a random boolean expression in variables $var_1$ through $var_n$.
**Output:** A boolean expression is returned.
**Argument options:** ([$var_1$, ..., $var_n$] to produce the same results. ♣ ({$var_1$, ..., $var_n$}, *MOD2*) to create an expression in modulo 2 canonical form. ♣ (exprb, {$var_1$, ..., $var_n$}, *CNF*) to create an expression in canonical conjunctive normal form. ♣ (exprb, {$var_1$,

..., var$_n$}, *DNF*) to create an expression in canonical disjunctive normal form. This option is the default.

### logic[satisfy](exprb)

Calculates a valuation of variables expression exprb such that boolean expression exprb is true.

**Output:** A set containing equations is returned.

**Argument options:** (exprb, {var$_1$, ..., var$_n$}) to include variable names var$_1$ through var$_n$ in the valuation. Typically, these variables are ones that do not occur in exprb.

**Additional information:** If exprb is not true under any valuation, NULL is returned.

**See also:** logic[bequal], logic[tautology]

### logic[tautology](exprb)

Determines whether boolean expression exprb is a tautology.

**Output:** A boolean value of either true or false is returned.

**Argument options:** (exprb, name) to assign to name a valuation of the variables in expr showing that it is not a tautology. name is assigned a set containing an equation for each variable if exprb is not a tautology, or NULL if it is.

**Additional information:** A boolean expression is a tautology if and only if it is true under every possible valuation. ♣ Without the third parameter name, logic[tautology] runs significantly faster for complicated expressions.

**See also:** logic[bequal], logic[satisfy]

### macro(name = expr)

Sets up an abbreviation of name for expression expr to be used in Maple procedures.

**Output:** A NULL expression is returned.

**Argument options:** (eqn$_1$, ..., eqn$_n$), where eqn$_i$ is of the form name$_i$ = expr$_1$, to define n macros. ♣ (name = name) to erase the macro for name.

**Additional information:** This command is primarily used to shorten a long name such as linalg[ffgausselim] to a more easily typed name like ffg. ♣ When a procedure containing macros is read in, all abbreviations are translated back into their long names. ♣ You cannot define one macro in terms of another.

**See also:** alias

*LA = 181–183* **LI = 132**

### makehelp(str, filename)

Reads in textual file filename and creates a TEXT object with the name *'help/text/str'*.

**Output:** A NULL expression is returned.

**Argument options:** (str, filename, libname) to store the resulting file under the library libname. ♣ (str$_1$/str$_2$, filename, libname) to assign the TEXT object to *'help/str$_1$/text/str$_2$'*.

**Additional information:** This command needs to be defined by readlib(makehelp) before being invoked. ♣ Each line in the textual file is converted to an individual string in the TEXT object. ♣ If the resulting files are placed in the correct directory, then corresponding help files are accessible with the ? syntax. ♣ For more information and an example, see the on-line help file.

See also: help, TEXT


### map(fnc, list)

Applies command fnc to each element of list.

**Output:** A list containing new values, in the appropriate order, is returned.

**Argument options:** (fnc, list, expr$_1$, expr$_2$, ..., expr$_n$) to apply fnc, with extra parameters expr$_1$ through expr$_n$, to the values of list. ♣ (fnc, expr), where expr is any type of expression, to apply fnc on each operand of expr and return a new expression of the same type.

**Additional information:** When map is called with two parameters, fnc must be a command that can take only one parameter. ♣ When extra parameters are specified, the operands of list or expr are substituted for the *first* parameter of fnc. There is no way to specify that the substitution occurs for a different parameter— this is a shortcoming of map.

See also: *op*, select, zip

**FL = 91–93, 212** *LA = 17, 79–80, 106, 114* **LI = 133**


### match(expr$_1$ = expr$_2$, var, name)

Performs rudimentary pattern matching on two expressions in var to determine if expr$_1$ can be matched to the pattern represented by expr$_2$. If a match is found, then a substitution set containing equations for the unknowns in expr$_2$ which make it equal to expr$_1$ is assigned to name.

**Output:** A boolean value of true or false is returned.

**Additional information:** match does not perform substitution on main variable var.

See also: *subs*, has

*LA = 94* **LI = 133**

### max(expr$_1$, ..., expr$_n$)

Determines which is the maximum value among expressions expr$_1$ through expr$_n$.

**Output:** If one expression can be determined to be the maximum, then that expression is returned. If max can disqualify one or more expressions, then an unevaluated max command with a reduced number of expressions is returned. Otherwise, the command is returned as entered.

**Additional information:** If certain expressions are extremely close in value, increasing the value of global variable Digits might help in determining the maximum.

**See also:** min, *maximize, minimize, extrema, Digits*
*LI = 134*


### member(expr, set)

Determines whether expr is an element of set.

**Output:** A boolean value of true or false is returned.

**Argument options:** (expr, list) to determine if expr is an element of list. ♣ (expr, list, name) to assign the position in list where expr is first encountered to name.

**Additional information:** Keep in mind that expr must be *exactly* equal to a first-level operand of set or list. It is not sufficient that expr be a subexpression of an element.

**See also:** has
*LA = 17, 51 LI = 137*


### min(expr$_1$, ..., expr$_n$)

Determines which is the minimum value among expressions expr$_1$ through expr$_n$.

**Output:** If one expression can be determined to be the minimum, then that expression is returned. If min can disqualify one or more expressions, an unevaluated min command with a reduced number of expressions is returned. Otherwise, the command is returned as entered.

**Additional information:** If certain expressions are extremely close in value, increasing the value of global variable Digits might help in determining the minimum.

**See also:** max, *minimize, maximize, extrema, Digits*
*LI = 134*


### set$_1$ minus set$_2$

Computes the difference between set$_1$ and set$_2$.
**Output:** A set is returned.

**Argument options:** 'minus'(set$_1$, set$_2$) to produce the same results. minus must be in backquotes because it is a keyword. ♣ name$_1$ minus name$_2$ to represent the difference between two unassigned names that may later be assigned to sets. The input is left unevaluated.

**Additional information:** The difference between set$_1$ and set$_2$ contains the elements of set$_1$ that are *not* found in set$_2$.
**See also:** intersect, union, symmdiff, member
**FL = 81** *LA = 17, 51*

### MOLS(int$_p$, posint$_1$, posint$_2$)

Creates a list containing posint$_2$ mutually orthogonal Latin squares of size int$_p$^posint$_1$.
**Output:** A list of arrays is returned.
**Additional information:** This command needs to be defined by readlib(MOLS) before being invoked. ♣ For a definition of what makes Latin squares orthogonal, see the on-line help page. ♣ Be warned that large values of the parameters can cause very lengthy calculations to be performed.
*LI = 303*

### npspin(A, var, list)

Computes the Newman-Penrose spin coefficients of array A, variable var, and list, using Debever's formalism.
**Output:** An array of spin coefficients is returned.
**Additional information:** This command needs to be defined by readlib(debever) before being invoked. ♣ The result of npspin can be simplified with simp and then used in curvature to calculate curvature components. ♣ printspin and printcurve print out the spin coefficients and the curvature components, respectively. ♣ For more information on all these commands, see the on-line help page for debever.
**See also:** cartan, tensor, petrov

### NPspinor package

Provides commands for manipulations of *Newman-Penrose/spinor* formalisms, including commutators and Pfaffian operators.
**Additional information:** To access the command fnc from the NPspinor package, use with(NPspinor, fnc) to load in a pointer for fnc only or with(NPspinor) to load in pointers to all the commands. Unfortunately, the long form NPspinor[fnc] does not work.
♣ This package updates the earlier np package, which is still available. Each command in np has a direct counterpart in NPspinor.

## NPspinor[basis](expr)

Computes the basis spinors present in expression expr.
**Output:** A set of spinors is returned.
**Additional information:** A spinor is typically of the form fnc[var].
♣ If no spinors are present in expr, an empty set is returned.
**See also:** NPspinor[del]

## NPspinor[checkvars]()

Checks to ensure that the global variables used within the NPspinor package are not assigned values.
**Output:** Warning messages telling which package variables have not been assigned values and which ones have are returned.
**Additional information:** The global variables are stored in the list 'NPspinor/globals'. ♣ If any of the global variables have assigned values, this could lead to trouble. Unassign them and run NPspinor[checkvars] again.
**See also:** NPspinor[eqns]

## NPspinor[conj](expr)

Computes the complex conjugate of expr in the Newman-Penrose/spinor formalism.
**Output:** An expression is returned.

## NPspinor[contract](expr)

Simplifies expr by evaluating contractions between basis spinors.
**Output:** An expression is returned.
**Additional information:** A spinor is typically of form fnc[var].
♣ For a list of possible contractions, see the on-line help page.
♣ Before NPspinor[contract] is fully effective, NPspinor[dyad] must be called on expr.
**See also:** NPspinor[dyad]

## NPspinor[D](expr$_{np}$)

Performs as a Newman-Penrose Pfaffian differential operator on expr$_{np}$, an expression or equation constructed of similar operators. NPspinor[D] corresponds to the null tetrad vector $l$.

**Output:** An expression or equation is returned.
**Additional information:** For more information on NPspinor[D] and examples of its use, see the help page for NPspinor[Pfaffian].
**See also:** NPspinor[V], NPspinor[X], NPspinor[Y], NPspinor[V_D], NPspinor[X_D], NPspinor[Y_D]


### NPspinor[del](expr, name$_1$, name$_2$)

Computes the covariant derivative of expr for free spinor indices name$_1$ and name$_2$.
**Output:** An expression is returned.
**Additional information:** The result is computed explicitly for scalars, unevaluated procedures calls, and basis 1-spinors. ♣ One of the indices name$_1$ or name$_2$ must be ''dotted'' (i.e., end in the letter c), while the other must not be ''dotted''. ♣ Before NPspinor[del] is fully effective, NPspinor[dyad] must be called on expr. ♣ For more information on this command, see the online help page.
**See also:** NPspinor[dyad], NPspinor[symm], NPspinor[basis]


### NPspinor[dyad](expr)

Expands any forms of the curvature spinors within expr with respect to the dyad spinors.
**Output:** An expression is returned.
**Additional information:** This command needs to be called before many other operations to get the expressions into workable form.
**See also:** NPspinor[contract], NPspinor[del]


### NPspinor[eqns]()

Initializes the 29 basic equations, eq1 through eq29, used in the NPspinor package.
**Output:** A NULL expression is returned.
**Argument options:** (int$_1$, ..., int$_n$) to display equations eq.int$_1$ through eq.int$_n$. All the int$_i$ must be between 1 and 29 inclusive.
**Additional information:** After this command has been called, the global variable neqs is set to 29. ♣ The formal name for these equations is *Newman-Penrose Ricci and Bianchi identities*.
**See also:** NPspinor[checkvars], NPspinor[suball]


### NPspinor[findsymm](expr, [name$_1$, ..., name$_n$])

Rewrites any symmetric collections of basic spinors in free indices name$_1$ through name$_2$ found in expr in terms of the unevaluated command Symm.

**Output:** An expression in unevaluated Symm commands is returned.
**Argument options:** (expr, [name$_1$, ..., name$_n$], *nice*) to present the results in a nicer-looking dummy notation.
**Additional information:** A maximum of six indices are allowed (i.e., n < 6). For more information, see the on-line help page.
**See also:** NPspinor[rewrite], NPspinor[symm]


### NPspinor[makeeqn](expr)

Converts expr into a format accepted by the *eqn* typesetting system.
**Output:** An string in *eqn* format is returned.
**Additional information:** This command is identical to eqn except that it treats certain Newman-Penrose quantities differently.
**See also:** eqn


### NPspinor[rewrite](expr, name$_1$, name$_2$)

Rewrites expr in terms of symbolic coefficients named name$_1$[1], name$_1$[2], etc. The coefficients that were originally in expr are assigned to the automatically created array named name$_2$.
**Output:** An expression is returned.
**Argument options:** (expr, name$_1$, name$_2$, [name$_3$, ..., name$_n$]) to treat expr as an expression only in indices name$_3$ through name$_n$.
**Additional information:** name$_1$ and name$_2$ must be different.
♣ To reset the coefficients, the assignment name$_1$ := name$_2$ can be done.
**See also:** NPspinor[symm], NPspinor[findsymm]


### NPspinor[suball](expr)

Substitutes for values in expr from the 29 basic equations, eq1 through eq29, used in the NPspinor package, and their conjugates.
**Output:** An expression is returned.
**Argument options:** (expr, int$_1$, int$_2$) to substitute using only the basic equations eq.int$_1$ through eq.int$_2$ and their conjugates.
**Additional information:** The form which uses *all* the basic equations can be quite slow. Try narrowing down the range of equations.
**See also:** NPspinor[eqns]


### NPspinor[symm](expr, [name$_1$, ..., name$_n$])

Computes the symmetric form of expr over free indices name$_1$ through name$_n$.
**Output:** An expression is returned.

**Additional information:** name$_1$ through name$_n$ must be either all "dotted" (i.e., end in the letter $c$), or all "undotted". ♣ For more information, see the on-line help page.
**See also:** NPspinor[del], NPspinor[rewrite], NPspinor[findsymm]


### NPspinor[V](expr$_{np}$)

Performs as a Newman-Penrose Pfaffian differential operator on expr$_{np}$, an expression or equation constructed of similar operators. NPspinor[V] corresponds to the null tetrad vector $n$.
**Output:** An expression or equation with the NPspinor[V] operator applied is returned.
**Additional information:** For more information on NPspinor[V] and examples of its use, see the help page for NPspinor[Pfaffian].
**See also:** NPspinor[D], NPspinor[X], NPspinor[Y], NPspinor[V_D], NPspinor[X_V], NPspinor[Y_V]


### NPspinor[V_D](expr)

Returns an equation in expr corresponding to the commutation relation which is associated with the Lie bracket $[V, D]$ for differential operators V and D.
**Output:** An equation whose left-hand side is the Lie bracket and whose right-hand side is its linear expansion in terms of the known differential operators is returned.
**See also:** NPspinor[V], NPspinor[D], NPspinor[X_D], NPspinor[X_V], NPspinor[Y_D], NPspinor[Y_V], NPspinor[Y_X]


### NPspinor[X](expr$_{np}$)

Performs as a Newman-Penrose Pfaffian differential operator on expr$_{np}$, an expression or equation constructed of similar operators. NPspinor[X] corresponds to the null tetrad vector $m$.
**Output:** An expression or equation with the NPspinor[X] operator applied is returned.
**Additional information:** For more information on NPspinor[X] and examples of its use, see the help page for NPspinor[Pfaffian].
**See also:** NPspinor[D], NPspinor[V], NPspinor[Y], NPspinor[X_D], NPspinor[X_V], NPspinor[Y_X]


### NPspinor[X_D](expr)

Returns an equation in expr corresponding to the commutation relation which is associated with the Lie bracket $[X, D]$ for differential operators X and D.

**Output:** An equation whose left-hand side is the Lie bracket and whose right-hand side is its linear expansion in terms of the known differential operators is returned.
**See also:** NPspinor[X], NPspinor[D], NPspinor[V_D], NPspinor[X_V], NPspinor[Y_D], NPspinor[Y_V], NPspinor[Y_X]

### NPspinor[X_V](expr)

Returns an equation in expr corresponding to the commutation relation which is associated with the Lie bracket $[X, V]$ for differential operators X and V.
**Output:** An equation whose left-hand side is the Lie bracket and whose right-hand side is its linear expansion in terms of the known differential operators is returned.
**See also:** NPspinor[X], NPspinor[V], NPspinor[V_D], NPspinor[X_D], NPspinor[Y_D], NPspinor[Y_V], NPspinor[Y_X]

### NPspinor[Y](expr$_{np}$)

Performs as a Newman-Penrose Pfaffian differential operator on expr$_{np}$, an expression or equation constructed of similar operators. NPspinor[Y] corresponds to the null tetrad vector *m*.
**Output:** An expression or equation with the NPspinor[Y] operator applied is returned.
**Additional information:** For more information on NPspinor[Y] and examples of its use, see the help page for NPspinor[Pfaffian].
**See also:** NPspinor[D], NPspinor[V], NPspinor[X], NPspinor[Y_D], NPspinor[Y_V], NPspinor[Y_X]

### NPspinor[Y_D](expr)

Returns an equation in expr corresponding to the commutation relation which is associated with the Lie bracket $[Y, D]$ for differential operators Y and D.
**Output:** An equation whose left-hand side is the Lie bracket and whose right-hand side is its linear expansion in terms of the known differential operators is returned.
**See also:** NPspinor[Y], NPspinor[D], NPspinor[V_D], NPspinor[X_D], NPspinor[X_V], NPspinor[Y_V], NPspinor[Y_X]

### NPspinor[Y_V](expr)

Returns an equation in expr corresponding to the commutation relation which is associated with the Lie bracket $[Y, V]$ for differential operators Y and V.
**Output:** An equation whose left-hand side is the Lie bracket and

## Miscellaneous

whose right-hand side is its linear expansion in terms of the known differential operators is returned.

**See also:** NPspinor[Y], NPspinor[V], NPspinor[V_D], NPspinor[X_D], NPspinor[X_V], NPspinor[Y_D], NPspinor[Y_X]

### NPspinor[Y_X](expr)

Returns an equation in expr corresponding to the commutation relation which is associated with the Lie bracket $[Y, X]$ for differential operators Y and X.

**Output:** An equation whose left-hand side is the Lie bracket and whose right-hand side is its linear expansion in terms of the known differential operators is returned.

**See also:** NPspinor[Y], NPspinor[X], NPspinor[V_D], NPspinor[X_D], NPspinor[X_V], NPspinor[Y_D], NPspinor[Y_V]

### numapprox package

Provides commands for numerical approximation of functions.

**Additional information:** To access the command fnc from the numapprox package, use the long form of numapprox[fnc], or with(numapprox, fnc) to load in a pointer for fnc only, or with(numapprox) to load in pointers to all the commands. ♣ This package contains three commands: numapprox[infnorm], numapprox[minimax], and numapprox[remez]. numapprox[minimax] is the central command and the only one detailed in a command listing. For more information on the other two commands, see their on-line help pages.

### numapprox[minimax](expr, var=a..b, [posint$_n$, posint$_d$])

Computes a numerical approximation to expr, an expression in var. The approximation is on interval [a, b] and has numerator and denominator that are polynomials in var of degree posint$_n$ and posint$_d$, respectively.

**Output:** A rational polynomial in var is returned.

**Argument options:** (expr, var=a..b, [posint$_n$, 0]) or (expr, var=a..b, posint) to return the best minimax polynomial approximation of degree posint. The result is a nested expression in var, not as a ratio of polynomials. ♣ (expr, var=a..b, [posint$_n$, posint$_d$], expr$_w$) to specify the positive weight function expr$_w$ as an expression in var. The default value is 1.

**Additional information:** The Remez algorithm is used to compute the approximation. For more information on this algorithm, see the on-line help page. ♣ If expr is nonzero on the entire interval, then

weight function 1/expr is used to minimize the relative error.
**See also:** approx[infnorm], *fsolve*

### numboccur(expr, expr$_s$)

Determines how many times subexpression expr$_s$ occurs in expression expr.
**Output:** A non-negative integer value is returned.
**Additional information:** expr can be *any* type of expression. ♣ expr is evaluated before the determination is done and numboccur can only find a subexpression of expr if it is a proper operand of that evaluated expression. See the entry for subs for more details.
**See also:** has, *subs, member, op*

### op(float)

Displays the first-level operands of floating-point value float.
**Output:** An expression sequence is returned.
**Argument options:** (*1*, float) to extract the mantissa of float. ♣ (*2*, float) to extract the exponent of float.
**See also:** *op*, Float

### optimize(expr)

Optimizes the representation of expr, an algebraic expression, through the use of common subexpressions.
**Output:** A computation sequence of equations representing expr is returned. The left-hand sides of the equations are global variables t1, t2, etc.
**Argument options:** (expr, A) to optimize the elements of an array of algebraic expressions. ♣ (expr, name=expr) or (expr, [name$_1$=expr$_1$, ..., name$_n$=expr$_n$]) to optimize equations whose right-hand sides are algebraic expressions.
**Additional information:** This command needs to be defined by readlib(optimize) before being invoked. ♣ 'optimize/makeproc' can be used to generate procedures from such a computation sequence. ♣ For more information, see the on-line help page.
**See also:** 'optimize/makeproc', cost, fortran, C
*LA = 26−27* ***LI = 306***

### 'optimize/makeproc'([name$_1$=expr$_1$, ..., name$_n$=expr$_n$])

Produces a procedure from the computation sequence create by optimize enclosed in list brackets.
**Output:** A procedure is returned.

**Argument options:** ([name$_1$=expr$_1$, ..., name$_n$=expr$_n$], *parameters*=[str$_1$, ..., str$_m$]) to specify that the procedure should have parameters str$_1$ through str$_m$. ♣ ([name$_1$=expr$_1$, ..., name$_n$=expr$_n$], *locals*=[str$_1$, ..., str$_m$]) to specify that the procedure should have str$_1$ through str$_m$ declared as local variables. ♣ ([name$_1$=expr$_1$, ..., name$_n$=expr$_n$], *globals*=[str$_1$, ..., str$_m$]) to specify that the procedure should have str$_1$ through str$_m$ declared as global variables.

**Additional information:** This command needs to be defined by readlib(optimize) before being invoked. ♣ Unless otherwise specified, name$_1$ through name$_m$ are declared as local variables. ♣ For more information, see the on-line help page.

**See also:** optimize

*LI = 306*

### petrov(A)

Computes the Petrov classification of the Weyl tensor for array A.
**Output:** A string, which details the classification, is returned.
**Additional information:** This command needs to be defined by readlib(petrov) before being invoked. ♣ Use nonzero to display any determined non-zero expressions. ♣ For more information on these commands, see the on-line help page for petrov.
**See also:** cartan, tensor, npspin

*LI = 307*

### polar(cmplx)

Converts complex value cmplx to its polar coordinates form.
**Output:** An unevaluated polar command is returned. The two parameters of the polar command are the modulus and argument of the complex value, respectively.
**Argument options:** (expr$_1$, expr$_2$), where expr$_1$ and expr$_2$ are real expressions, to represent the appropriate polar value.
**Additional information:** This command needs to be defined by readlib(polar) before being invoked. ♣ The appropriate transformations are done to ensure that the first parameter is positive.
**See also:** evalc, *convert(polar)*

### protect(name)

Prevents the name name from being modified by the user.
**Output:** No output is returned.
**Argument options:** (name$_1$, ..., name$_n$) to protect all the names name$_1$ through name$_n$.
**Additional information:** Protected names cannot be modified by interactive commands or library code. ♣ Most built-in Maple

names are protected by default. ♣ To remove the protection from a name, use unprotect.
**See also:** unprotect

### rand()

Creates a random 12-digit non-negative integer.
**Output:** An integer is returned.
**Argument options:** ($int_1..int_2$) to return a procedure that returns a random integer in the range $int_1$ to $int_2$. ♣ (int) to return a procedure that returns a random integer in the range 0 to $int-1$.
**Additional information:** You can have more than one random number generator running at the same time. ♣ To alter the seeding value, assign a new non-zero integer to the global variable, _seed.
**See also:** *randpoly, linalg[randmatrix], linalg[randvector], geometry[randpoint]*
*LI = 171*

### Re(cmplx)

Extracts the real component of complex value cmplx.
**Output:** A real value is returned.
**Additional information:** If it is not possible to determine the real component of cmplx, then Re is left unevaluated.
**See also:** Im, argument, evalc

### related(name)

Displays the *SEE ALSO* section of the on-line help page for Maple command or structure name.
**Output:** A NULL expression is returned.
**Argument options:** ('name') if name is a keyword such as quit. ♣ ($name_1, name_2$) to display information from a more specific help page, where $name_2$ is a subtopic of $name_1$ or $name_2$ is a command in the package $name_1$.
**Additional information:** The *SEE ALSO* section of a help page contains the names of other commands and structures similar to name.
**See also:** help, usage, example, info

### round(expr)

Rounds real expression expr to the nearest integer value.
**Output:** If expr is a numeric value, then an integer is returned. Otherwise, round returns unevaluated.
**Argument options:** (cmplx) to round a complex valued expression. The result is equivalent to round(Re(expr)) + I * round(Im

(expr)). ♣ (1, expr) to represent the derivative of the round command at expr.
**Additional information:** The derivative of the round command is 0 wherever it is defined.
**See also:** trunc, frac, ceil, floor
*LI = 217*

### search(str$_1$, str$_2$)

Determines whether string str$_1$ contains an occurrence of string str$_2$.
**Output:** A boolean value of true or false is returned.
**Argument options:** (str$_1$, str$_2$, name) to assign the index of the first character of substring str$_2$ in str$_1$, if it is present, to name.
**Additional information:** This command needs to be defined by readlib(search) before being invoked. ♣ To extract a substring, use substring.
**See also:** searchtext, SearchText, substring, length, cat
*LI = 317*

### searchtext(str$_1$, str$_2$)

Determines the position of the first occurrence of str$_2$ in string str$_1$.
**Output:** An integer value representing the character position in str$_2$ where str$_1$ occurs is returned.
**Additional information:** If str$_2$ is not found in str$_1$, then a value of 0 is returned. ♣ This command is *not* case-sensitive. For a case-sensitive search, use SearchText. ♣ To extract a substring, use substring.
**See also:** SearchText, search, substring

### SearchText(str$_1$, str$_2$)

Determines the position of the first occurrence of str$_2$ in string str$_1$.
**Output:** An integer value representing the character position in str$_2$ where str$_1$ occurs is returned.
**Additional information:** If str$_2$ is not found in str$_1$, then a value of 0 is returned. ♣ This command is case-sensitive. ♣ To extract a substring, use substring.
**See also:** searchtext, search, substring

### select(fnc$_b$, list)

Creates a new list by selecting the elements of list which satisfy boolean command fnc$_b$.
**Output:** A list containing the retained values, in the appropriate

order, is returned.
**Argument options:** (fnc$_b$, set) to select elements from set. ♣ (fnc, list, expr$_1$, expr$_2$, ..., expr$_n$) to apply boolean function fnc, with extra parameters expr$_1$ through expr$_n$, to the values of list. ♣ (fnc, expr), where expr is sum or a product, to apply fnc on each term of expr and return a new expression of the same type containing just those terms that passed the test.
**Additional information:** When select is called with two parameters, fnc must be a command taking only one parameter. ♣ When extra parameters are specified, the operands of list or expr are substituted for the *first* parameter of fnc. There is no way to specify that substitution occurs for a different parameter—this is a shortcoming of select.
**See also:** op, map
*LI = 184*


### shake(expr)

Creates a floating-point range approximation for the value of expr.
**Output:** A list containing a floating-point range is returned.
**Argument options:** (expr, n) to compute the range approximation to an accuracy controlled by n. The default is the value of Digits.
**Additional information:** This command needs to be defined by readlib(shake) before being invoked. ♣ For more information on how shake works, see the on-line help page.
**See also:** evalr
*LI = 282*


### simplify(expr, '&*')

Simplifies expr, an expression containing non-commutative &* operators.
**Output:** An expression in operator notation is returned.
**Additional information:** This command needs to be defined by readlib(commutat) before being invoked.
**See also:** commutat, c, simplify(c), convert('&*'), convert(c)


### simplify(expr, c)

Simplifies expr, an expression containing non-commutative mutiplication expressions in terms of unevaluated c calls.
**Output:** An expression is returned.
**Additional information:** This command needs to be defined by readlib(commutat) before being invoked.
**See also:** commutat, c, simplify('&*'), convert('&*'), convert(c)

## spline([exprx$_1$, ..., exprx$_n$], [expry$_1$, ..., expry$_n$], name)

Computes spline segment polynomials (in variable name) corresponding to the spline of degree 3 on knot sequence represented by independent values exprx$_1$ through exprx$_n$ and dependent values expry$_1$ through expry$_n$.

**Output:** An unevaluated If structure representing the ranges of name and the segment polynomials is returned.

**Argument options:** ([exprx$_1$, ..., exprx$_n$], [expry$_1$, ..., expry$_n$], name, posint) to create splines of degree posint. ♣ ([exprx$_1$, ..., exprx$_n$], [expry$_1$, ..., expry$_n$], name, option), where option is *linear*, *quadratic*, *cubic*, or *quartic*, to create splines of appropriate degree.

**Additional information:** This command needs to be defined by readlib(spline) before being invoked. ♣ For more information on this command, see the on-line help page.

**See also:** bspline, interp


## substring(str, m..n)

Extracts the substring of str including the characters at positions m through n inclusive.

**Output:** A string is returned.

**Additional information:** If m is less than 1, the substring starting at position 1 is extracted. ♣ If n is greater than length(str), the substring ending at position length(str) is extracted. ♣ If n is greater than int$_1$, the null string is returned.

**See also:** searchtext, SearchText, search, length, cat

*LI = 208*


## symmdiff(set$_1$, ..., set$_n$)

Computes the symmetric difference of sets set$_1$ through set$_n$.

**Output:** A set is returned.

**Argument options:** (expr$_1$, ..., expr$_n$) to represent the symmetric difference of expressions expr$_1$ through expr$_n$. An expression in minus, union, and intersect operators is returned.

**Additional information:** This command needs to be defined by readlib(symmdiff) before being invoked. ♣ The symmetric difference of sets set$_1$ and set$_2$ is defined as (set$_1$ union set$_2$) minus (set$_1$ intersect set$_2$). This formula is expanded when more than two parameters are given.

**See also:** union, intersect, minus

### tensor()

Computes curvature tensors in a coordinate basis.

**Output:** A NULL expression is returned.

**Additional information:** This command needs to be defined by readlib(tensor) before being invoked. ♣ tensor differs in usage from most other Maple commands. In order to alter the coordinates for the $n$ dimensions, you must assign new values to the names x1, ..., xn. You *cannot* choose your own variable names. Likewise, the metric tensors involved are assigned to the names g.i.j, where i and j range from 1 to n. The fact that these variable names are predefined is the reason that the tensor command call needs no parameters. ♣ To change the number of dimensions from the default value of 4, assign a new value to Ndim. ♣ There are other no-parameter commands that are activated when tensor is read. They include invmetric, Riemann, Einstein, and Weyl. The results of calls to these commands are assigned to predefined variable names, such as C132 and R4144. ♣ For more information on these commands, see the on-line help page for tensor.

**See also:** npspin, cartan, petrov

*LI = 321*

### TEXT(str$_1$, ..., str$_n$)

Prints the sequence of strings str$_1$ through str$_n$, starting each string on a new line.

**Output:** A sequence of strings, each displayed on a new line, is returned.

**Additional information:** This command is used to create the on-line Maple help pages. ♣ For a brief example of how to create your own help pages, see the on-line help page for TEXT.

**See also:** makehelp, search, length, cat, *type(TEXT)*

**FL = 139–142**

### thiele([exprx$_1$, ..., exprx$_n$], [expry$_1$, ..., expry$_n$], var)

Computes, using Thiele's interpolation formula, an expression in var representing the continued fraction interpolation of independent values exprx$_1$ through exprx$_n$ and dependent values expry$_1$ through expry$_n$.

**Output:** A continued fraction in var is returned.

**Additional information:** This command needs to be defined by readlib(thiele) before being invoked. ♣ If two or more of the exprx$_i$ values are equal, an error message is returned. ♣ For more information on Thiele's interpolation formula, see the on-line help page.

**See also:** *interp, sinterp*

*LI = 324*

## totorder package

Provides commands to compute total ordering of names.
**Additional information:** To access the command fnc from the totorder package, use the long form of the name totorder[fnc], or with(totorder, fnc) to load in a pointer for fnc only, or with(totorder) to load in pointers to all the commands. ♣ To define a total ordering, use totorder[tassume].

## totorder[forget](var)

Removes relationships involving variable var from the current total ordering.
**Output:** A sequence whose first element is *forgotten* and whose second element is var is returned.
**Argument options:** (*everything*) to remove *all* relationships from the total ordering.
**Additional information:** var must be totally separable from the ordering for this command to work. ♣ totorder[tassume] is used to add relationships to the total ordering.
**See also:** totorder[tassume], totorder[tis], totorder[ordering]

## totorder[ordering]()

Displays the current total ordering.
**Output:** A sequence of inequalities is returned.
**See also:** totorder[tassume], totorder[forget], totorder[tis]

## totorder[tassume](ineq$_1$, ..., ineq$_n$)

Defines a total ordering on the variables in inequalities ineq$_1$ through ineq$_n$, specified by the inequalities themselves.
**Output:** A sequence whose first element is *assumed* and whose subsequent elements are inequalities defining the relationships is returned.
**Additional information:** If a total ordering already exists, calling totorder[tassume] adds the new relationships to the total ordering. ♣ Inequalities can use any of the operators <, >, and =. ♣ Inequalities must be given in either ascending or descending order. ♣ totorder[forget] is used to remove relationships from the total ordering.
**See also:** totorder[forget], totorder[tis], totorder[ordering]

## totorder[tis](ineq)

Determines whether inequality ineq is true in the current total ordering.

## Miscellaneous

**Output:** A boolean value of either true or false is returned.
**Additional information:** ineq does not need to look exactly like one of the inequalities that defined the total ordering. ♣ Either side of ineq can contain expressions in the variables of the total ordering. ♣ If totorder[tis] cannot determine the validity of ineq, an error message is returned.
**See also:** totorder[tassume], totorder[forget], totorder[ordering]

### trunc(expr)

Truncates real expression expr to an integer value by removing the fractional part.
**Output:** If expr is a numeric value, then an integer is returned. Otherwise, trunc returns unevaluated.
**Argument options:** (cmplx) to truncate a complex valued expression. The value returned equals trunc(Re(expr)) + I *trunc(Im(expr)). ♣ (1, expr) to represent the derivative of the trunc command at expr.
**Additional information:** The derivative of the trunc command is 0 wherever it is defined.
**See also:** frac, round, ceil, floor
*LI = 217*

### tutorial()

Runs an on-line tutorial covering the basics of Maple.
**Output:** Not applicable.
**Argument options:** (n), where n is an integer between 1 and 15 to start the tutorial at a specific section.
**Additional information:** The on-line tutorial automatically creates a worksheet with input, output, and text regions. All the user has to do is hit specific keys to control the flow of information. ♣ Because the tutorial was written in Maple's ownlanguage, it can be easily updated. For more information, see the help page for tutorial.
**See also:** help, example

### type(expr, 'and')

Tests whether expr is a logical expression whose main operator is and.
**Output:** A boolean value of either true or false is returned.
**Additional information:** The parameter *and* must be in backquotes because it is a keyword.
**See also:** minus, type('or'), type('not'), type('logical')

Miscellaneous 511

**type(expr,** *'intersect'*)

Tests whether expr is an unevaluated intersection of two sets.
**Output:** A boolean value of either true or false is returned.
**Additional information:** The parameter *intersect* must be in backquotes because it is a keyword.
**See also:** intersect, type(*'minus'*), type(*'union'*)


**type(expr,** *logical*)

Determines whether expr is a logical expression.
**Output:** A boolean value of either true or false is returned.
**Additional information:** expr is a logical if it tests positive as type *'and'*, *'or'*, or *'not'*.
**See also:** type(*'and'*), type(*'or'*), type(*'not'*)
*LI = 225*


**type(expr,** *'minus'*)

Tests whether expr is an unevaluated difference of two sets.
**Output:** A boolean value of either true or false is returned.
**Additional information:** The parameter *minus* must be in backquotes because it is a keyword.
**See also:** minus, type(*'intersect'*), type(*'union'*)


**type(expr,** *'not'*)

Tests whether expr is a logical expression whose main operator is the negation not.
**Output:** A boolean value of either true or false is returned.
**Additional information:** The parameter *not* must be in backquotes because it is a keyword.
**See also:** minus, type(*'and'*), type(*'or'*), type(*'logical'*)


**type(expr,** *'or'*)

Tests whether expr is a logical expression whose main operator is or.
**Output:** A boolean value of either true or false is returned.
**Additional information:** The parameter *or* must be in backquotes because it is a keyword.
**See also:** minus, type(*'and'*), type(*'not'*), type(*'logical'*)


**type(expr,** *'union'*)

Tests whether expr is an unevaluated union of two sets.
**Output:** A boolean value of either true or false is returned.

**Additional information:** The parameter *union* must be in backquotes because it is a keyword.
**See also:** union, type(*'minus'*), type(*'intersect'*)

### unapply(expr, var)

Converts expr, an expression in var, to a functional operator taking one parameter.
**Output:** A functional operator is returned.
**Argument options:** unapply(expr, $var_1$, $var_2$, ..., $var_n$) to convert expr to a functional operator in n parameters.
**Additional information:** Whenever a functional operator is to be made from an already existing expression, use unapply.
**See also:** *student[procmake]*
**FL = 37, 220**  *LI = 257*

### $set_1$ union $set_2$

Computes the union of $set_1$ and $set_2$.
**Output:** A set is returned.
**Argument options:** 'union'($set_1$, $set_2$) to produce the same results. union must be in backquotes because it is a keyword. ♣ $name_1$ union $name_2$ to represent the union of two unassigned names that may later be assigned to sets. The input is left unevaluated.
**Additional information:** The union of $set_1$ and $set_2$ contains the elements found in *either* $set_1$ *or* $set_2$.
**See also:** minus, intersect, member
**FL = 81** *LA = 17, 51*

### unprotect(name)

Removes the name protection from name, allowing it to be modified.
**Output:** No output is returned.
**Argument options:** ($name_1$, ..., $name_n$) to remove name protection from all the names $name_1$ through $name_n$.
**Additional information:** Protected names cannot be modified by interactive commands or library code. ♣ Most built-in Maple names are protected by default. ♣ You cannot unprotect Maple *keywords*. ♣ To protect a name, use protect.
**See also:** protect

## usage(name)

Displays the *CALLING SEQUENCE* section of the on-line help page for Maple command or structure name.

**Output:** A NULL expression is returned.

**Argument options:** ('name') if name is a keyword such as quit. ♣ (name$_1$,name$_2$) to display information from a more specific help page, where name$_2$ is a subtopic of name$_1$ or name$_2$ is a command in the package name$_1$.

**Additional information:** The ??name syntax is much easier to use, and does not require a semicolon at the end of the statement as usage(name) does. ♣ The calling sequence provides you with a template for the parameters to a command.

**See also:** help, info, example, related


## zip(fnc, [expra$_1$, ..., expra$_n$], [exprb$_1$, ..., exprb$_n$])

Creates a new list by applying two-parameter command fnc to value pairs expra$_1$ and exprb$_1$ through expra$_n$ and exprb$_n$.

**Output:** A list containing n values is returned.

**Argument options:** (fnc, [expra$_1$, ..., expra$_m$], [exprb$_1$, ..., exprb$_n$], expr) to use default value expr to fill up empty elements at the end of the shorter of the two lists.

**Additional information:** If the two input lists are of different lengths and no fourth parameter is given, the resulting list is the same length as the shorter list. ♣ There is no way to use zip with commands that take three or more parameters.

**See also:** map

**FL = 93** *LI = 262*

# Index

!, 19
" " ", 30
" ", 30
", 30
<, 30
<=, 30
<>, 30
', 32
( ), 24
(..), 20
*, 19
+, 19
-, 19
., 18
/, 14
:=, 28
:, 12
;, 12
=, 29
???, 473
??, 513
?, 486
?, 6
@, 24
[], 22
$, 20
&^
    in difforms, 41
    in liesymm, 127
&and, 463
&iff, 463
&implies, 464
&mod, in liesymm, 127
&nand, 464
&nor, 464
&not, 464
&or, 464
&xor, 465

^, 19
{}, 21
', 31

about, 465
abs, 334
acycpoly, in networks, 245
add
    in linalg, 77
    in powseries, 52
addcol, in linalg, 77
addedge, in networks, 245
addressof, 422
addrow, in linalg, 78
addvertex, in networks, 246
adj, in linalg, 78
adjacency, in networks, 246
adjoint, in linalg, 78
Ai, 334
alias, 465
allpairs, in networks, 246
allvalues, 148
altitude, in geometry, 203
amortization, 465
anames, 422
ancestor, in networks, 246
angle
    in geom3d, 194
    in linalg, 78
animate, in plots, 399
animate3d, in plots, 400
annul, in liesymm, 128
appendto, 422
apply, in transform, 325
Appolonius, in geometry, 204
arccos, 335
arccosh, 335

# Index

arccoshp, in padic, 290
arccosp, in padic, 290
arccot, 335
arccoth, 336
arccothp, in padic, 290
arccotp, in padic, 290
arccsc, 336
arccsch, 336
arccschp, in padic, 290
arccscp, in padic, 291
arcsec, 336
arcsech, 337
arcsechp, in padic, 291
arcsecp, in padic, 291
arcsin, 337
arcsinh, 337
arcsinhp, in padic, 291
arcsinp, in padic, 291
arctan, 338
arctanh, 338
arctanhp, in padic, 292
arctanp, in padic, 292
area
    in geom3d, 194
    in geometry, 204
are_collinear
    in geom3d, 195
    in geometry, 204
are_concurrent
    in geom3d, 195
    in geometry, 204
areconjugate, in group, 480
are_harmonic, in geometry, 205
are_orthogonal, in geometry, 205
are_parallel
    in geom3d, 195
    in geometry, 205
are_perpendicular
    in geom3d, 195
    in geometry, 205
are_similar, in geometry, 206
are_tangent
    in geom3d, 196
    in geometry, 206
args, 422
argument, 466
array, 73
arrivals, in networks, 247
assemble, 423
assign, 118
assigned, 423
assume, 466
asubs, 366
asympt, 38
augment, in linalg, 78

B, in numtheory, 279
backsub, in linalg, 79
band, in linalg, 79
basis
    in linalg, 79
    in NPspinor, 496
    in simplex, 136
bell, in combinat, 237
bequal, in logic, 489
bernoulli, 268
bernstein, 149
BesselI, 338
BesselJ, 338
BesselK, 339
BesselY, 339
beta, in random, 312
Beta, 339
bezout, in linalg, 80
Bi, 339
bicomponents, in networks, 247
binomial, in combinat, 237
binomial, 236
binomiald, in random, 313
bisector, in geometry, 206
blackscholes, 467
BlockDiagonal, in linalg, 80
blockmatrix, in linalg, 80
boxplot, in statplots, 321
break, 423
bsimp, in logic, 490
bspline, 467
by, 426

c, 467
C, 468
canon, in logic, 490
cartan, 468
cartprod, in combinat, 237
Catalan, 340

Index  517

cauchy, in random, 313
cdf, in statevalf, 319
ceil, 468
center
    in geometry, 206
    in group, 480
centralizer, in group, 481
centroid
    in geom3d, 196
    in geometry, 206
cfrac, in numtheory, 279
cfracpol, in numtheory, 280
changevar, in student, 58
character, in combinat, 237
charmat, in linalg, 80
charpoly
    in linalg, 80
    in networks, 247
chebyshev, 39
checkvars, in NPspinor, 496
Chi, 340
    in combinat, 238
chisquare, in random, 313
choose, in combinat, 238
chrem, 269
chrompoly, in networks, 248
Ci, 340
circle
    in geometry, 207
    in networks, 250
circumcircle, in geometry, 207
classi, 119
classmark, in transform, 325
close, 423
    in liesymm, 128
coeff, 149
coefficientofvariation, in describe, 303
coeffs, 149
coeftayl, 39
col, in linalg, 81
coldim, in linalg, 81
collect, 366
collinear, in projgeom, 224
colspace, in linalg, 81
colspan, in linalg, 81
combinat package, 236-244
combine, 39, 367

commutat, 469
companion, in linalg, 82
complement, in networks, 248
complete, in networks, 248
completesquare, in student, 58
compoly, 150
components, in networks, 248
compose, in powseries, 52
concat, in linalg, 82
concur, in projgeom, 225
concyclic, in geometry, 207
cond, in linalg, 82
conformal, in plots, 400
conic
    in geometry, 208
    in projgeom, 225
conj, in NPspinor, 496
conjpart, in combinat, 238
conjugate, in projgeom, 225
conjugate, 469
connect, in networks, 249
connectivity, in networks, 249
constants, 340
content, 150
Content, 150
contourplot, in plots, 401
contract
    in networks, 249
    in NPspinor, 496
convergs, 469
convert('*'), 368
convert('+'), 367
convert('&*'), 469
convert('and'), 470
convert(base), 368
convert(binary), 368
convert(binomial), 244
convert(c), 470
convert(confrac), 368
convert(D), 119
convert(decimal), 369
convert(degrees), 369
convert(diff), 119
convert(double), 369
convert(Ei), 341
convert(equality), 119
convert(erf), 341
convert(erfc), 341

convert(*exp*), 341
convert(*expln*), 342
convert(*expsincos*), 342
convert(*factorial*), 370
convert(*float*), 370
convert(*fraction*), 370
convert(*GAMMA*), 342
convert(*hex*), 370
convert(*horner*), 151
convert(*hostfile*), 424
convert(*hypergeom*), 40
convert(*lessequal*), 119
convert(*lessthan*), 120
convert(*list*), 371
convert(*listlist*), 371
convert(*ln*), 342
convert(*mathorner*), 151
convert(*matrix*), 73
convert(*metric*), 371
convert(*mod2*), 372
convert(*multiset*), 372
convert(*name*), 372
convert(*octal*), 372
convert(*'or'*), 470
convert(*parfrac*), 372
convert(*polar*), 373
convert(*polynom*), 40
convert(*radians*), 373
convert(*radical*), 373
convert(*rational*), 373
convert(*ratpoly*), 40
convert(*RoofOf*), 374
convert(*set*), 374
convert(*sincos*), 343
convert(*sqrfree*), 151
convert(*std*), in simplex, 136
convert(*stdle*), in simplex, 137
convert(*string*), 374
convert(*tan*), 343
convert(*trig*), 343
convert(*vector*), 74
convert, in logic, 490
convexhull
    in geometry, 208
    in simplex, 137
coordinates
    in geom3d, 197
    in geometry, 208
coplanar, in geom3d, 197

copy, 74
copyinto, in linalg, 82
core, in group, 481
cos, 343
cosets, in group, 481
cosh, 344
coshp, in padic, 292
cosp, in padic, 292
cosrep, in group, 481
cost, 424
cot, 344
coth, 344
cothp, in padic, 292
cotp, in padic, 292
count, in describe, 304
countcuts, in networks, 250
countmissing, in describe, 304
counttrees, in networks, 250
covariance, in describe, 304
crossprod, in linalg, 83
csc, 345
csch, 345
cschp, in padic, 293
cscp, in padic, 293
csgn, 345
ctangent, in projgeom, 225
cterm, in simplex, 137
cumulativefrequency, in transform, 326
curl, in linalg, 83
cyclebase, in networks, 250
cyclotomic, in numtheory, 280
cylinderplot, in plots, 401


d
    in difforms, 42
    in liesymm, 128
D, 40
    in NPspinor, 496
D(*fnc*), 120
daughter, in networks, 250
dawson, 346
dcdf, in statevalf, 319
Dchangevar, in DEtools, 121
decile, in describe, 305
decodepart, in combinat, 238
defform, in difforms, 42
define, 424

Index   519

definite, in linalg, 83
degree, 151
degreeseq, in networks, 251
del, in NPspinor, 497
delcols, in linalg, 84
delete, in networks, 251
deletemissing, in transform, 326
delrows, in linalg, 84
denom, 374
densityplot, in plots, 401
departures, in networks, 251
DEplot, in DEtools, 121
DEplot1, in DEtools, 122
DEplot2, in DEtools, 122
depvars, in liesymm, 129
derived, in group, 482
DerivedS, in group, 482
describe subpackage, 303-311
DESol, 120
det, in linalg, 84
Det, 74
detailf, in geometry, 208
determine, in liesymm, 129
DEtools package, 121-124
device, 412
dfieldplot, in DEtools, 123
diag, in linalg, 84
diameter
    in geometry, 209
    in networks, 252
diff, 41
Diff, 41, 124
diff(fnc(var)), 124
difforms package, 41-44
Digits, 470
dilog, 346
dinic, in networks, 252
dinterp, 152
Dirac, 346
disassemble, 425
discreteuniform, in random, 314
discrim, 152
Discrim, 152
display, in plots, 402
display3d, in plots, 402
distance
    in geom3d, 197
    in geometry, 209
    in student, 59

DistDeg, 152
distrib, in logic, 491
diverge, in linalg, 85
divide, 153
Divide, 153
divideby, in transform, 326
divisors, in numtheory, 280
djspantree, in networks, 252
do, 426
done, 425
dotprod, in linalg, 85
Doubleint, in student, 59
dsolve, 124
dsolve(*numeric*), 125
dual
    in logic, 491
    in simplex, 137
duplicate, in networks, 252
dvalue, in liesymm, 129
dyad, in NPspinor, 497


E, 346
    in numtheory, 280
edges, in networks, 253
Ei, 347
eigenvals, in linalg, 85
Eigenvals, 74
eigenvects, in linalg, 86
elif, 430
ellipse, in geometry, 209
ellipsoid, 194
else, 430
empirical, in random, 314
encodepart, in combinat, 239
end, 439
ends, in networks, 253
entermatrix, in linalg, 86
entries, 75
environ, in logic, 491
eqn, 471
eqns, in NPspinor, 497
equal, in linalg, 86
equate, in student, 59
erf, 347
erfc, 347
ERROR, 425
Eta, in liesymm, 129
euler, 269

Eulercircle, in geometry, 210
Eulerline, in geometry, 210
eulermac, 44
eval, 471
Eval, 153
evala, 153
eval(array), 75
evalb, 471
evalc, 472
evalf, 472
evalf(*int*), 44
'evalf/int', 45
evalf(*Int*), 44
evalgf, 154
evalhf, 472
evalm, 76
evaln, 473
evalp, in padic, 293
evalpow, in powseries, 52
evalr, 473
eval(table), 76
eweight, in networks, 253
example, 473
excircle, in geometry, 210
exp, 348
expand, 375
Expand, 375
expandoff, 375
expandon, 376
exponential
    in linalg, 87
    in random, 314
expp, in padic, 293
extend, in linalg, 87
extract, 425
extrema, 45
extvars, in liesymm, 130

F, in numtheory, 280
factor, 155
Factor, 155
factorEQ, in numtheory, 280
factors, 156
Factors, 156
factorset, in numtheory, 280
FAIL, 426
false, 348
feasible, in simplex, 138

fermat, in numtheory, 281
ffgausselim, in linalg, 87
FFT, 156
fi, 430
fibonacci
    in combinat, 239
    in linalg, 88
fieldplot, in plots, 403
fieldplot3d, in plots, 403
finance, 474
find_angle, in geometry, 210
findsymm, in NPspinor, 497
finduni, in grobner, 159
finite, in grobner, 159
firstpart, in combinat, 239
fit subpackage, 311-312
fixdiv, 157
Float, 474
floor, 474
flow, in networks, 253
flowpoly, in networks, 254
fnormal, 475
for/from/by/to/do/od, 426
for/from/by/while/do/od, 427
for/in/do/od, 427
for, 426
forget, 428
    in totorder, 509
formpart, in difforms, 42
fortran, 475
fourier, 157
fpconic, in projgeom, 226
frac, 475
fratio, in random, 315
freeze, 428
frequency, in transform, 326
FresnelC, 348
Fresnelf, 348
Fresnelg, 349
FresnelS, 349
frobenius, in linalg, 88
from, 426
frontend, 428
fsolve, 126
fundcyc, in networks, 254

G, in orthopoly, 172
galois, 157

gamma, 349
    in random, 315
GAMMA, 349
Gauss, 429
gausselim, in linalg, 88
GaussInt package, 269-274
gaussjord, in linalg, 88
gbasis, in grobner, 160
gc, 429
gcd, 157
Gcd, 158
gcdex, 158
Gcdex, 158
genfunc package, 476-479
genmatrix, in linalg, 89
genpoly, 158
geom3d package, 194-203
geometricmean, in describe, 305
geometry package, 203-219
Gergonnepoint, in geometry, 210
getcoeff, in liesymm, 130
getform, in liesymm, 130
getlabel, in networks, 254
GF, 479
GIbasis, in GaussInt, 270
GIchrem, in GaussInt, 270
GIdivisor, in GaussInt, 270
GIfacset, in GaussInt, 270
GIfactor, in GaussInt, 270
GIfactors, in GaussInt, 271
GIgcd
    in GaussInt, 271
    in numtheory, 281
GIgcdex, in GaussInt, 271
GIhermite, in GaussInt, 271
GIissqr, in GaussInt, 271
GIlcm, in GaussInt, 272
GImcmbine, in GaussInt, 272
GInearest, in GaussInt, 272
GInodiv, in GaussInt, 272
GInorm, in GaussInt, 272
GIorder, in GaussInt, 272
GIphi, in GaussInt, 273
GIprime, in GaussInt, 273
GIquadres, in GaussInt, 273
GIquo, in GaussInt, 273
GIrem, in GaussInt, 273
GIroots, in GaussInt, 274

girth, in networks, 254
GIsieve, in GaussInt, 274
GIsmith, in GaussInt, 274
GIsqrt, in GaussInt, 274
global, 439
grad, in linalg, 89
gradplot, in plots, 404
gradplot3d, in plots, 404
GramSchmidt, in linalg, 90
graphical, in networks, 254
graycode, in combinat, 239
grelgroup, in group, 482
grobner package, 159-162
group package, 480-486
groupmember, in group, 482
grouporder, in group, 482
gsimp, in networks, 255
gsolve, in grobner, 160
gunion, in networks, 255


H, in orthopoly, 172
hadamard, in linalg, 90
harmonic
    in geometry, 211
    in projgeom, 226
harmonic, 350
harmonicmean, in describe, 305
has, 486
hasclosure, in liesymm, 130
hastype, 486
head, in networks, 255
heap, 429
Heaviside, 350
help, 486
hermite, in linalg, 90
Hermite, 162
hessian, in linalg, 91
hilbert, in linalg, 91
histogram, in statplots, 322
history, 430
hook, in liesymm, 131
htranspose, in linalg, 91
hypergeom, 350


I, 351
icdf, in statevalf, 320
icontent, 162

# Index

idcdf, in statevalf, 320
if, 430
ifactor, 275
ifactors, 275
iFFT, 162
if/then/elif/then/else/fi, 430
igcd, 275
igcdex, 276
ihermite, in linalg, 91
ilcm, 276
ilog, 351
ilog10, 351
ilog[expr], 351
Im, 487
imagunit, in numtheory, 281
implicitplot, in plots, 405
implicitplot3d, in plots, 405
importdata, in stats, 321
in, 427
incidence, in networks, 255
incident, in networks, 256
incircle, in geometry, 211
indegree, in networks, 256
indepvars, in liesymm, 131
indets, 430
index, in numtheory, 281
indexfunc, in linalg, 92
indices, 76
induce, in networks, 256
infinity, 351
info, 487
infolevel, 431
initialize, 431
innerprod, in linalg, 92
insert, 431
int, 45
Int, 46
intbasis, in linalg, 92
integrand, in student, 59
inter
    in geom3d, 197
    in geometry, 211
    in projgeom, 226
    in group, 483
intercept, in student, 60
interface(*echo*), 432
interface(*indentamount*), 432
interface(*iris*), 432
interface(*labeling*), 432
interface(*labelwidth*), 433
interface(*plotdevice*), 393
interface(*plotoutput*), 394
interface(*postplot*), 394
interface(*preplot*), 394
interface(*prettyprint*), 433
interface(*prompt*), 433
interface(*quiet*), 433
interface(*screenheight*), 434
interface(*screenwidth*), 434
interface(*verboseproc*), 434
interface(*version*), 435
interface(*wordsize*), 435
interp, 163
Interp, 163
intersect, 487
intparts, in student, 60
inttovec, in combinat, 240
inverse
    in linalg, 93
    in powseries, 53
inversion, in geometry, 211
invfourier, 163
invlaplace, 164
invperm, in group, 483
invphi, in numtheory, 281
invztrans, 164
iquo, 276
iratrecon, 276
irem, 277
iroot, 277
irreduc, 164
Irreduc, 164
is, 488
isabelian, in group, 483
iscont, 46
is_equilateral, in geometry, 211
isgiven, 488
ismith, in linalg, 93
isnormal, in group, 483
isolate, 376
isolve, 126
isprime, 277
isqrfree, 277
isqrt, 278
is_right, in geometry, 212
issqr, 278
issqrfree, in numtheory, 282
issubgroup, in group, 483

iszero, in linalg, 93
ithprime, 278

J, in numtheory, 282
jacobi, in numtheory, 282
jacobian, in linalg, 93
join, in projgeom, 226
jordan, in linalg, 94
JordanBlock, in linalg, 94

kernel, in linalg, 94
kronecker, in numtheory, 282
kurtosis, in describe, 306

L
    in numtheory, 282
    in orthopoly, 172
lambda, in numtheory, 283
laplace, 165
laplaced, in random, 315
laplacian, in linalg, 94
lasterror, 435
lastpart, in combinat, 240
latex, 488
lattice, 76
lccutc, in projgeom, 226
lccutr, in projgeom, 227
lccutr2p, in projgeom, 227
lcm, 165
lcoeff, 165
LCS, in group, 484
ldegree, 165
leadmon, in grobner, 160
leastsqrs, in linalg, 95
leastsquare, in fit, 311
leftbox, in student, 60
leftsum, in student, 61
legendre, in numtheory, 283
length, 489
lexorder, 489
lhs, 376
Li, 352
libname, 435
Lie, in liesymm, 131
liesymm package, 127-135
limit, 46
Limit, 47

linalg package, 77-106
line
    in geometry, 212
    in projgeom, 227
line3d, in geom3d, 198
linearcorrelation, in describe, 306
linemeet, in projgeom, 227
linsolve, in linalg, 95
ln, 352
lnGAMMA, 352
local, 439
log, 352
log10, 353
log[expr], 353
logic package, 489-492
logistic, in random, 316
loglogplot, in plots, 405
lognormal, in random, 316
logp, in padic, 293
logplot, in plots, 406
lprint, 436
Lrank, in liesymm, 131

M, in numtheory, 283
macro, 492
makeeqn, in NPspinor, 498
makeforms, in liesymm, 132
makehelp, 493
makeproc, in student, 61
make_square, in geometry, 212
map, 493
match, 493
matrix, in linalg, 95
matrixplot, in plots, 406
max, 494
maxdegree, in networks, 257
maximize, 47
    in simplex, 138
maxnorm, 166
maxorder, 166
mcombine, in numtheory, 283
mean, in describe, 306
meandeviation, in describe, 307
median
    in describe, 307
    in geometry, 213
MeijerG, 353

mellin, 166
mellintable, 167
member, 494
mersenne, in numtheory, 283
middlebox, in student, 61
middlesum, in student, 61
midpoint
    in geom3d, 198
    in geometry, 213
    in projgeom, 228
    in student, 62
min, 494
mincut, in networks, 257
mindegree, in networks, 257
minimax, in numapprox, 501
minimize, in simplex, 139
minimize, 47
minkowski, in numtheory, 284
minor, in linalg, 96
minpoly, in linalg, 96
minpoly, 167
minus, 494
mipolys, in numtheory, 284
mixpar
    in difforms, 43
    in liesymm, 132
mlog, in numtheory, 284
mobius, in numtheory, 284
mod, 167
mode, in describe, 308
modp1, 168
modp1(*Constant*), 169
modp1(*ConvertIn*), 169
modp1(*ConvertOut*), 169
modp1(*One*), 170
modp1(*Randpoly*), 170
modp1(*Zero*), 170
modpol, 170
MOLS, 495
moment, in describe, 308
moving, in transform, 327
mroot, in numtheory, 285
msolve, 135
msqrt, in numtheory, 285
mtaylor, 48
mulcol, in linalg, 96
mulperms, in group, 484
mulrow, in linalg, 97
multconst, in powseries, 53

multiapply, in transform, 327
multinomial, in combinat, 240
multiply
    in linalg, 97
    in powseries, 53

Nagelpoint, in geometry, 213
nargs, 436
nearestp, in numtheory, 285
negative, in powseries, 53
negativebinomial, in random, 317
neighbors, in networks, 257
networks package, 245-262
new, in networks, 257
next, 436
nextpart, in combinat, 240
nextprime, 278
Nextprime, 171
nops, 376
norm, in linalg, 97
norm, 171
normal, 377
Normal, 377
NormalClosure, in group, 484
normald, in random, 317
normalf, in grobner, 161
normalize, in linalg, 98
normalizer, in group, 484
notchedbox, in statplots, 322
npspin, 495
NPspinor package, 495-501
nthcover, in numtheory, 285
nthdenom, in numtheory, 286
nthnumer, in numtheory, 286
nthpow, in numtheory, 286
NULL, 437
nullspace, in linalg, 98
Nullspace, 106
numapprox package, 501-502
numbcomb, in combinat, 241
numboccur, 502
numbpart, in combinat, 241
numbperm, in combinat, 241
numer, 377
numtheory package, 278-289

od, 426
odeplot, in plots, 406
on_circle, in geometry, 213
on_line, in geometry, 214
on_plane, in geom3d, 198
onsegment
    in geom3d, 198
    in geometry, 213
    in projgeom, 228
on_sphere, in geom3d, 199
op, 378
op(2DPlot), 395
op(3DPlot), 395
op(circle), 219
op(conic), 220
op(conicpj), 220
op(*Diff*), 48
op(ellipse), 220
op(*eval*(array)), 106
op(*eval*(table)), 106
op(expr$_p$), 289
op(float), 502
op(graph), 262
op(*Int*), 48
op(*Limit*), 49
op(line), 221
op(line3d), 221
op(linepj), 221
op(plane), 221
op(point), 222
op(point3d), 222
op(pointpj), 222
op(polynom), 171
op(*Product*), 49
op(pseries), 49
op(ratpoly), 171
op(series), 50
op(sphere), 223
op(square), 223
op(*Sum*), 50
op(tetrahedron), 223
op(triangle), 223
op(triangle3d), 224
open, 437
operator, 437
optimize, 502
'optimize/makeproc', 502
options, 439
orbit, in group, 484

order, 51
    in numtheory, 286
Order, 470, 51
ordering, in totorder, 509
ordp, in padic, 294
orthocenter, in geometry, 214
orthog, in linalg, 98
orthopoly package, 172-173
outdegree, in networks, 258

P, in orthopoly, 173
padic package, 289-296
parallel
    in geom3d, 199
    in geometry, 214
parity, in difforms, 43
parse, 437
partition, in combinat, 242
path, networks, 258
PDEplot, in DEtools, 123
pdf, in statevalf, 320
percentile, in describe, 308
permanent, in linalg, 98
permgroup, in group, 485
permrep, in group, 485
permute, in combinat, 242
perpen_bisector, in geometry, 215
perpendicular
    in geom3d, 199
    in geometry, 214
petersen, in networks, 258
petrov, 503
pf, in statevalf, 320
phaseportrait, in DEtools, 123
phi, in numtheory, 286
Pi, 353
piecewise, 354
pivot
    in linalg, 98
    in simplex, 139
pivoteqn, in simplex, 140
pivotvar, in simplex, 140
plane, in geom3d, 199
plot, 396
plot3d, 397
plots package, 399-412

# Index

point
    in geometry, 215
    in projgeom, 228
point3d, in geom3d, 200
pointplot, in plots, 407
pointto, 438
poisson, in random, 317
poisson, 51
polar, 503
polar_point, in geometry, 215
polarp, in projgeom, 228
polarplot, in plots, 407
pole_line, in geometry, 215
poleline, in projgeom, 228
polygonplot, in plots, 408
polygonplot3d, in plots, 408
polyhedraplot, in plots, 408
potential, in linalg, 99
powcreate, in powseries, 53
powdiff, in powseries, 54
Power, 173
powerpc, in geometry, 216
powerps, in geom3d, 200
powerset, in combinat, 242
powexp, in powseries, 54
powint, in powseries, 54
powlog, in powseries, 54
Powmod, 174
powpoly, in powseries, 54
powseries[powsolve], 135
powseries package, 52-56
powsolve, in powseries, 135
powsubs, in student, 62
pprimroot, in numtheory, 287
prem, 174
Prem, 174
pres, in group, 485
prevpart, in combinat, 242
prevprime, 296
Prevprime, 174
Prime, 175, 296
Primfield, 175
Primitive, 175
primpart, 175
Primpart, 176
primroot, in numtheory, 287
print, 438
printf, 438
printlevel, 439

proc, 439
procbody, 440
proc/local/global/options/end, 439
procmake, 440
product, 56
Product, 56
projection
    in geom3d, 200
    in geometry, 216
projgeom package, 224-229
prolong, in liesymm, 132
proot, 176
protect, 503
Psi, 353
psqrt, 176
ptangent, in projgeom, 229

quadraticmean, in describe, 309
quantile
    in describe, 309
    in statplots, 323
quantile2, in statplots, 323
quartile, in describe, 309
quit, 440
quo, 176
Quo, 177
quotient, in powseries, 55

rad_axis, in geometry, 216
rad_center, in geometry, 216
radius
    in geom3d, 201
    in geometry, 216
radnormal, 378
rad_plane, in geom3d, 201
radsimp, 378
rand, 504
randbool, in logic, 491
randcomb, in combinat, 243
RandElement, in group, 485
randmatrix, in linalg, 99
random, in networks, 258
random subpackage, 312-319
randpart, in combinat, 243
randperm, in combinat, 243
randpoint, in geometry, 217

# Index 527

randpoly, 177
Randpoly, 177
Randprime, 177
randvector, in linalg, 99
range
    in describe, 310
    in linalg, 99
rank
    in linalg, 100
    in networks, 259
rankpoly, in networks, 259
ratform, in linalg, 100
ratio, in simplex, 140
rationalize, 378
ratrecon, 178
Ratrecon, 178
ratvaluep, in padic, 294
Re, 504
read, 441
readdata, 441
readlib, 441
readline, 442
readstat, 442
realroot, 178
recipoly, 179
reduce, in liesymm, 133
reflect
    in geom3d, 201
    in geometry, 217
related, 504
rem, 179
Rem, 179
remember, 442
remove, in transform, 327
replot, in plots, 409
residue, 57
restart, 443
resultant, 179
Resultant, 180
RETURN, 443
reversion, in powseries, 55
rewrite, in NPspinor, 498
rgf_charseq, in genfunc, 476
rgf_encode, in genfunc, 476
rgf_expand, in genfunc, 477
rgf_findrecur, in genfunc, 477
rgf_hybrid, in genfunc, 477
rgf_norm, in genfunc, 477
rgf_pfrac, in genfunc, 477

rgf_relate, in genfunc, 478
rgf_sequence, in genfunc, 478
rgf_simp, in genfunc, 478
rgf_term, in genfunc, 479
rhs, 379
rightbox, in student, 62
rightsum, in student, 63
RootOf, 180
rootp, in padic, 294
roots, 180
Roots, 181
rootsunity, in numtheory, 287
rotate, in geometry, 217
round, 504
row, in linalg, 100
rowdim, in linalg, 100
rowspace, in linalg, 100
rowspan, in linalg, 100
rref, in linalg, 101
rsolve, 135
rtangent, in projgeom, 229


safeprime, in numtheory, 288
satisfy, in logic, 492
save, 443
scalarmul, in linalg, 101
scalarpart, in difforms, 43
scaleweight, in transform, 328
scatter1d, in statplots, 323
scatter2d, in statplots, 324
search, 505
searchtext, 505
SearchText, 505
sec, 354
sech, 355
sechp, in padic, 294
secp, in padic, 295
select, 505
seq, 444
series, 57
setoptions, in plots, 409
setoptions3d, in plots, 409
setup
    in liesymm, 133
    in simplex, 141
shake, 506
Shi, 355
shortpathtree, in networks, 259

show, in networks, 259
showtangent, in student, 63
showtime, 444
shrink, in networks, 260
Si, 355
sides, in geometry, 217
sigma, in numtheory, 288
sign, 181
signum, 355
similitude, in geometry, 217
simpform, in difforms, 43
simplex package, 136-141
simplify, 379
simplify('@'), 379
simplify('&*'), 506
simplify(*atsign*), 380
simplify(*c*), 506
simplify(*Ei*), 356
simplify(*GAMMA*), 356
simplify(*hypergeom*), 356
simplify(*ln*), 356
simplify(*polar*), 380
simplify(*power*), 380
simplify(*radical*), 380
simplify(*RootOf*), 380
simplify(*sqrt*), 357
simplify(*trig*), 357
simpson, in student, 63
Simsonline, in geometry, 218
sin, 357
singular, 57
singularvals, in linalg, 101
sinh, 357
sinhp, in padic, 295
sinp, in padic, 295
sinterp, 181
skewness, in describe, 310
slope, in student, 64
smith, in linalg, 101
Smith, 181
solvable, in grobner, 161
solve, 141
sort(list), 381
sort(poly), 182
spacecurve, in plots, 410
span, in networks, 260
spanpoly, in networks, 260
spantree, in networks, 260
sparsematrixplot, in plots, 410

sphere, in geom3d, 201
sphereplot, in plots, 410
spline, 507
split, 182
    in transform, 328
spoly, in grobner, 162
sprem, 182
Sprem, 183
sq2factor, in numtheory, 288
sqrfree, 183
Sqrfree, 183
sqrt, 358
sqrtp, in padic, 295
square, in geometry, 218
sscanf, 444
Ssi, 358
ssystem, 445
stack, in linalg, 102
standarddeviation, in describe, 310
standardize, in simplex, 141
standardscore, in transform, 328
statevalf subpackage, 319-320
statplots subpackage, 321-325
stats package, 321-329
statsort, in transform, 328
status, 445
statvalue, in transform, 329
stirling1, in combinat, 243
stirling2, in combinat, 244
stop, 445
student package, 58-64
studentst, in random, 318
sturm, 183
sturmseq, 184
suball, in NPspinor, 498
subgrel, in group, 485
submatrix, in linalg, 102
subs, 381
subsets, in combinat, 244
subsop, 381
substring, 507
subtract, in powseries, 55
subvector, in linalg, 102
sum, 65
Sum, 65
sumbasis, in linalg, 103
surfdata, in plots, 411

Svd, 107
swapcol, in linalg, 103
swaprow, in linalg, 103
Sylow, in group, 486
sylvester, in linalg, 103
symm, in NPspinor, 498
symmdiff, 507
symmetric
    in geom3d, 202
    in geometry, 218
symmetry, in statplots, 324
system, 445
system(str), 446

T, in orthopoly, 173
table, 107
tail, in networks, 261
tally, in transform, 329
tallyinto, in transform, 329
tan, 359
tangent
    in geom3d, 202
    in geometry, 218
tangentpc, in geometry, 219
tangentte, in projgeom, 229
tanh, 358
tanhp, in padic, 295
tanp, in padic, 296
tassume, in totorder, 509
tau, in numtheory, 288
tautology, in logic, 492
taylor, 65
tcoeff, 184
TD, in liesymm, 133
tensor, 508
termscale, in genfunc, 479
testeq, 142
tetrahedron, in geom3d, 202
TEXT, 508
textplot, in plots, 411
textplot3d, in plots, 411
tharmonic, in projgeom, 229
thaw, 446
then, 430
thiele, 508
thue, in numtheory, 288
time, 446
tis, in totorder, 509

to, 426
toeplitz, in linalg, 104
totorder package, 509-510
tpsform, in powseries, 55
trace, 446
    in linalg, 104
trace(fnc), 447
transform subpackage, 325-329
translate, 184
    in liesymm, 134
transpose, in linalg, 104
traperror, 447
trapezoid, in student, 64
triangle, in geometry, 219
triangle3d, in geom3d, 202
trigsubs, 382
Tripleint, in student, 64
true, 359
trunc, 510
tubeplot, in plots, 412
tutorial, 510
tuttepoly, in networks, 261
type
    in geom3d, 203
    in geometry, 219
type('!'), 449
type('\*\*'), 448
type('\*'), 447
type('+'), 448
type('..'), 448
type('.'), 448
type('<='), 142
type('<>'), 142
type('<'), 142
type('='), 142
type('^'), 448
type('algebraic'), 449
type(algext), 449
type(algfun), 449
type(algnum), 450
type(algnumext), 450
type('and'), 510
type(anything), 450
type(arctrig), 359
type(array), 107
type(boolean), 450
type(complex), 450
type(complex(integer)), 296
type(const), 66

type(*constant*), 451
type(*cubic*), 184
type(*equation*), 143
type(*even*), 451
type(*evenfunc*), 359
type(*expanded*), 185
type(*facint*), 296
type(*float*), 451
type(*form*), 66
type(*fraction*), 451
type(*function*), 359
type(*graph*), 263
type(*heap*), 451
type(*indexed*), 108
type(*integer*), 452
type(*'intersect'*), 511
type(*laurent*), 66
type(*linear*), 185
type(*list*), 452
type(*listlist*), 452
type(*logical*), 511
type(*mathfunc*), 360
type(*matrix*), 108
type(*'minus'*), 511
type(*monomial*), 185
type(*name*), 452
type(*negative*), 452
type(*negint*), 453
type(*nonneg*), 453
type(*nonnegint*), 453
type(*'not'*), 511
type(*nothing*), 453
type(*numeric*), 453
type(*odd*), 453
type(*oddfunc*), 360
type(*operator*), 453
type(*'or'*), 511
type(*PLOT*), 413
type(*PLOT3D*), 413
type(*point*), 454
type(*polynom*), 185
type(*posint*), 454
type(*positive*), 454
type(*primeint*), 297
type(*procedure*), 454
type(*quadratic*), 186
type(*quartic*), 186
type(*radext*), 454
type(*radfun*), 455

type(*radfunext*), 455
type(*radical*), 455
type(*radnum*), 455
type(*radnumext*), 456
type(*range*), 456
type(*rational*), 456
type(*ratpoly*), 186
type(*realcons*), 456
type(*relation*), 456
type(*RootOf*), 456
type(*scalar*), 67, 108
type(*series*), 67
type(*set*), 457
type(*sqrt*), 457
type(*square*), 457
type(*string*), 457
type(*table*), 109
type(*taylor*), 67
type(*TEXT*), 457
type(*trig*), 360
type(*type*), 457
type(*uneval*), 458
type(*'union'*), 511
type(*vector*), 109

U, in orthopoly, 173
unames, 458
unapply, 512
unassign, 458
uniform, in random, 318
union, 512
unprotect, 512
untrace, 458
usage, 513
userinfo, 458

V, in NPspinor, 499
value, 67
valuep, in padic, 296
vandermonde, in linalg, 104
variance, in describe, 311
V_D, in NPspinor, 499
vdegree, in networks, 261
vecpotent, in linalg, 105
vectdim, in linalg, 105
vectoint, in combinat, 244
vector, in linalg, 105
vertices, in networks, 261

void, in networks, 262
volume, in geom3d, 203
vweight, in networks, 262


W, 360
wdegree
    in difforms, 44
    in liesymm, 134
wedgeset, in liesymm, 134
weibull, in random, 318
while, 427
with, 459
words, 459
write, 459
writeln, 460
writeto, 460
Wronskian, in linalg, 105
wsubs, in liesymm, 134


X, in NPspinor, 499
X_D, in NPspinor, 499
xscale, in statplots, 324
xshift, in statplots, 325
X_V, in NPspinor, 500
xyexchange, in statplots, 325


Y, in NPspinor, 500
Y_D, in NPspinor, 500
Y_V, in NPspinor, 500
Y_X, in NPspinor, 501


Zeta, 361
zip, 513
ztrans, 186